W0225691

ATOMS IN
ASTROPHYSICS

PHYSICS OF ATOMS AND MOLECULES

Series Editors:

P. G. Burke, *The Queen's University of Belfast, Northern Ireland*
H. Kleinpoppen, *Institute of Atomic Physics, University of Stirling, Scotland*

R. B. Bernstein *(New York, U.S.A.)*
J. C. Cohen-Tannoudji *(Paris, France)*
R. W. Crompton *(Canberra, Australia)*
J. N. Dodd *(Dunedin, New Zealand)*
G. F. Drukarev *(Leningrad, U.S.S.R.)*
W. Hanle *(Giessen, Germany)*

Editorial Advisory Board:

C. J. Joachain *(Brussels, Belgium)*
W. E. Lamb, Jr. *(Tucson, U.S.A.)*
P.-O. Löwdin *(Gainesville, U.S.A.)*
H. O. Lutz *(Bielefeld, Germany)*
M. R. C. McDowell *(London, U.K.)*
K. Takayanagi *(Tokyo, Japan)*

1979:
ATOM – MOLECULE COLLISION THEORY: A Guide for the Experimentalist
Edited by Richard B. Bernstein

1980:
COHERENCE AND CORRELATION IN ATOMIC COLLISIONS
Edited by H. Kleinpoppen and J. F. Williams

VARIATIONAL METHODS IN ELECTRON – ATOM SCATTERING THEORY
R. K. Nesbet

1981:
DENSITY MATRIX THEORY AND APPLICATIONS
Karl Blum

INNER-SHELL AND X-RAY PHYSICS OF ATOMS AND SOLIDS
Edited by Derek J. Fabian, Hans Kleinpoppen, and Lewis M. Watson

1982:
INTRODUCTION TO THE THEORY OF LASER – ATOM INTERACTIONS
Marvin H. Mittleman

1983:
ATOMS IN ASTROPHYSICS
Edited by P. G. Burke, W. B. Eissner, D. G. Hummer, and I. C. Percival

1983:
ELECTRON – ATOM AND ELECTRON – MOLECULE COLLISIONS
Edited by Juergen Hinze

1983:
PROGRESS IN ATOMIC SPECTROSCOPY, Part C
Edited by H. J. Beyer and Hans Kleinpoppen

A Continuation Order Plan is available for this series. A continuation order will bring delivery of each new volume immediately upon publication. Volumes are billed only upon actual shipment. For further information please contact the publisher.

ATOMS IN ASTROPHYSICS

Edited by

P. G. BURKE
The Queen's University of Belfast
Belfast, Northern Ireland

W. B. EISSNER
Science and Engineering Research Council
Daresbury, England

D. G. HUMMER
Joint Institute for Laboratory Astrophysics
University of Colorado
Boulder, Colorado

and

I. C. PERCIVAL
University of London
London, England

PLENUM PRESS • NEW YORK AND LONDON

Library of Congress Cataloging in Publication Data

Main entry under title:

Atoms in astrophysics.

   (Physics of atoms and molecules)
   Includes bibliographical references and index.
   1. Astrophysics. 2. Atoms. 3. Seaton, M. J. I. Burke, P. G. II. Series
QB463.A86   1983   523.01'97   82-22517
ISBN-13:978-1-4613-3538-2   e-ISBN-13:978-1-4613-3536-8
DOI: 10.1007/978-1-4613-3536-8

© 1983 Plenum Press, New York
Softcover reprint of the hardcover 1st edition 1983

A Division of Plenum Publishing Corporation
233 Spring Street, New York, N.Y. 10013

All rights reserved

No part of this book may be reproduced, stored in a retrieval system, or transmitted in any form
or by any means, electronic, mechanical, photocopying, microfilming, recording, or otherwise,
without written permission from the Publisher

CONTRIBUTORS

D. R. BATES • *Department of Applied Mathematics and Theoretical Physics, The Queen's University of Belfast, Belfast BT7 1NN, Northern Ireland*

P. G. BURKE • *Department of Applied Mathematics and Theoretical Physics, The Queen's University of Belfast, Belfast BT7 1NN, Northern Ireland*

A. DALGARNO • *Harvard–Smithsonian Center for Astrophysics, Cambridge, Massachusetts 02138*

JACQUES DUBAU • *Observatoire de Paris–Meudon, 92190 Meudon, France*

W. EISSNER • *Daresbury Laboratory, Science and Engineering Research Council, Daresbury, Warrington WA4 4AD, England*

D. R. FLOWER • *Department of Physics, University of Durham, Durham DH1 3LE, England*

DAVID L. MOORES • *University College London, Gower Street, London WC1 E6BT, England*

D. W. NORCROSS • *Joint Institute for Laboratory Astrophysics, University of Colorado and National Bureau of Standards, Boulder, Colorado 80309*

H. NUSSBAUMER • *Institute of Astronomy, ETH Zentrum, CH 3092 Zurich, Switzerland*

G. PEACH • *Department of Physics and Astronomy, University College London, Gower Street, London WC1 E6BT, England*

IAN PERCIVAL • *Department of Applied Mathematics, Queen Mary College, University of London, Mile End Road, London E1 4NS, England*

HANNELORE E. SARAPH • *University College London, Gower Street, London WC1 E6BT, England*

P. J. STOREY • *Department of Physics and Astronomy, University College London, Gower Street, London WC1 E6BT, England*

HENRI VAN REGEMORTER • *Observatoire de Paris–Meudon, 92190 Meudon, France*

PREFACE

It is hard to appreciate but nevertheless true that Michael John Seaton, known internationally for the enthusiasm and skill with which he pursues his research in atomic physics and astrophysics, will be sixty years old on the 16th of January 1983. To mark this occasion some of his colleagues and former students have prepared this volume. It contains articles that describe some of the topics that have attracted his attention since he first started his research work at University College London so many years ago.

Seaton's association with University College London has now stretched over a period of some 37 years, first as an undergraduate student, then as a research student, and then, successively, as Assistant Lecturer, Lecturer, Reader, and Professor. Seaton arrived at University College London in 1946 to become an undergraduate in the Physics Department, having just left the Royal Air Force in which he had served as a navigator in the Pathfinder Force of Bomber Command. There are a number of stories of how his skill with instruments and the precision of his calculations, later to be so evident in his research, saved his crew from enemy action, and on one occasion, on a flight through the Alps, from a collision with Mount Blanc that at the time was shrouded in clouds. Only when the clouds suddenly lifted to reveal the mountain in all its glory close by, did the crew have cause to be grateful that, as many of the readers of this book will be well aware, a Seaton calculation is carried out as if his life depended on it.

The period immediately after the Second World War marked the start of a new era, both for atomic physics and astrophysics at University College London, and for the influnce that UCL was to have internationally in these fields. Having worked on the Manhattan Project, H. S. W. Massey had just returned to his appointment as Goldsmid Professor of Applied Mathematics, and he was joined in the Department of Mathematics by a number of colleagues, including D. R. Bates, R. L. F. Boyd, R. A. Buckingham, E. H. S. Burhop, J. C. Gunn and J. B. Hasted. A. Dalgarno, B. L. Moiseiwitsch, E. A. Power and A. L. Stewart were among the students who joined this group soon afterwards.

Seaton graduated with a first-class honors degree in Physics in 1948, and started his research under the direction of Bates in the Mathematics Department. Bates, of course, took his role as supervisor very seriously, and

vii

on one occasion gave a formal series of lectures that only one student, Seaton, attended. This clearly provides a strong argument against the present overwhelming emphasis that is put upon staff–student ratios and the undesirability of small classes. Between 1949 and 1951, during the course of the work for his Ph.D. thesis on *Quantal Calculations of Certain Reaction Rates with Applications to Astrophysical and Geophysical Problems*, Seaton published six papers, and significantly these were equally distributed between *Monthly Notices of the Royal Astronomical Society* and physics journals. The theme of this thesis was to become the basis of his subsequent scientific work.

In October 1950 Massey moved to the Physics Department to take up his new appointment as Quain Professor of Physics. He was accompanied by Bates, Buckingham, and Burhop who became Readers in Physics, L. Castillejo and Seaton who became Assistant Lecturers, Boyd and Hasted who were appointed to ICI Fellowships, and Dalgarno and Moiseiwitsch who were then completing their Ph.D. research. At about this time, Seaton started his classic paper on *The Hartree–Fock Equations for Continuous States with Applications to Electron Excitation of the Gound State Configuration Terms of OI.* How this work laid the foundation for the later developments is shown in the review on *Low-Energy Electron Collisions with Complex Atoms and Ions* by two co-editors of this book, P. G. Burke who owes his introduction to atomic collision physics to Seaton in 1956 and W. Eissner whom Seaton introduced in 1967 to methods for complex atoms feasible with the advent of the new generation of computers. In addition, in his chapter on *Forbidden Atomic Lines in Auroral Spectra*, Bates discusses the important influence this paper has had on certain astrophysical problems.

In 1957, together with I. C. Percival, another co-editor of this book, Seaton gave the first rigorous discussion of the partial wave theory of electron–hydrogen collisions, and in collaboration with Castillejo, also discussed the theory of the long-range interactions between electrons and hydrogen atoms. Seaton's longstanding interest in this topic is highlighted in the chapter on *Long-Range Interactions in Atoms and Diatomic Molecules* by G. Peach. In 1958, Percival and Seaton published the first general algebraic formulation of the problem of the polarization of line radiation by electron impact, and this paper now forms the basis for the interpretation of electron–photon coincidence experiments.

Until the early 1960's the detailed calculation of wave functions for complex atomic systems was still a very tedious and formidable task, even for the electron–hydrogen system, and theoreticians could not hope to provide astrophysicists with the large amounts of atomic data that they required. Simpler methods were explored, and Seaton applied the semiclassical impact parameter method to the excitation of allowed transitions between states of high principal quantum number. Percival's chapter on *Collisions between Charged Particles and Highly Excited Atoms* links up with this early work.

Seaton took a leave of absence from UCL for the academic year 1954-5 and went to France as Chargé de Recherche at the Institut d'Astrophysique in Paris. This visit marked the start not only of his long association with H. van Regemorter, but also of the close collaboration that has subsequently developed between University College and the Observatoire de Paris at Meudon. More recently, this collaboration has been extended to include the Observatoire de Nice. The chapter by J. Dubau and H. van Regemorter on *Electron–Ion Processes in Hot Plasmas* represents one area that has been strongly influenced by this association. In recognition of his contributions to atomic physics and astrophysics, Seaton was awarded the degree of Docteur honoris causa by the Observatoire de Paris in 1976.

Also in 1954, Seaton realized that the quantum defect theory provided a powerful tool for the interpretation and production of atomic data, and his first publication on this subject appeared in *Comptes Rendus* in 1955. The enormous developments in this field are described in this book by D. L. Moores and H. E. Saraph in their chapter on *Applications of Quantum Defect Theory*.

Finally, during this period, Seaton carried out the work that established the importance of proton collisions with atomic ions in laboratory and astrophysical plasmas. These processes are discussed by Dalgarno in the chapter on *Proton Impact Excitation of Positive Ions.*

In 1961 Seaton visited the University of Colorado at Boulder. At this time the Joint Institute for Laboratory Astrophyics was being established, and in 1964 Seaton was appointed a Fellow-Adjoint of the Institute in recognition of his contribution to its scientific activities. A very active collaboration between UCL and JILA in both atomic physics and astrophysics followed. Three former postgraduate students of Seaton's are now staff members of JILA, and five others have visited Boulder on postdoctoral or visiting fellowships. One of these students, D. G. Hummer, another co-editor, began his career in astrophysics with Seaton in 1959, and during the period 1961–4, published several papers with him on the structure of planetary nebulae. Although for many years Hummer has been a senior member of JILA, he continues to be a regular and welcome visitor to UCL.

In the 1960's Seaton also became interested in the properties of the central stars of planetary nebulae, and in a joint paper with the late R. J. Harman, managed for the first time to obtain a well-defined track for these objects on the Hertzsprung–Russell diagram. They showed that these exceptionally hot stars in fact represented a natural stage in stellar evolution. In his chapter on *Planetary Nebulae* D. R. Flower reviews the present state of our knowledge of these astrophysical plasmas.

Seaton became well-known as one of the few people who understood both the detailed methods for the calculation of atomic data, and the requirements for these data in astrophysics. He strove to teach astrophysicists about the calculation of atomic data and he encouraged the atomic physicists to produce accurate data of direct relevance to astrophysical problems. In fact, it was a problem in the interpretation of processes in the solar corona, pointed out to Seaton in a letter from A. Unsöld, that led to the discovery in 1964, by his former student and colleague A. Burgess, of the importance of dielectronic recombination. In the book published in honor of Bates' 60th birthday, Seaton and P. J. Storey devote a chapter to the complex processes involved. He also became very keen to make astronomical observations himself. He made a pact with the astronomer D. E. Osterbrock that they would swap jobs for a year. Osterbrock would come to UCL to calculate atomic data and he would spend a year carrying out observations. Osterbrock fulfilled his pledge in 1968, but it was not until 1978 that Seaton started to make observations with the International Ultraviolet Explorer. Such was his enthusiasm, that between 1978 and 1981 he collaborated on ten papers that combined observations of novae and nebulae with interpretative theory. During much of this period he also had the honor to be president of the Royal Astronomical Society, and was deeply involved in the Society's affairs.

In the late 1960's he decided to exploit the new generation of computers that were then becoming available to obtain accurate data on atomic structure and on atomic collisions that could be used with confidence in astrophysics. Typically, he became completely immersed in the use of this new technology, doing a lot of programming himself and developing new computational methods. As a result, a package of programs was developed at University College that is now widely used throughout the world. Some of these programs are described in the chapter by H. Nussbaumer and P. J. Storey on *The University College Computer Package for the Calculation of Atomic Data: Aspects of Development and Application.* Other aspects of the programs for electron–atom collisions are described by D. W. Norcross in his chapter on *Numerical Methods for Asymptotic Solutions of Scattering Equations.*

In 1968 Seaton played a major role in setting up and supporting the international journal *Computer Physics Communications*, both as an Advisory Editor and as an active contributor of programs. He was also involved in launching the *Journal of Physics* series in 1968; he was the first Honorary

Editor of *Journal of Physics B: Atomic and Molecular Physics* and helped to establish it as the leading international journal in its field.

In recognition of his great contribution to atomic physics and astrophysics Seaton was elected a Fellow of the Royal Society in 1967. The most recent honor bestowed on him was the award, in 1982, of an Honorary D. Sc. degree by the Queen's University of Belfast.

We hope that this volume will reveal something of Seaton's prodigious influence on the subject of atoms in astrophyics, and at the same time will provide an up-to-date survey of many important aspects of this field.

P. G. Burke
W. Eissner
D. G. Hummer
I. C. Percival

CONTENTS

3. Collisions between Charged Particles and Highly Excited Atoms

Ian Percival

4. Proton Impact Excitation of Positive Ions

A. Dalgarno

5. Long-Range Interactions in Atoms and Diatomic Molecules

G. Peach

7. Electron–Ion Processes in Hot Plasmas

Jacques Dubau and Henri van Regemorter

8. The University College Computer Package for the
Calculation of Atomic Data: Aspects of Development
and Application

H. Nussbaumer and P. J. Storey

9. Planetary Nebulae

D. R. Flower

10. Forbidden Atomic Lines in Auroral Spectra

D. R. Bates

LOW-ENERGY ELECTRON COLLISIONS WITH COMPLEX ATOMS AND IONS

P. G. BURKE AND W. EISSNER

1. Introduction

In this chapter we review the theory and numerical methods and give some illustrative results for low-energy elastic scattering and excitation of atoms and positive ions by electron impact. We also describe briefly the major computer program packages that are available and widely used in this field. We restrict our considerations mainly to low electron impact energies where the velocity of the incident or scattered electron is of the same order as the velocity of the electrons taking an active part in the collision. We will, however, consider briefly some extensions of the theory to intermediate energies above the ionization threshold but not so high that perturbation theories can be used with reliability. We will mainly be concerned with collisions involving complex atoms and ions which for our purposes here means when the target contains more than two electrons. Nevertheless, the methods and programs which have been developed have been widely used to study collisions with hydrogenlike and heliumlike ions and we will mention some of this work.

The modern theory of this subject can be traced to the classic paper by Seaton[1] in 1953 in which he generalized the Hartree–Fock equations for continuum states and then applied these equations to calculate transitions between the ^3P, ^1D, and ^1S terms of the ground-state configuration of atomic oxygen. As Seaton pointed out, although Hartree–Fock wave functions had previously been used in various continuum state problems,[2-9]

P. G. BURKE ● Department of Applied Mathematics and Theoretical Physics, The Queen's University of Belfast, Belfast BT7 1NN, Northern Ireland   W. EISSNER ● Daresbury Laboratory, Science & Engineering Research Council, Daresbury, Warrington WA4 4AD, England.

no complete justification for the procedure had been given, and the equations used, derived largely by analogy with bound-state equations, did not include the coupling terms required in collision theory. The calculations on atomic oxygen were rapidly extended by Seaton to O II and N II[10] and to N I,[11] providing for the first time quantitative estimates of the electron impact excitation and de-excitation of the forbidden lines of the ground-state configurations for these systems. These results were of fundamental importance in many applications, for example, in the physics of the upper atmosphere[12] and gaseous nebulae[13] which are discussed in other chapters of this book.

The basic theory described by Seaton required the solution of sets of coupled integro-differential equations for the motion of the scattered electron. In the early 1950s, computers were still in their infancy, and it was not possible to solve these equations exactly. Consequently, approximate methods of solution were developed by Seaton. In spite of these restrictions, the results obtained by Seaton in these early papers have stood the test of time remarkably well and in most cases are very close to the exact numerical results obtained in recent years. Because of the limitation of space here, we will concentrate on methods which have been developed for the exact solution of these equations; however, we must not forget that approximate methods of solution, such as the Coulomb–Born approximation and the distorted wave approximation, can give accurate results when the charge on the target ion becomes large or the energy becomes high and are widely used in these situations.

In our later discussion of these coupled integro-differential equations, we will concentrate on developments made in this subject since the review of the numerical solution of the coupled integro-differential equations by Burke and Seaton in 1971.[14] However, it may be of interest to present at this point a brief review of the development of this subject following Seaton's 1953 paper. The first numerical solution of coupled equations in this field was made by Marriott,[15] who used a hand operated desk calculator to solve the two coupled s-wave equations resulting from the inclusion of the $1s$ and $2s$ states in e–H scattering. He developed a noniterative treatment of the exchange integrals suggested by Percival[16] which was analogous to methods used for the treatment of exchange in atomic structure calculations by Hartree.[17] Following this work, Percival and Seaton[18] developed the general theory of e–H excitation, and it became clear that an intensive effort was needed to solve the resultant equations.

Seaton initiated this effort at University College London in late 1956, in which one of the authors of this chapter (P.G.B.) was fortunate to be involved for the first time in atomic collision work, and a start was made on programming these equations for the electronic computers which were then becoming more widely available. This led to two independent general computer programs for e–H scattering. The first on the English Electric Pilot ACE and DEUCE computers in the United Kingdom by Burke, Burke, Percival, and McCarroll,[19,20] which was first used by John[21] to obtain e–H static exchange phase shifts, and the second on the IBM 704 computers in the United States by Burke and Schey[22] who first used this program to predict resonances lying below the $n = 2$ thresholds. At about the same time, Smith[23] had independently generalized and programmed the two-channel calculation of Marriott for arbitrary angular momentum, and he then joined forces with Burke and Schey to obtain $1s$–$2s$ and $1s$–$2p$ excitation cross sections for e–H and e–He$^+$, where the $1s$, $2s$, and $2p$ states were included in the expansion of the total wave function.[24-27] This work involved the solution of sets of four coupled integro-differential equations. Shortly afterward, similar and independent calculations were carried out for e–H scattering by Damburg and Gailitis,[28] Omdivar,[29] and McEachran and Fraser,[30,31] all giving essentially the same results and confirming the correctness of five independent computer programs.

Although work on e–H scattering and on e–H-like ion scattering continued throughout the 1960s using these programs (and continues today using the new generation of programs), attention in the mid-1960s turned increasingly toward the development of programs to solve exactly the equations resulting from an electron incident on an atom or ion with one open $2p$ or $3p$ shell. This work was largely stimulated by unpublished lecture notes by Seaton[32] and by the experience gained in developing programs to solve the e–H problem mentioned above. This led to a program to solve this still rather specialized problem by Smith, Henry, Burke, and Sinfailam,[33-35] while complementary work was being carried out by the University College group led by Seaton.[36,37] Many calculations on first- and second-row atoms and ions involving just one configuration of the target followed from this work.

By the late 1960s it was realized by a number of groups that the time was right for the development of programs that would enable the calculation of low-energy collisions of electrons with any complex atom or ion provided relativistic effects were not too important.[38-42] The combination of new theoretical approaches, new computational methods, and the probability that computers capable of solving at least 20 coupled integrodifferential equations would soon be available all served to stimulate this work. This work has now been brought to fruition and a number of program packages are generally available. These are principally packages developed at University College, London under the direction of Seaton,[43-48] at Queen's University, Belfast,[49-54] at Baton Rouge, Louisiana,[55,56] and at IBM, San Jose.[57–59] All of these packages have now been widely used for many complex atoms and ions.

The next steps in the continuous development of programs, numerical methods, and theory are now well under way. On the programming side, this takes the form of how to make optimum use of the new generation of vector and parallel processing computers such as the CRAY-1.[60] On the method side there is a continual search for numerical approaches that optimize the solution of the coupled integro-differential equations arising in this problem,[61] while on the theoretical side the impetus is in two directions. Firstly, to extend the theory to enable heavy atoms where relativistic effects are important to be considered.[62–65] Secondly, to extend the approaches that have been successful at low energies, where only a few channels are open, to intermediate energies where an infinite number of bound states as well as continuum states can be excited.[66,67]

The rest of this chapter is arranged as follows. In Section 2 we discuss the basic theory and derive from a variational principle the coupled integro-differential equations describing the collision of an electron with a complex atom or ion. We also derive an expression for the cross section and collision strength and discuss recent methods used to include relativistic effects. In Section 3 we describe methods of solving these coupled integro-differential equations. We will limit ourselves in this discussion to the region of configuration space where exchange effects are important, remembering that the solution in the asymptotic region is discussed elsewhere in this book.[68] Finally, in Section 4 we summarize the major computer program packages developed over the last decade to solve this problem.

2. Theory of Electron Collisions with Atoms and Ions

2.1. Expansion of the Collision Wave Function

We commence by considering the scattering of electrons by atoms and ions where relativistic effects can be neglected. We must then solve the Schrödinger equation

$$(H_{N+1} - E)\Psi = 0, \tag{1}$$

for an N-electron target atom or ion where the Hamiltonian, in atomic units, takes the form

$$H_{N+1} = -\frac{1}{2} \sum_{i=1}^{N+1} \left(\nabla_i^2 + \frac{2Z}{r} \right) + \sum_{i>j=1}^{N+1} \frac{1}{r_{ij}}, \tag{2}$$

where Z is the charge on the nucleus. This Hamiltonian is diagonal in both the total orbital angular momentum L and the total spin S, their components in some preferred direction as well as the total parity π. This allows us to write down solutions of Eq. (1) which are eigenstates of these quantum numbers as well as of the energy. If we are not considering ionization, we can expand this wave function in the form

$$\Psi^\Gamma(X_{N+1}) = \mathcal{Q} \sum_{i=1}^{n} \overline{\Phi}_i^\Gamma(x_1 \cdots x_N \hat{\mathbf{r}}_{N+1} \sigma_{N+1}) r_{N+1}^{-1} F_i^\Gamma(r_{N+1})$$

$$+ \sum_{i=1}^{m} \chi_i^\Gamma(x_1 \cdots x_{N+1}) a_i^\Gamma, \tag{3}$$

where $\Gamma \equiv LSM_L M_S \pi$ and will usually be omitted for notational convenience, X_{N+1} represents the space and spin coordinates of all $N+1$ electrons, $\mathcal{Q}$ is an operator that antisymmetrizes the first summation with respect to the interchange of the coordinates of any pair of electrons, $x_i \equiv \mathbf{r}_i \sigma_i$ represents the space and spin coordinates of the ith electron, and the $\overline{\Phi}_i$ are channel functions formed by coupling a target wave function $\Phi_i(x_1 \cdots x_N)$ with the angular and spin parts of the wave function for the $(N+1)$th electron whose reduced radial wave function is $F_i(r_{N+1})$. Finally the χ_i are a set of L^2 integrable antisymmetrized functions. Given a set of basis functions Φ_i and χ_i, our problem is to derive and solve the equations for the unknown functions F_i and the coefficients a_i.

Before deriving these equations we first make some general comments on the physical and mathematical basis for Eq. (3). We include in the first expansion on the right-hand side all target states of interest. By this we mean that we include both the initial and final states in the collision process of interest, as well as other states which are expected to be closely coupled to them during the collision. We may also include some pseudostates in this expansion which approximately represent continuum states of the target which are not included explicitly in this expansion. These pseudostates may be chosen either so that the polarizabilities of the initial and/or final states are accurately included,[69] which is particularly important at low impact energies, or on some more general but less well defined grounds such as making partial allowance for intermediate continuum states during the collision. These states are expected to be particularly important for incident electron energies above the ionization threshold.[70–73]

The second expansion on the right-hand side of Eq. (3), over what we call correlation functions χ_i, performs two roles. Firstly, for computational convenience the radial functions F_i are usually chosen to be orthogonal to the atomic orbitals in the target states Φ_i with the same angular symmetry;

e.g., in e–O scattering the p-wave continuum functions are taken to be orthogonal to the $2p$ orbital in the target. However, this means that the intermediate collision complex with the configuration $1s^2 2s^2 2p^5$, which is expected to play an important role in $^2P^\circ$ scattering, is excluded from the first expansion. This configuration is then one of those that must be included in the second expansion to ensure completeness. The second role of the second expansion is to represent short-range correlation effects. An example of this is in the work of Burke and Taylor[74,75] on e–H and e–He$^+$ scattering where up to 20 Hylleraas-type terms were included in the expansion. For complex atoms and ions it is more convenient to introduce additional contracted pseudo-orbitals with the same range (but more nodes) as the Hartree–Fock orbitals of the target and to construct additional correlation configurations from these.[76–78] Again, like the pseudostates included in the first expansion, these correlation functions represent part of the continuum omitted from the first expansion.

One final point associated with Eq. (3) concerns the role of the antisymmetrization operator $\mathcal{A}$. This clearly imposes the requirements of the Pauli exclusion principle on the wave function. However, it was also pointed out by Castillejo, Percival, and Seaton[79] that if it had not been included electron exchange effects would require continuum terms, involving a singularity in the energy integral, to be included in the first summation over the target states. As well as avoiding this difficulty it can be shown that with the inclusion of $\mathcal{A}$ Eq. (3) satisfies the exact boundary conditions, provided all target states that can be energetically excited are included. This in turn leads to the important property that in this low-energy region, provided that accurate target states are used, the sum of eigenphases converges to its exact value from below as the basis of short-range correlation terms is expanded to completeness.[80–83] This enables accurate results to be systematically obtained using expansion (3) at energies where only a few channels are open.

2.2. Expansion of the Target Wave Function

We have seen that the expansion of the total wave function in Eq. (3) involves an expansion over a set of target states and pseudostates. For complex atoms and ions these states are not known exactly and they are thus usually expanded in the form

$$\Phi_i(X_N) = \sum_j \phi_j(x_1 \cdots x_N) c_{ij}, \tag{4}$$

where the ϕ_j are a given set of antisymmetrized single configuration basis functions belonging to a particular $LS\pi$ state of the target. These basis functions are similar to the basis functions χ_i in Eq. (3) and are usually constructed from the same one-electron radial orbitals $P_{nl}(r)$, the essential difference being that they describe a system of N rather than $N + 1$ electrons.

The radial orbitals $P_{nl}(r)$ used in the construction of the basis functions ϕ_j and χ_j satisfy the orthonormality relation

$$\int_0^\infty P_{nl}(r)P_{n'l}(r)\,dr = \delta_{nn'}, \tag{5}$$

where n labels the principal quantum number and l the orbital angular momentum quantum number of the orbital. These orbitals often include the Hartree–Fock orbitals of the target as well as additional pseudo-orbitals as discussed above. Alternatively, orbitals may be constructed from solutions in a statistical model potential.[43,84] They may be tabulated at a grid of points or they may be expressed in analytical form as a sum of Slater type orbitals (STOs)

$$P_{nl}(r) = \sum_j b_j \frac{(2\xi_j)^{k_j + 1/2}}{\sqrt{(2k_j)!}}\, r^{k_j} e^{-\xi_j r}, \tag{6}$$

where $k_j \geqslant l + 1$ and the coefficients b_j, k_j, and ξ_j depend on n and l.

The configurations ϕ_j can then be constructed from these one-electron radial orbitals and their corresponding angular and spin components. The coefficients c_{ij} in Eq. (4) are obtained by diagonalizing the N electron Hamiltonian H_N defined by Eq. (2) with $N + 1$ replaced by N, so that

$$\langle \Phi_i | H_N | \Phi_j \rangle = \delta_{ij} E_i^N. \tag{7}$$

Usually only a few states with the lowest energies for each target symmetry $L_i S_i \pi_i$ of interest are retained in Eq. (3). States with higher energies are unphysical, representing in some average way the continuous spectrum of H_N, and are usually omitted.

As already mentioned, some pseudostates may also be included in Eq. (3). The construction of pseudostates representing polarizabilities of a complex atom or ion has been discussed by Vo Ky Lan, Le Dourneuf, Burke, and Mitchell.[85,86]

The success of any practical collision calculation for a complex atom or ion depends to a very large extent on the choice of the radial orbitals P_{nl} and the configurations included in the expansion of the target states and pseudostates. In Section 4 we will discuss programs that allow these orbitals and states to be calculated.

2.3. Variational Principles and the Derivation of the Coupled Integro-Differential Equations

In this section we derive the Kohn variatonal principle[87,88] satisfied by the trial wave function defined by Eq. (3). We then show that coupled integro-differential equations satisfied by the functions $F_i(r)$ follow from this principle.

We start by considering the boundary conditions satisfied by the functions $F_i(r)$. In order to do this we must introduce a second index j on these functions which defines the channel in which the wave is incident. Another way of explaining the need for this additional index is [see Eq. (27)] that these radial functions satisfy n coupled second-order integro-differential equations which from the general theory of differential equations have n linearly independent solutions that are zero at the origin. There are several representations used for these solutions, corresponding to taking different linearly independent combinations of these solutions. The S matrix is defined by the following boundary conditions:

$$F_{ij}(r) \underset{r \to 0}{\sim} r^{l_i + 1} \tag{8}$$

and

$$\left. \begin{aligned} F_{ij}(r) &\underset{r \to \infty}{\sim} k_i^{-1/2} \left(e^{-i\theta_i(r)} \delta_{ij} - e^{i\theta_i(r)} S_{ij} \right), && k_i^2 > 0 \\ F_{ij}(r) &\underset{r \to \infty}{\sim} 0, && k_i^2 < 0 \end{aligned} \right\} \quad j = 1, \ldots, n_a \tag{9}$$

where

$$k_i^2 = 2(E - E_i^N). \tag{10}$$

E and E_i^N are defined by Eqs. (1) and (7), respectively, and

$$\theta_i(r) = k_i r - \tfrac{1}{2} l_i \pi - \eta_i \ln 2 k_i r + \sigma_i,$$

$$\eta_i = -(Z - N)/k_i, \tag{11}$$

$$\sigma_i = \arg \Gamma(l_i + 1 + i\eta_i),$$

l_i being the orbital angular momentum of the scattered electron in the ith channel. In general, not all the channels will be open (i.e., have $k_i^2 > 0$). We will assume that of the n channels defined by Eq. (3), n_a are open and they are ordered so that they occur first. It is easy to see that of the n linearly independent solutions which are zero at the origin, n_a satisfy the physical requirement that the closed-channel solutions vanish asymptotically as in Eq. (9). The remaining $n_b = n - n_a$ solutions have nonphysical exponentially increasing solutions in the closed channels and are not considered here.

In numerical applications it is convenient to use wave functions that are real rather than complex. We can achieve this by taking linear combinations of the independent solutions defined by Eq. (9) giving solutions that satisfy the real asymptotic boundary conditions

$$
\left.
\begin{aligned}
F_{ij}(r) &\underset{r\to\infty}{\sim} k_i^{-1/2}(\sin\theta_i(r)\delta_{ij} + \cos\theta_i(r)K_{ij}), \quad k_i^2 > 0 \\
F_{ij}(r) &\underset{r\to\infty}{\sim} 0, \quad k_i^2 < 0,
\end{aligned}
\right\} \quad j = 1, \ldots, n_a
$$

$$(12)$$

where for notational convenience we have used the same function F to denote both asymptotic forms of Eqs. (9) and (12). By comparing Eqs. (9) and (12) we find that the two $n_a \times n_a$ matrices S_{ij} and K_{ij} are related by the matrix equation

$$
\mathbf{S} = \frac{1 + i\mathbf{K}}{1 - i\mathbf{K}}.
$$

$$(13)$$

It is easy to show from flux conservation and time-reversal invariance that the S matrix is unitary and symmetric and thus the K matrix is real and symmetric. The symmetric K matrices for each $LS\pi$ combination contains the complete information on the collision processes in the space spanned by the basis functions Φ_i and χ_i, and it is these quantities that we wish to calculate and from which the cross sections can be obtained.

Of course, the asymptotic forms (9) and (12) are not the only ones that can be written down. Indeed, by replacing $\theta_i(r)$ by $\theta_i(r) + \tau$ in Eq. (12), where τ is some constant,[89] we can define a new τ-dependent K matrix. In circumstances where the incident energy is such that the K matrix is close to a pole when $\tau = 0$, this feature can be used to define a new $K(\tau)$ matrix which is no longer nearly singular. This has some advantages in certain numerical methods used for calculating the K matrices. Of course, it is clear that, given the K matrix with $\tau = 0$, we can in principle calculate the K matrix for any nonzero value of τ by just recombining the solutions defined by Eq. (12).

We are now in a position to derive the required variational principle. We consider the following integral over the space and spin coordinates of all $N + 1$ electrons:

$$
I_{ij} = \int \Psi_i^{\Gamma*}(X_{N+1})(H_{N+1} - E)\Psi_j^{\Gamma}(X_{N+1})\,dX_{N+1},
$$

$$(14)$$

where both Ψ_i^{Γ} and Ψ_j^{Γ} are defined by Eq. (3) and where the additional indices i and j correspond to the second index we have introduced on the functions F_{ij} [and also by implication on the coefficient a_i in Eq. (3)]. Equation (14) can be reduced using Eq. (7). We find after some considerable manipulation that

$$I_{ij} = \int_0^{\infty} \sum_{kl} F_{ki}(r) \left\{ -\frac{1}{2} \left[\frac{d^2}{dr^2} - \frac{l_k(l_k+1)}{r^2} + \frac{2Z}{r} + k_k^2 \right] F_{kj}(r) \delta_{kl} \right.$$

$$\left. + V_{kl}(r) F_{lj}(r) + \int_0^{\infty} W_{kl}(r,r') F_{lj}(r') \, dr' \right\} dr, \quad (15)$$

which can be written in matrix notation as

$$\mathbf{I} = \int_0^{\infty} \mathbf{F}^T \mathbf{L} \mathbf{F} \, dr, \quad (16)$$

where $\mathbf{L}$ is the integro-differential operator in curly brackets in (15). In Eq. (15) we have defined the direct potential $V_{kl}(r)$ by

$$V_{kl}(r_{N+1}) = \int \overline{\Phi}_k^*(x_1 \cdots x_N \hat{\mathbf{r}}_{N+1} \sigma_{N+1}) \sum_{i=1}^{N} \frac{1}{r_{iN+1}},$$

$$\times \overline{\Phi}_l(x_1 \cdots x_N \hat{\mathbf{r}}_{N+1} \sigma_{N+1}) \, dx_1 \cdots dx_N \, d\hat{\mathbf{r}}_{N+1} \, d\sigma_{N+1}, \quad (17)$$

while the nonlocal potential operator $W_{kl}(r,r')$ arises from two types of terms. The first corresponds to the exchange term in the first expansion in Eq. (3), which gives

$$-N \int \overline{\Phi}_k^*(x_1 \cdots x_N \hat{\mathbf{r}}_{N+1} \sigma_{N+1}) r_{N+1}^{-1} r_{N\,N+1}^{-1} \overline{\Phi}_l(x_1 \cdots x_{N-1} x_{N+1} \hat{\mathbf{r}}_N \sigma_N)$$

$$\times r_N^{-1} F_{lj}(r_N) \, dx_1 \cdots dx_N \, r_{N+1}^2 \, d\hat{\mathbf{r}}_{N+1} \, d\sigma_{N+1}. \quad (18)$$

The second arises from the correlation functions χ_i in Eq. (3). We can choose these functions to diagonalize H_{N+1} with eigenvalues ε_i, and we define the radial functions

$$U_{ks}(r_{N+1}) = \int \overline{\Phi}_k^*(x_1 \cdots x_N \hat{\mathbf{r}}_{N+1} \sigma_{N+1}) r_{N+1}^{-1}$$

$$\times (H_{N+1} - E) \chi_s(X_{N+1}) \, dx_1 \cdots dx_N \, r_{N+1}^2 \, d\hat{\mathbf{r}}_{N+1} \, d\sigma_{N+1}. \quad (19)$$

The second expansion in Eq. (3) then gives

$$-\sum_s \int_0^\infty U_{ks}(r_{N+1}) \frac{1}{\epsilon_s - E} U_{ls}(r_N) F_{lj}(r_N) dr_N. \qquad (20)$$

Clearly both Eqs. (18) and (20), which together form the last term in Eq. (15), are integral operators acting on the radial function $F_{lj}(r)$.

The explicit form of Eqs. (17), (18), and (20) have been written down in a few simple cases, e.g., for e–H scattering[18] and for electron scattering by first-row atoms and ions involving a single $1s^2 2s^2 2p^q$ configuration.[33–35] However, for complex atoms and ions of most interest it is only practical to construct these potentials using a computer. Nevertheless, certain general statements can be made about the form of these potentials. It follows from Eq. (17) that the direct potential behaves asymptotically as

$$V_{kl}(r) \underset{r \to \infty}{\sim} \frac{N}{r} \delta_{kl} + \sum_{\lambda=1}^{\lambda_{\max}} a_{kl}^\lambda r^{-\lambda-1}. \qquad (21)$$

The first term corresponds to the screening of the nuclear charge by the target electrons. It is this term when combined with the term $2Z/r$ in Eq. (15) which leads to the logarithmic term in θ_i in Eq. (11). The second terms are long-range dipole, quadrupole, and higher multipole potentials coupling the channels. These potentials contribute particularly to allowed transitions and can cause severe difficulties in the solution of the coupled integro-differential equations. This problem is taken up in another chapter in this book by Norcross.[68] On the other hand, the exchange potential while severely complicating the solution of these equations vanishes exponentially at infinity, being limited by the range of the target states Φ_i and the correlation functions χ_i included in expansion (3).

We now consider variations of the integral I due to arbitrary variations of the functions F about the exact solution satisfying the boundary conditions defined by Eqs. (8) and (12). These variations satisfy

$$\delta F(0) = 0, \qquad (22)$$

$$\delta F(r) \underset{r \to \infty}{\sim} k^{-1/2} \cos \theta(r) \delta K$$

where k and θ are diagonal matrices with non-zero elements k_i and θ_i. The corresponding variation in I is to first order

$$\delta I = \int_0^\infty (\delta F^T L F + F^T L \delta F) dr. \qquad (23)$$

Using Green's theorem and the symmetry of the direct and exchange potentials, we find that

$$\delta \mathbf{I} = 2 \int_0^\infty \delta \mathbf{F}^T \mathbf{L} \mathbf{F} \, dr - \tfrac{1}{2} \left(\mathbf{F}^T \frac{d}{dr} \delta \mathbf{F} - \delta \mathbf{F}^T \frac{d}{dr} \mathbf{F} \right) \Bigg|_{r=0}^{r=\infty}. \tag{24}$$

The surface term can be simply evaluated using the boundary condition satisfied by the function $\mathbf{F}$ and its variation $\delta \mathbf{F}$. We then obtain

$$\delta \mathbf{I} = 2 \int_0^\infty \delta \mathbf{F}^T \mathbf{L} \mathbf{F} \, dr + \tfrac{1}{2} \delta \mathbf{K}. \tag{25}$$

It follows that for arbitrary variations $\delta \mathbf{F}$ the variational principle[88]

$$\delta(\mathbf{I} - \tfrac{1}{2} \mathbf{K}) = 0 \tag{26}$$

leads immediately to the requirement that $\mathbf{L}\mathbf{F} = 0$. Written out explicitly using Eqs. (15) and (16), these equations become, after dropping the second index on $\mathbf{F}$

$$\left[\frac{d^2}{dr^2} - \frac{l_i(l_i + 1)}{r^2} + \frac{2Z}{r} + k_i^2 \right] F_i(r)$$

$$= 2 \sum_{j=1}^n \left[V_{ij}(r) F_j(r) + \int_0^\infty W_{ij}(r, r') F_j(r') \, dr' \right]$$

$$+ \sum_{nl} \lambda_{i \, nl} P_{nl}(r) \delta_{ll_i}, \qquad i = 1, \ldots, n, \tag{27}$$

where in order to impose the constraint mentioned earlier—that the function $F_i(r)$ should be orthogonal to all the target orbitals $P_{nl}(r)$ with the same symmetry—we have included a term in Eq. (27) involving Lagrange multipliers $\lambda_{i \, nl}$ which are chosen so that

$$\int_0^\infty P_{nl}(r) F_i(r) \, dr = 0 \qquad \text{for} \quad l = l_i. \tag{28}$$

Equations (27) are the required coupled integro-differential equations that must be solved.

The variational principle expressed in Eq. (26) clearly depends on the asymptotic form (12) chosen for the functions $F_i(r)$. However, we have seen that this asymptotic form is not unique, indeed an infinity of different variational principles can be constructed by taking different linear combinations of the n_a linearly independent solutions of Eq. (27). However, all these variational principles are satisfied by the exact solution of Eq. (27). Their main interest lies in constructing different approximate solutions. It follows from Eq. (26) that if $\mathbf{F}^t$ is an approximate solution to Eq. (27), and $\mathbf{K}^t$ is the corresponding approximate K matrix, then an improved K matrix, correct to second order in the error in the wave function, is given by the Kohn corrected K matrix[88]

$$\mathbf{K}^{Kohn} = \mathbf{K}^t - 2\mathbf{I}^t, \tag{29}$$

where $\mathbf{I}^t$ is calculated using the approximate solutions $\mathbf{F}^t$.

2.4. Derivation of the Cross Section

Once the K matrices and the corresponding S matrices are calculated by solving Eq. (27) for all $LS\pi$ combinations of importance, the cross section can easily be obtained using the method of Blatt and Biedenharn[90] and Lane and Thomas.[91]

We first expand the initial state, which we assume for simplicity is an electron moving in the field of a neutral target, in partial waves. Now for a plane wave, propagating in the z direction so that $kz = kr\cos\vartheta$

$$e^{ikz} = (4\pi)^{1/2} \sum_{l=0}^{\infty} i^l (2l + 1) \left(\frac{\pi}{2kr} \right)^{1/2} J_{l+1/2}(kr) Y_{l0}(\vartheta, \varphi)$$

$$\underset{r\to\infty}{\sim} \frac{(4\pi)^{1/2}}{2ikr} \sum_{l=0}^{\infty} i^l (2l + 1)^{1/2} (e^{i\theta(r)} - e^{-i\theta(r)}) Y_{l0}(\vartheta, \varphi), \tag{30}$$

where $J_n(\rho)$ are Bessel functions and $Y_{lm}(\vartheta, \varphi)$ are spherical harmonics. Our initial state, consisting of an electron in a plane-wave state incident on an atomic state denoted by the quantum numbers $\alpha_i L_i S_i M_{L_i} M_{S_i} \pi_i$ is then

$$\Phi_i(x_1 \cdots x_N) e^{ik_i z_{N+1}} \chi_{(1/2)\mu_i}(\sigma_{N+1})$$

$$\underset{r_{N+1}\to\infty}{\sim} \frac{i\pi^{1/2}}{k_i r_{N+1}} \sum_{LSl_i} (L_i M_{L_i} l_i 0 | LM_L)(S_i M_{S_i} \tfrac{1}{2} \mu_i | SM_S) i^{l_i} (2l_i + 1)^{1/2}$$

$$\times \overline{\Phi}_i(x_1 \cdots x_N \hat{\mathbf{r}}_{N+1} \sigma_{N+1})(e^{-i\theta_i(r)} - e^{i\theta_i(r)}), \tag{31}$$

where $\chi_{(1/2)\mu_i}(\sigma_{N+1})$ is the spin function of the incident electron and the quantities $(L_i M_{L_i} l_i 0 | LM_L)$ and $(S_i M_{S_i} \tfrac{1}{2} \mu_i | SM_S)$ are the usual Clebsch–Gordan coefficients.

We now equate the ingoing waves in Eqs. (3) and (30), where we assume that the radial functions F_i have the asymptotic form defined by Eq. (9). If we write our total wave function as

$$\Psi = \Psi_{\text{inc}} + \Psi_{\text{reac}}, \tag{32}$$

where Ψ_{inc} is defined by Eq. (30), then

$$\Psi_{\text{reac}} \underset{r_{N+1} \to \infty}{\sim} \sum_{\substack{LS\pi l_i l_i \\ \alpha_i L_i S_i}} \frac{i\pi^{1/2}}{k_i^{1/2} k_j^{1/2} r_{N+1}} (L_i M_{L_i} l_i 0 \,|\, L M_L)(S_i M_{S_i} \tfrac{1}{2} \mu_i \,|\, S M_S)$$

$$\times i^{l_i}(2l_i+1)^{1/2} \overline{\Phi}_j(x_1 \cdots x_N \hat{\mathbf{r}}_{N+1} \sigma_{N+1}) e^{i\theta_j} \big[\delta_{ji} - S_{ji}^{LS\pi} \big], \tag{33}$$

where we have now shown explicitly the dependence of the S matrix elements on L, S, and π. We now calculate the scattering amplitude for exciting a particular atomic state $\alpha_j L_j S_j M_{L_j} M_{S_j} \pi_j$ with the scattered electron moving in the direction ϑ, φ by expanding the channel functions $\overline{\Phi}_j$ in Eq. (33) in terms of target states Φ_j. We obtain

$$\Psi_{\text{reac}} \underset{r_{N+1} \to \infty}{\sim} \sum_{\alpha_j L_j S_j M_{L_j} M_{S_j} \mu_i} \Phi_j(x_1 \cdots x_N) \chi_{(1/2)\mu_j}(\sigma_{N+1}) f_{ji}(\hat{\mathbf{r}}_{N+1}) e^{ik_j r_{N+1}}, \tag{34}$$

where the scattering amplitude f_{ji} is defined by

$$f_{ji}(\vartheta, \varphi) = -i\left(\frac{\pi}{k_i k_j}\right)^{1/2} \sum_{\substack{LS\pi \\ l_i l_j m}} i^{l_i - l_j}(2l_i+1)^{1/2}(L_i M_{L_i} l_i m_i \,|\, L M_L)$$

$$\times (S_i M_{S_i} \tfrac{1}{2} \mu_i \,|\, S M_S)(L_j M_{L_j} l_j m \,|\, L M_L)$$

$$\times (S_j M_{S_j} \tfrac{1}{2} \mu_j \,|\, S M_S) T_{ji}^{LS\pi} Y_{l_j m}(\vartheta, \varphi), \tag{35}$$

where we have introduced the T matrix by the equation

$$T_{ji}^{LS\pi} = S_{ji}^{LS\pi} - \delta_{ji}. \tag{36}$$

The differential cross section for a transition from state $\alpha_i L_i S_i$ to state $\alpha_j L_j S_j$ is then

$$\frac{d\sigma_{ji}}{d\Omega} = \frac{k_j}{k_i} |f_{ji}(\vartheta, \varphi)|^2, \tag{37}$$

while the total cross section is obtained by averaging over initial spin states, summing over final spin states and integrating over all directions of the scattered electron giving

$$\sigma_{tot}(\alpha_i L_i S_i \rightarrow \alpha_j L_j S_j) = \frac{\pi}{k_i^2} \sum_{\substack{LS\pi \\ l_i l_j}} \frac{(2L+1)(2S+1)}{2(2L_i+1)(2S_i+1)} |T_{ji}^{LS\pi}|^2 \qquad (38)$$

in units of a_0^2.

This result is also valid for ions for inelastic transitions although, as is well known, the elastic scattering cross section in this case is infinite due to the Coulomb singularity in the angular distribution in the forward direction.

A further quantity, called collision strength $\Omega(i, j)$, first introduced by Hebb and Menzel,[92] is often used in applications. In terms of the cross section it is defined by the equation

$$\Omega(i, j) = \frac{(2L_i+1)(2S_i+1)k_i^2}{\pi} \sigma_{tot}(i \rightarrow j) = \frac{\omega_i k_i^2}{\pi} \sigma_{tot}(i \rightarrow j) \qquad (39)$$

where ω_i is the statistical weight of the target state. The collision strength has the advantage of being both dimensionless and symmetric in i and j. The modifications to these formulae, which occur when fine-structure effects are considered, are given in Section 2.5.1.

2.5. Inclusion of Relativistic Effects

As the atomic number of the target increases, relativistic effects become progressively more important. In this section we consider how these effects can be included in low-energy electron collisions. There are two main ways in which relativistic effects play a role. Firstly, there is a direct effect which may be considered as a distortion which the strong nuclear potential, partially screened by the electronic charge distribution of the target, induces in the wave function describing the motion of the scattered electron. Secondly, there is an indirect effect caused by the change in the charge distribution of the target due to relativity. Our basic problem is to show how these effects can be included as well as electron exchange, correlation, and channel coupling effects which have already been considered.

There are several ways in which relativistic effects can be included in the electron–atom collision problem. For intermediate weight atoms and ions, relativistic effects can be accounted for by adding the additional terms from the Breit–Pauli Hamiltonian to the nonrelativistic Hamiltonian defined by Eq. (2). If these terms are sufficiently small they can be included using perturbation theory; however, as they become more important for heavier targets the direct solution of the Schrödinger equation with the modified Hamiltonian must be carried out. This approach has been considered by Jones,[62] Scott and Burke,[63] and Sinfailam and Baylis.[93,94] A fundamental question which must be asked about this approach is whether or not it can give reliable results for heavy atoms such as Hg.

The alternative approach, which is clearly appropriate for heavy atoms, is to base the collision problem on the Dirac equation. Carse and Walker[95,96] have derived the coupled integro-differential equations for e–H scattering using this approach, and Walker[96] has obtained results for hydrogenlike ions with the nuclear charge lying between 2 and 100. In the case of complex atoms and ions, the theory was developed by Chang,[97] within an R-matrix framework (see Section 3.3), and applied to e–Ne⁺ scattering.[98] However, no further results were reported, so recently Norrington and Grant[65] have reconsidered this problem and developed a general program package that has given results in good agreement with earlier nonrelativistic calculations on Ne⁺.

Finally, we mention methods that can be used when relativistic effects are small so that the fine-structure intervals between the levels of the target can be neglected in a first approximation. In this case, the calculation is carried out in LS coupling using the nonrelativistic Hamiltonian. The corresponding nonrelativistic K matrices are then recoupled to give K matrices and the corresponding cross sections between the fine-structure levels. This is the basis of a program written by Saraph[99,100] which has had wide use in electron–ion collisions. This approach, combined with an effective range expansion given by Ross and Shaw,[101] was used by Burke and Mitchell[102] to give results close to fine-structure thresholds for e–Cs scattering.

2.5.1. Use of the Breit–Pauli Hamiltonian. The Breit–Pauli Hamiltonian has been widely discussed in the literature (e.g., by Bethe and Salpeter[103] and by Akhiezer and Berestetsky[104]). Here we will just present the basic formulas. The Hamiltonian for an electron incident on an N-electron target is given by

$$H_{N+1}^{BP} = H_{N+1}^{NR} + H_{N+1}^{REL} \qquad (40)$$

where the nonrelativistic Hamiltonian H_{N+1}^{NR} is defined by Eq. (2) and where H_{N+1}^{REL} consists of one- and two-body relativistic terms resulting from the reduction of the Dirac equation and the Breit interaction to Pauli form.

The one-body operators are

$$H_{N+1}^{\text{mass}} = -\frac{1}{8}\alpha^2 \sum_{i=1}^{N+1} \nabla_i^4, \qquad \text{mass-correction term} \qquad (41a)$$

$$H_{N+1}^{\text{D}_1} = -\frac{1}{8}\alpha^2 Z \sum_{i=1}^{N+1} \nabla_i^2\left(\frac{1}{r_i}\right), \qquad \text{one-body Darwin term} \qquad (41b)$$

$$H_{N+1}^{\text{so}} = \frac{1}{2}\alpha^2 Z \sum_{i=1}^{N+1} r_i^{-3}(\mathbf{l}_i \cdot \mathbf{s}_i), \qquad \text{spin–orbit interaction} \qquad (41c)$$

The two-body operators are

$$H_{N+1}^{\text{soo}} = -\frac{1}{2}\alpha^2 \sum_{i\neq j}^{N+1} \left(\frac{\mathbf{r}_{ij}}{r_{ij}^3} \times \mathbf{p}_i\right) \cdot (\mathbf{s}_i + 2\mathbf{s}_j), \qquad \text{mutual spin–orbit and}$$

$$\text{spin–other orbit terms}$$

$$(42a)$$

$$H_{N+1}^{\text{ss}} = -\alpha^2 \sum_{i<j=1}^{N+1} \frac{1}{r_{ij}^3}\left[\mathbf{s}_i \cdot \mathbf{s}_j - \frac{3(\mathbf{s}_i \cdot \mathbf{r}_{ij})(\mathbf{s}_j \cdot \mathbf{r}_{ij})}{r_{ij}^2}\right], \qquad \text{spin–spin term} \qquad (42b)$$

$$H_{N+1}^{\text{oo}} = -\frac{1}{2}\alpha^2 \sum_{i<j=1}^{N+1}\left[\frac{\mathbf{p}_i \cdot \mathbf{p}_j}{r_{ij}} + \frac{\mathbf{r}_{ij}(\mathbf{r}_{ij} \cdot \mathbf{p}_i) \cdot \mathbf{p}_j}{r_{ij}^3}\right], \qquad \text{orbit–orbit term}$$

$$(42c)$$

$$H_{N+1}^{\text{D}_2} = \frac{1}{4}\alpha^2 \sum_{i<j=1}^{N+1} \nabla_i^2\left(\frac{1}{r_{ij}}\right), \qquad \text{two-body Darwin term}$$

$$(42d)$$

$$H_{N+1}^{\text{ssc}} = -\frac{8\pi\alpha^2}{3} \sum_{i<j=1}^{N+1} (\mathbf{s}_i \cdot \mathbf{s}_j)\delta(\mathbf{r}_{ij}), \qquad \text{spin–contact term}$$

$$(42e)$$

We must now solve the modified Schrödinger equation defined by Eq. (1) with H_{N+1} replaced by H_{N+1}^{BP}. We expand the total wave function, as in

Eq. (3), in terms of an expansion over target states plus an expansion over correlation functions; however, L and S and their z components are no longer conserved. Instead, it is often convenient[63,99,100] to adopt the pair coupling scheme introduced by Racah.[105] In this scheme we have

$$\mathbf{J}_i + \mathbf{l} = \mathbf{K}, \qquad \mathbf{K} + \mathbf{s} = \mathbf{J}, \tag{43}$$

where $\mathbf{J}_i$ is the total angular momentum of the target, $\mathbf{l}$ is the orbital angular momentum of the scattered electron, $\mathbf{s}$ is the spin of the scattered electron, and $\mathbf{J}$ is the total angular momentum. The conserved quantities in the collision are now JM_J and π. This scheme is particularly useful for intermediate weight atoms and ions where the relativistic effects are not too large since the transformation from LS coupling involves the relatively simple recoupling coefficient

$$\langle ((L_iS_i)J_i,l)K,\tfrac{1}{2};JM_J \,|\, (L_il)L,(S_i\tfrac{1}{2})S;JM_J \rangle$$

$$= \left[(2J_i + 1)(2L + 1)(2K + 1)(2S + 1) \right]^{1/2}$$

$$\times W(LlS_iJ_i;L_iK)W(LJS_i\tfrac{1}{2};SK), \tag{44}$$

where the W coefficients are standard Racah coefficients.

An alternative coupling scheme which has been considered[62] is the intermediate coupling scheme defined by

$$\mathbf{l} + \mathbf{s} = \mathbf{j}, \qquad \mathbf{J}_i + \mathbf{j} = \mathbf{J}. \tag{45}$$

In this case the transformation coefficient from LS coupling is

$$\langle (S_iL_i)J_i,(\tfrac{1}{2}l)j;JM_J \,|\, (S_i\tfrac{1}{2})S,(L_il)L;JM_J \rangle$$

$$= \left[(2J_i + 1)(2j + 1)(2S + 1)(2L + 1) \right]^{1/2}$$

$$\times (SM_SLM_L|JM_J) \begin{Bmatrix} S_i & L_i & J_i \\ \tfrac{1}{2} & l & j \\ S & L & J \end{Bmatrix} \tag{46}$$

which now involves the Wigner 9-j coefficient.[106]

As in Section 2.3, we can now derive a set of coupled integro-differential equations, analogous to Eq. (27), from the Kohn variational principle.[88] One essential difference from Eq. (27) is that since L and S are no longer conserved, the number of coupled equations for the same number of target states increases substantially. This, coupled with the fact that the atoms and ions considered are more complex and must also be treated relativistically, means that the solution of these equations is far more complex.

A further problem arises from the form of the potential near the origin. The variational principle leads to a term of the form $\alpha^2 V^2 r$ arising from the mass-correction term. This term also occurs in the second-order equation form of Dirac's equation given by Mott and Massey[107] and Brown and Bauer.[108] The relativistic correction to the static potential in this case is

$$V_{\text{rel}} = -\frac{\alpha^2}{2}\left[V^2(r) - \frac{\kappa V'(r)}{r\left[\gamma + 1 + \alpha^2 V(r)\right]} \right.$$

$$\left. + \frac{1}{2}\frac{V''(r)}{\left[\gamma + 1 + \alpha^2 V(r)\right]} - \frac{3}{4}\frac{\alpha^2 V'^2(r)}{\left[\gamma + 1 + \alpha^2 V(r)\right]^2} \right], \quad (47)$$

where $\gamma = (1 - \alpha^2 k^2)^{-1/2}$, the symbol κ is defined by

$$\kappa = j + \tfrac{1}{2}, \qquad \text{when} \quad l = j + \tfrac{1}{2}$$

$$\kappa = -j - \tfrac{1}{2}, \qquad \text{when} \quad l = j - \tfrac{1}{2} \qquad (48)$$

and where the primes denote the derivative with respect to r. Solutions of the radial equations including this term do not have the usual r^{l+1} behavior at the origin as in Eq. (8), but instead behave as

$$F_\kappa(r) \underset{r\to 0}{\sim} r^\lambda, \qquad (49)$$

where

$$\lambda = \tfrac{1}{2} + (\kappa^2 - \alpha^2 Z^2)^{1/2}. \qquad (50)$$

Thus the indicial equation is not the same for spin-up as for spin-down. In practical calculations, however, this behavior has not caused difficulty.

We conclude this section by stating the modified form of the cross sections and collision strengths in the new coupling schemes. In the pair-coupling scheme we have

$$\sigma_{\text{tot}}\left(\alpha_i J_i \to \alpha_j J_j\right) = \frac{\pi}{2k_i^2(2J_i + 1)} \sum_{\substack{J\pi \\ \kappa_i \kappa_j l_i l_j}} (2J + 1)\left|T_{ji}^{J\pi}\right|^2 \qquad (51)$$

in units of a_0^2, and

$$\Omega(i, j) = \frac{(2J_i + 1)k_i^2}{\pi} \sigma_{\text{tot}}(i \to j), \tag{52}$$

while in the intermediate coupling scheme we have

$$\sigma_{\text{tot}}(\alpha_i J_i \to \alpha_j J_j) = \frac{\pi}{2k_i^2(2J_i + 1)} \sum_{\substack{J\pi \\ j_i j_j l_i l_j}} (2J + 1)|T_{ji}^{J\pi}|^2 \tag{53}$$

and $\Omega(i, j)$ is again given by Eq. (52). Although Eqs. (51) and (53) are formally similar, the respective T matrices must be calculated in the appropriate representations corresponding to Eqs. (43) and (45).

  2.5.2. *Use of the Dirac Hamiltonian.* The relativistic calculations of atomic structure based on the Dirac Hamiltonian have been described in considerable detail by Grant.[109,110] Here we will limit ourselves to discussing the significant differences that occur in treating the electron collision problem. This problem has been discussed by Chang.[97,98]

  The Dirac Hamiltonian for the collision of an electron with an N-electron target is defined by

$$H_{N+1}^D = \sum_{i=1}^{N+1} \left(c\boldsymbol{\alpha} \cdot \mathbf{p}_i + \beta' c^2 - \frac{Z}{r_i} \right) + \sum_{i>j=1}^{N+1} \frac{1}{r_{ij}}, \tag{54}$$

where $\beta' = \beta - 1$, α and β are the usual Dirac operators, and c is the velocity of light in vacuum. Quantum electrodynamic (QED) corrections to the electron–electron interaction considered by Breit, as well as other QED corrections, are often included in atomic structure calculations[111] but have not yet been considered in collision work.

  The expansion of the total wave function for a particular $JM_J\pi$ combination takes the general form of Eq. (3). However, now both the bound orbitals in the target and correlation functions and the orbitals representing the scattered electron are given by Dirac orbitals. These are defined in terms of large and small components $P(r)$ and $Q(r)$ by

$$\phi(\mathbf{r}, \sigma) = \frac{1}{r} \left[\begin{array}{c} P_a(r)\chi_{\kappa m}(\hat{\mathbf{f}}, \sigma) \\ Q_a(r)\chi_{-\kappa m}(\hat{\mathbf{f}}, \sigma) \end{array} \right] \tag{55}$$

for the bound orbitals and

$$F(\mathbf{r}, \sigma) = \frac{1}{r} \left[\begin{array}{c} P_c(r)\chi_{\kappa m}(\hat{\mathbf{f}}, \sigma) \\ Q_c(r)\chi_{-\kappa m}(\hat{\mathbf{f}}, \sigma) \end{array} \right] \tag{56}$$

for the continuum orbitals, where $a = n\kappa m$, $c = k\kappa m$, and the spin-angle function $\chi_{\kappa m}$ is defined by

$$\chi_{\kappa m}(\hat{\mathbf{r}}, \sigma) = \sum_{m_l m_i} (l m_l \tfrac{1}{2} m_i \mid jm) Y_{l m_l}(\vartheta, \varphi) \chi_{(1/2) m_i}(\sigma). \tag{57}$$

The angular state is defined by the κ quantum number, already defined by Eq. (48), or alternatively by the quantum numbers (l, j). The bound orbitals, which we assume are known from a previous atomic structure calculation, form an orthonormal set so that

$$\int_0^\infty \left[P_{a_i}(r) P_{a_j}(r) + Q_{a_i}(r) Q_{a_j}(r) \right] dr = \delta_{ij}, \tag{58}$$

where a_i and a_j differ only in the n quantum number, while the continuum orbitals, which we wish to calculate, are assumed orthogonal to all the bound orbitals

$$\int_0^\infty \left[P_{a_i}(r) P_{c_j}(r) + Q_{a_i}(r) Q_{c_j}(r) \right] dr = 0 \tag{59}$$

again where a_i and c_j differ only in the radial quantum numbers. These orbitals are then combined to define the total wave function using j-j coupling.

We can now derive coupled integro-differential equations for the unknown radial functions $P_c(r)$ and $Q_c(r)$ in a similar way to our procedure in Section 2.3, except that we now use the Dirac Hamiltonian rather than the nonrelativistic Hamiltonian, and our coupling scheme is diagonal in $JM_J\pi$ rather than in $LM_L SM_S\pi$. These equations are coupled first-order equations instead of the coupled second-order equations obtained nonrelativistically.

As in the nonrelativistic case we can neglect the exchange and correlation terms beyond a certain radius, and the coupled equations then simplify to the form

$$P_i' + \frac{\kappa_i}{r} P_i - \frac{1}{c}\left(2c^2 + \epsilon_i + \frac{Z}{r} \right) Q_i = -\frac{1}{c} \sum_{j=1}^{n} \sum_{\lambda=1}^{\lambda_{\max}} a_{ij}^\lambda r^{-\lambda-1} Q_j,$$

$$Q_i' - \frac{\kappa_i}{r} Q_i + \frac{1}{c}\left(\epsilon_i + \frac{Z}{r} \right) P_i = \frac{1}{c} \sum_{j=1}^{n} \sum_{\lambda=1}^{\lambda_{\max}} a_{ij}^\lambda r^{-\lambda-1} P_i, \tag{60}$$

where ϵ_i is the channel energy and we have omitted the subscript c since we are now only concerned with the continuum orbitals. The long-range potential coefficients a_{ij}^λ are analogous to those considered in Eq. (21). For low-energy scattering and small residual charge $Z - N$, these equations can be approximated by their nonrelativistic limit given by

$$P_i'' - \frac{\kappa_i(\kappa_i + 1)}{r^2} P_i + 2\left(\epsilon_i + \frac{Z}{r}\right)P_i$$

$$= 2\sum_{j=1}^{n}\sum_{\lambda=1}^{\lambda_{max}} a_{ij}^\lambda r^{-\lambda-1}P_j(r), \qquad i = 1, \ldots, n, \qquad (61)$$

which have the asymptotic form

$$\left.\begin{array}{ll} P_{ij}(r) \underset{r\to\infty}{\sim} \left[\frac{1}{k_i}\left(1 + \frac{\epsilon_i}{2c}\right)\right]^{1/2}\left[\sin\theta_i(r)\delta_{ij} + \cos\theta_i(r)K_{ij}\right] & k_i^2 > 0 \\ \\ P_{ij}(r) \underset{r\to\infty}{\sim} 0, & k_i^2 < 0 \end{array}\right\} \; j = 1, \ldots, n_a$$

$$(62)$$

where n_a is the number of open channels and $k_i^2 = 2\epsilon_i + \epsilon_i^2/c^2$. These equations are identical in form to Eq. (12) obtained in the nonrelativistic case and also to those obtained using the Breit–Pauli Hamiltonian, and similar numerical methods can be used in their solution.[68] The main difficulty in practice arises from the fine-structure splitting of the atomic energy levels which can be small, giving rise to acute numerical problems. In conclusion, we note that the derivation of the S matrix and T matrix follows, using the standard formulas from the K matrix defined by Eq. (62). The cross sections and collision strengths then follow as in Section 2.5.1.

3. Numerical Solutions of the Coupled Integro-Differential Equations

We review in this section the numerical methods developed to solve the coupled integro-differential equations (27) or the corresponding equations discussed in Section 2.5 when relativistic effects are included. Our main objective is to obtain the K matrix defined by (12), but in many applications the radial wave functions $F_{ij}(r)$, for all r lying within the radius of the target, are required. We commence with a short discussion of early methods, i.e., those developed prior to 1968. We then review the main methods in use today.

3.1. Early Work

This can nearly all be classified under two headings: firstly, methods using an iterative treatment of the nonlocal exchange potential, and secondly, methods that reduce the coupled integro-differential equations to systems of coupled differential equations. We will see that both of these methods have difficulties in certain circumstances, but nevertheless they are still used with advantage, in certain circumstances.

   3.1.1. *Iterative Methods.* Following Burke and Seaton[14] we write Eq. (27) in the form

$$\mathbf{LF} = 0, \tag{63}$$

where the operator L is divided into two parts

$$\mathbf{L} = \mathbf{L}_0 - \mathbf{W} \tag{64}$$

and usually, but not always, $\mathbf{L}_0$ contains all the terms in Eq. (27) except the nonlocal exchange potential operator and $\mathbf{W}$ represents this operator. Equation (63) can then be written as

$$\mathbf{L}_0\mathbf{F} = \mathbf{WF} \tag{65}$$

which can be solved iteratively by writing

$$\mathbf{L}_0\mathbf{F}_0 = 0,$$
$$\mathbf{L}_0\mathbf{F}_s = \mathbf{WF}_{s-1}, \qquad s \geqslant 1. \tag{66}$$

The boundary conditions corresponding to Eq. (12) are enforced at each stage of the iteration yielding $\mathbf{K}_s$. Clearly the equations for $s = 0$ are a set of n coupled differential equations, while those for $s \geqslant 1$ have additional but known inhomogeneous terms. The Langrange orthogonalization terms are usually retained in L_0 and are enforced at each state of the iteration.

   A procedure for the solution of Eq. (27) by iterating the Born sequence was introduced for e–H scattering by Smith, McEachran, and Fraser[112] and further studied by McEachran and Fraser.[30] Convergence of this method is often poor for s-wave scattering where the potential interaction has the most effect, but it can be a very useful method for high angular momenta or high energies.

   The basic problem with the iterative approach is that convergence may be slow or nonexistent, particularly in the neighborhood of resonances, which are important in low-energy electron collisions.[22] At an energy near a resonance, the K matrix goes through a pole and thus $\mathbf{K}_s$ and $\mathbf{K}_{s+1}$ may differ substantially, making convergence difficult to achieve.

   Many ways of speeding the convergence have been proposed, perhaps the most interesting and successful being the iterative-variational methods discussed by Saraph and Seaton[113,114] and by McEachran et al.[31] and Sheorey[115]. These methods are based on the use of the Kohn variational principle defined by Eq. (29), where the integral $\mathbf{I}^t$ is calculated by taking a trial function consisting of a linear combination of the iterated functions $\mathbf{F}_s$ defined by Eq. (66).

3.1.2. *Reduction to a System of Coupled Differential Equations.* This method, first used in e–H scattering by Marriott[15] on the suggestion of Percival,[16] depends on the special form taken by the nonlocal exchange operator defined by Eq. (18). We expand the electron–electron interaction in the form

$$\frac{1}{r_{ij}} = \sum_{\lambda=0}^{\infty} \frac{r_<^\lambda}{r_>^{\lambda+1}} P_\lambda(\cos\theta_{ij}), \tag{67}$$

where $r_<$ and $r_>$ are the smaller and greater of r_i and r_j, and θ_{ij} is the angle between the vectors $\hat{\mathbf{r}}_1$ and $\hat{\mathbf{r}}_2$. The total contribution of Eq. (18) to the exchange operator in Eq. (27) is then

$$\sum_i \sum_j \int_0^\infty W_{ij}(r,r') F_j(r')\, dr' = \sum_{k=1}^{n_e} a_k\, y_{\lambda_k}(P_{1_k} F_{l_k}; r) P_{2_k}(r), \tag{68}$$

where $P_{1_k}(r)$ and $P_{2_k}(r)$ are known target orbitals, the a_k are constants, which depend on i and j, and are obtained by carrying out the remaining integrals in Eq. (18) over known orbitals, F_{l_k} is one of the continuum orbitals F_j, $j = 1, \ldots, n$, λ_k is defined by Eq. (67), and n_e is the total number of exchange terms. The functions $y_\lambda(PF; r)$ are defined by

$$y_\lambda(PF; r) = r^{-\lambda-1} \int_0^r P(r') F(r') r'^\lambda\, dr' + r^\lambda \int_r^\infty P(r') F(r') r'^{-\lambda-1}\, dr'. \tag{69}$$

It is now easy to show that the function $Y_\lambda(r) = r y_\lambda(PF; r)$ satisfies the second-order differential equation

$$\frac{d^2 Y_\lambda(r)}{dr^2} = \frac{\lambda(\lambda+1)}{r^2} Y_\lambda(r) - (2\lambda+1)\frac{P(r)F(r)}{r} \tag{70}$$

subject to the boundary conditions

$$Y_\lambda(r) \underset{r\to 0}{\sim} a r^{\lambda+1} \tag{71}$$

and

$$Y_\lambda(r) \underset{r\to\infty}{\sim} b r^{-\lambda}, \tag{72}$$

where the constants a and b are defined by the integrals

$$a = \int_0^\infty P(r)F(r)r^{-\lambda-1}\,dr, \qquad b = \int_0^\infty P(r)F(r)r^\lambda\,dr. \qquad (73)$$

It follows that the n_e exchange terms in Eq. (68), together with the original n coupled equations (27), are reduced to $n + n_e$ coupled differential equations subject to the boundary conditions defined by Eqs. (8), (12), (71), and (72). These equations can be solved using one of the standard techniques involving inward and outward integration and matching at some intermediate value of r to given n_a linearly independent solutions.[14]

The inclusion of correlation terms χ_i in the original expansion (3) gives rise to an additional nonlocal separable potential defined by Eq. (20). This introduces an additional complication in the solution of the differential equations, but a noniterative method can still be used as discussed by Burke and Taylor.[74,75]

The main difficulty with this approach is just the very large number of additional differential equations which have to be retained for realistic calculations for complex atoms or ions. For typical calculations carried out at the present time, involving say between 10 and 20 channels, where each target state may involve between 5 and 10 configurations, the value of n_e may be several thousand. The corresponding computing time required then becomes completely prohibitive. However, for light atoms involving a few channels or for one-channel calculations on complex atoms the method is still useful.

3.2.   Reduction to a System of Linear Algebraic Equations

This method introduced into electron–atom collisions by Seaton[42,46] forms the basis of IMPACT, one of the major computer program packages in this field.[48] It overcomes the difficulties of both methods discussed in Section 3.1 by being noniterative and by its efficiency being only weakly dependent on the number of exchange terms. In this method, the coupled integro-differential equations (27) are reduced to a system of linear algebraic equations for the values of the functions $F_i(r)$, $i = 1, \ldots, n$, which are tabulated at a set of mesh points in the region where exchange and correlation effects are important. The method was not completely new, similar approaches having been used by Robertson[116] for nuclear scattering problems and by Thompson[117] for electron scattering, but Seaton, by introducing a number of new ideas in the numerical and programming techniques used, revolutionized its use for calculating the collision of electrons by complex atoms and ions.

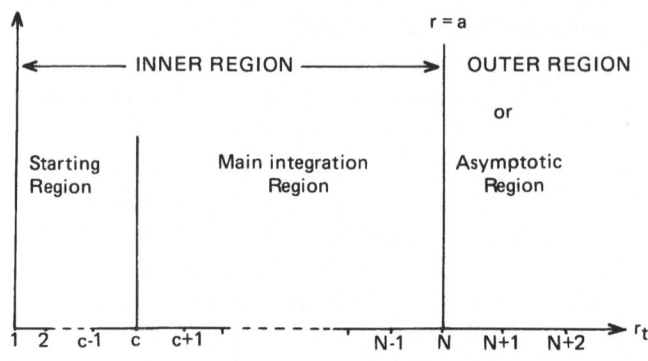

Radial Integration Regions

FIGURE 1. The radial integration regions used in the solution of Eq. (27).

The region of integration is divided into subregions as illustrated in Figure 1. The inner region $r \leqslant a$ is where exchange and correlation effects are important, and the outer region $r > a$ is where the coupled integro-differential equations reduce to coupled differential equations, where the potential is defined by Eq. (21). The method described in this section is concerned with the solution in the inner region. First a mesh of N points is used to span this region where

$$r_1 = 0, \qquad r_{i-1} < r_i, \qquad i = 1, \ldots, N, \qquad r_N = a. \tag{74}$$

In addition, two further points r_{N+1} and r_{N+2} are introduced which are used to match the solution in the inner region to the solution in the asymptotic region.[68] It follows that the total number of points used in the tabulation of each radial function is $N + 2$.

In order to keep the set of algebraic equations which must be solved as small as possible, the mesh points and the numerical methods for representing the differential and integral operators in Eq. (27) must be chosen with care. The starting region in Fig. 1 requires special treatment. Here the $F_i(r)$ are expanded as

$$F_i(r) = r^{l_i+1}\left[\left(1 - \frac{Zr}{l_i + 1}\right)A_i + r^2 g_i(r)\right], \tag{75}$$

where the unknown functions $g_i(r)$, which remain finite at threshold, are represented by their value at the mesh mpoints $r_1 r_2 \cdots r_c$. The number of mesh points c in the region may be 3, 4, or 5.

In the main integration region the tabular points are not equidistant but tend to get further apart as r increases. The object is to include about 4 or 5 points in each half-wavelength of the radial functions, which from experience is found to give good accuracy in the final solution. However, care must be taken in increasing the interval since it is found that the integration becomes unstable if the intervals $r_i - r_{i-1}$ change too rapidly. In practice

$$\frac{r_i - r_{i-1}}{r_{i-1} - r_{i-2}} = 1.2 \tag{76}$$

is found to be a suitable criterion to apply.

The finite-difference formulas used in the representation of the differential and integral operators in Eq. (27) have been described in detail by Seaton[46] and will not be reproduced here. However, it is clear that both operators can be expressed in terms of linear combinations of the values of the functions $F_i(r)$ at the mesh points. The effectiveness of these formulas in reducing the number of tabular points $N + 2$ is clearly of paramount importance.

Finally, we consider the form of the linear algebraic equations which must be solved. Clearly representing the n functions $F_i(r)$ gives rise to $n \times (N + 2)$ unknowns. We obtain further m unknowns corresponding to the coefficients a_i multiplying the correlation functions in Eq. (3). Finally, we have to impose the orthogonality constraints—Eq. (28). This gives rise to further n_λ Lagrange multipliers. Thus the total number of linear algebraic equations is

$$n_{\text{tot}} = n \times (N + 2) + m + n_\lambda. \tag{77}$$

A complete set of linearly independent solutions, satisfying the boundary conditions defined by Eq. (12) requires n_a solutions of these equations.

In the outer region, the coupled differential equations are solved numerically[68] to yield solutions with the asymptotic forms

$$S_{ij}(r) \underset{r \to \infty}{\sim} k_i^{-1/2} \sin \theta_i \delta_{ij}$$
$$C_{ij}(r) \underset{r \to \infty}{\sim} k_i^{-1/2} \cos \theta_i \delta_{ij} \qquad i = 1, \ldots, n; \quad j = 1, \ldots, n_a \tag{78}$$

and

$$E_{ij}(r) \underset{r \to \infty}{\sim} \kappa_i^{-1/2} \exp\left[-\kappa_i r + \frac{Z - N}{\kappa_i} \ln (2\kappa_i r) \right] \delta_{ij},$$

$$i = 1, \ldots, n; \quad j = n_a + 1, \ldots, n \tag{79}$$

where $\kappa_i = |k_i|$. The solution vectors must then satisfy

$$F_{ij}(r) = \sum_{l=1}^{n_a} \left[S_{il}(r)a_{lj} + C_{il}(r)b_{lj} \right] + \sum_{l=n_a+1}^{n} E_{il}(r)c_{lj},$$

$$i = 1, \ldots, n; \quad j = 1, \ldots, n_a, \quad (80)$$

where a_{lj}, b_{lj}, and c_{lj} are constants determined by fitting the solutions at the points r_{N+1} and r_{N+2}. In principle, these matrices are not all independent, but equivalent linearly independent solutions can be obtained by making different choices of a_{lj}.

In the original approach, the linear algebraic equations were defined by the matrix equations

$$\mathbf{C} \cdot \mathbf{G} = \mathbf{X} \cdot \mathbf{A}, \quad (81)$$

where C has dimensions ($n_{tot} \times n_{tot}$), G and X have dimensions $n_{tot} \times n_a$ and A is the matrix with elements a_{lj} in Eq. (80) which has dimensions ($n_a \times n_a$). For a given choice of A, which was taken to be the unit matrix, these equations can be solved to yield the unknown matrix G. As the work progressed it was found that good linear independence of the solutions was not always obtained with this choice of A. Consequently, Eq. (81) was repartitioned giving

$$(\mathbf{C}, -\mathbf{X}) \cdot \begin{pmatrix} \mathbf{G} \\ \mathbf{A} \end{pmatrix} = 0 \quad (82)$$

and row and column interchanges made so that the most nonsingular submatrix of dimension ($n_{tot} \times n_{tot}$) in $(\mathbf{C}, -\mathbf{X})$ is used in the Gauss elimination method. Further details of this procedure have been given by Seaton[46] and by Crees et al.[48]

We conclude the brief discussion of this method by mentioning that it has recently been applied with considerable success in electron–molecule scattering by Crees and Moores[118] and by Schneider and Collins.[119,120]

3.3. R-Matrix Method

This method, introduced into electron–atom collisions by Burke, Hibbert, and Robb,[40,49] forms the basis of RMATRX, a computer program package that enables electron and photon collisions with complex atoms and ions as well as atomic polarizabilities to be calculated.[52,53] This approach was first developed in nuclear physics by Wigner[121,122] and Wigner and Eisenbud[123] and later applications and developments in nuclear physics were reviewed by, in particular, Lane and Thomas,[91] Lane and Robson,[124,125] and Mahaux and Weidenmüller.[126] Its applications in nonrelativistic electron–atom collisions were reviewed by Burke and Robb[50] and recent relativistic developments have been made by Chang,[97,98] Scott and Burke,[63] and Norrington and Grant[65] (see Section 2.5).

As in the linear algebraic equations method, configuration space is divided into two regions, an inner and an outer region, as illustrated in Fig. 1. Again the radius $r = a$ of the inner region is chosen so that electron exchange and correlation effects can be neglected for $r \geqslant a$. The objective of the method is then to calculate the R matrix, which is defined by the equation

$$F_i(a) = \sum_{j=1}^{n} R_{ij}(E)\left(a\frac{dF_j}{dr} - b_j F_j\right)_{r=a} , \qquad i = 1, \ldots, n \qquad (83)$$

by solving the collision problem in the inner region. Here F_i and dF_i/dr are the solutions of Eq. (27) defined on the boundary $r = a$, and the b_j are parameters whose choice is usually arbitrary.

We solve the collision problem in the inner region by expanding the total wave function, in analogy with Eq. (3), in a discrete basis defined for each $LS\pi$ combination, by

$$\Psi_k = \mathcal{Q} \sum_{i=1}^{n} \sum_{j=1}^{p} \bar{\Phi}_i r_{N+1}^{-1} v_j(r_{N+1})a_{ijk} + \sum_{i=1}^{m} \chi_i b_{ik} , \qquad (84)$$

where $\bar{\Phi}_i$ and χ_i are defined as in Eq. (3), and we have expanded the radial function F_i in terms of radial basis functions v_j defined over the range $0 \leqslant r \leqslant a$ as described below. For radial basis functions satisfying arbitrary boundary conditions at $r = a$, the Hamiltonian defined by Eq. (2) is not Hermitian in the inner region due to the nonvanishing surface terms. These terms, however, can be cancelled by introducing the Bloch operator[127]

$$L_b = \sum_{i=1}^{n} |\bar{\Phi}_i\rangle \frac{1}{2}\delta(r - a)\left(\frac{d}{dr} - \frac{b_i}{r}\right)\langle\bar{\Phi}_i|. \qquad (85)$$

We now rewrite the Schrödinger equation (1) as

$$(H_{N+1} + L_b - E)\Psi = L_b\Psi, \qquad (86)$$

which can be formally solved giving

$$\Psi = (H_{N+1} + L_b - E)^{-1}L_b\Psi \qquad (87)$$

and we expand the Green's function $(H + L_b - E)^{-1}$ in the inner region in terms of the discrete basis Ψ_k. The coefficients a_{ijk} and b_{ik} are obtained by diagonalizing $H_{N+1} + L_b$ giving

$$\langle \Psi_k | H_{N+1} + L_b | \Psi_{k'} \rangle_I = E_k \delta_{kk'}, \qquad (88)$$

where the integral in this equation is over the inner region. Since $H_{N+1} + L_b$ is Hermitian in this region, the eigenvalues E_k are real. Equation (87) can then be written as

$$|\Psi\rangle = \frac{1}{2} \sum_{kj} \frac{|\Psi_k\rangle\langle\Psi_k|\overline{\Phi}_j\rangle}{E_k - E} \left(\frac{d}{dr} - \frac{b_j}{r} \right) \langle\overline{\Phi}_j | \Psi\rangle. \qquad (89)$$

Projecting this equation onto the channel functions $\overline{\Phi}_i$, assuming that Ψ is defined by Eq. (3), and evaluating it at $r = a$ yields immediately Eq. (83) where the R matrix is given by

$$R_{ij}(E) = \frac{1}{2a} \sum_k \frac{w_{ik}(a)w_{jk}(a)}{E_k - E} \qquad (90)$$

and we have introduced the surface amplitudes

$$w_{ik}(a) = \sum_j v_j(a)a_{ijk}. \qquad (91)$$

A major fraction of the computer time goes into calculating the matrix elements involved in Eq. (88) and carrying out the diagonalization. However, this needs to be carried out only once to determine the R matrix for all low energies. If the asymptotic solutions are known, Eq. (80) is then substituted into Eq. (83) to determine the asymptotic coefficients a_{ij} and b_{ij} and consequently the K and S matrices.

In principle, the summation over k in Eq. (90) is infinite in extent. However, in practice, only a finite number of terms can be retained, corresponding to the number of terms retained on the right-hand side of Eq. (84). It is then necessary to choose the basis functions $v_j(r)$ with care to ensure rapid convergence. In the work on electron–atom and electron–ion scattering, the basis functions are chosen to be eigensolutions of a model Hamiltonian defined by the following radial equations for each angular momentum:

$$\left[\frac{d^2}{dr^2} - \frac{l(l+1)}{r^2} + V(r) + k_i^2 \right] v_i(r) = \sum_n \lambda_{i\,nl} P_{nl}(r) \qquad (92)$$

subject to the boundary conditions

$$v_i(0) = 0, \qquad \frac{a}{v_i} \left.\frac{dv_i}{dr}\right|_{r=a} = b \qquad (93)$$

and the orthonormality condition

$$\int_0^a v_i(r)v_j(r)\,dr = \delta_{ij} \qquad (94)$$

and where the Langrange multipliers $\lambda_{i\,nl}$ are chosen so that the $v_i(r)$ are orthogonal to the target and pseudo-orbitals $P_{nl}(r)$ with the same angular symmetry. This ensures that the radial functions $F_i(r)$, which are expanded in terms of these $v_i(r)$, satisfy Eq. (28). If the potential $V(r)$ is chosen to be a suitable local potential approximately representing the charge distribution of the target, then rapid convergence is achieved provided that a correction, first considered by Buttle,[128] is added to the diagonal elements of Eq. (90). This correction is easily calculated by solving Eq. (92) at each energy under consideration.

Other choices of these basis functions have been made. In the work of Fano and Lee[129] and Lee,[130] using a related eigenchannel formalism,[131] STOs were used. Analytical representation of these radial orbitals are also appropriate in electron–molecule scattering, since the integrals in Eq. (88) can then be evaluated with only small modifications of standard bound-state computer programs. This is the basis of work on electron–molecule scattering using the R-matrix method by Schneider and collaborators[132–137] and by Burke and co-workers.[138,139] However, evidence is accumulating that numerical basis functions may be more appropriate even in this case if a large range of energies is being considered,[140] since it is difficult in this case to construct sufficient linearly independent analytical basis functions.

In comparing the R-matrix method with the linear algebraic equations method, one criterion is the number of functions v_i required to represent each F_i in the former method compared with the number of tabulation points N in Eq. (74). For a given energy this ratio is normally about $1:2$. Another criterion also in favor of the R-matrix method is that only one diagonalization of Eq. (88) is required for all energies considered, while Eq. (82) must be solved at each energy. However, it must be pointed out that many of the operations in Eq. (82) are energy independent and thus need only be carried out once. Finally, we note that Eq. (82) involves the solution of simultaneous equations, while Eq. (88) involves the computationally longer process of matrix diagonalization. In conclusion, it is found that for first-row atoms it is better to use the linear algebraic equations method if only a few energy points are required; however, if the calculation is required at many energies, the R-matrix method is preferable. For heavier atoms and ions involving very many channels, the extra space required to store the linear algebraic equations defined by Eq. (82) as opposed to the symmetric Hamiltonian matrix in the R-matrix method favors the latter method. .

3.4. Matrix Variational Method

The Hulthén and Kohn[87,88] variational methods were first used by Massey and Moiseiwitsch[141] in electron–atom scattering to solve the static exchange equations for e–H scattering obtained when just the ground state is retained in Eq. (3). Later Schwartz[142,143] showed how the Kohn variation method could be used to obtain almost exact results in the elastic scattering region, for e–H scattering. He also pointed out difficulties that arise from certain anomalous singularities in the method. Now we review how this method has been generalized to enable low-energy inelastic electron scattering by complex atoms to be accurately calculated. This work has been carried out largely by Nesbet and collaborators and the method and applications have been reviewed in a number of recent publications.[57–59]

As in the linear algebraic equations method and the R-matrix method, the matrix variational method again starts from expansion (3) for the collision wave function. However, unlike these previous methods configuration space is not divided into two regions, but instead a trial wave function is defined over all configuration space and is substituted into the multichannel Kohn variational principle defined by Eq. (26), or into one of the equivalent variational principles obtained using different asymptotic forms. The trial wave function itself has two parts, as in Eq. (3). One part represents the asymptotic form of the wave function, and a second short-range part consists of a sum of L^2 integrable correlation functions.

The part that represents the asymptotic form of the wave function consists in approximating the open-channel functions in Eq. (3) as follows:

$$F_i(r) = w_{0i}(r)\alpha_{0i} + w_{1i}(r)\alpha_{1i}, \qquad i = 1, \ldots, n_a \tag{95}$$

where the functions $w_{0i}(r)$ and $w_{1i}(r)$ satisfy the boundary conditions

$$w_{0i}(r) \underset{r \to \infty}{\sim} k_i^{-1/2} \sin\left(k_i r - \tfrac{1}{2} l_i \pi\right),$$

$$w_{1i}(r) \underset{r \to \infty}{\sim} k_i^{-1/2} \cos\left(k_i r - \tfrac{1}{2} l_i \pi\right),$$
$$\tag{96}$$

so that Eq. (95) is consistent with Eq. (12). The functions $w_{0i}(r)$ and $w_{1i}(r)$ are cut off at the origin so that they satisfy the correct boundary conditions there. If Coulomb potentials, or certain other long-range potentials, are present, such that analytical solutions can be found, these can be used instead of $w_{0i}(r)$ and $w_{1i}(r)$ defined by Eq. (96) in the asymptotic region. In practice, for numerical convenience, these functions are also chosen to be orthogonal to the target and pseudo-orbitals $P_{nl}(r)$ with the same angular symmetry. Using Eq. (95) for the open-channel functions in Eq. (3), we find that the integral

$$I = \int \Psi^{\Gamma*}(X_{N+1})(H_{N+1} - E)\Psi^{\Gamma}(X_{N+1})\,dX_{N+1} \equiv \langle \Psi^{\Gamma}|H_{N+1} - E|\Psi^{\Gamma}\rangle \quad (97)$$

reduces to the quadratic expression

$$I = \sum_{pq}\sum_{ij} \alpha_{pi} m_{pq}^{ij} \alpha_{qj}, \quad (98)$$

where

$$m_{pq}^{ij} = M_{pq}^{ij} - \sum_{kl} M_{pi,k}(M^{-1})_{kl} M_{l,qj}. \quad (99)$$

The matrix terms are the free–free matrix elements

$$M_{pq}^{ij} = \langle \mathcal{A}\overline{\Phi}_i w_{pi}|H_{N+1} - E|\mathcal{A}\overline{\Phi}_j w_{qj}\rangle, \quad (100)$$

the bound–free matrix elements

$$M_{pi,k} = \langle \mathcal{A}\overline{\Phi}_i w_{pi}|H_{N+1} - E|\chi_k\rangle, \quad (101)$$

and the bound–bound matrix elements

$$M_{kl} = \langle \chi_k|H_{N+1} - E|\chi_l\rangle. \quad (102)$$

It follows from Eqs. (97) and (98) that exact solutions of the Schrödinger equation require

$$\sum_{qj} m_{pq}^{ij}\alpha_{qj} = 0 \quad (103)$$

which should have n_a linearly independent solutions, corresponding to the n_a linearly independent solutions of the coupled integro-differential equations (27). However, unlike Eq. (27), Eq. (103) is a linear homogeneous algebraic equation which does not in general have solutions at arbitrary energies, and we must therefore look for approximate variational solutions. This can be achieved by using the Kohn variational principle defined by Eq. (26). The matrix elements m_{pq}^{ij} define $\mathbf{I}$ in this equation and we can obtain various forms of the variational principle by making different choices of the asymptotic boundary conditions as discussed earlier. Equation (26) then leads to a set of linear simultaneous equations for the K-matrix elements as well as giving an approximate collision wave function. A K matrix, correct to second-order in the errors in the wave function, is then given by Eq. (29).

As pointed out earlier, anomalous singularities arise in this formalism. They can be shown to be due to singularities in the matrices m_{00} or m_{11} where the rows and columns of these matrices are defined by the indices i and j in Eq. (99). The energies at which these singularities occur, however, can be changed either by changing the basis or by using a different form of the variational principle. An anomaly-free method, based on the Kohn variational principle, has been discussed by Nesbet and Oberoi[144] which overcomes these singularity problems.

In comparing this method with the previous two methods, discussed in Sections 3.2 and 3.3, we note that while both the former methods attempt to calculate the functions F_i accurately in the inner region, the representation given by Eq. (95) is arbitrarily determined by the cutoff used in this region. Thus the role of the second expansion in Eq. (3) assumes more significance in the matrix variational method since it has to represent the discrepancy between the exact F_i and the functions $w_{pi}(r)$ in this region as well as additional correlation effects. A further difference occurs in the outer region where both the previous methods use exact numerical solutions defined by Eqs. (78) and (79), while, in principle, Eq. (95) is only exact in this region if long-range multipole potentials are not present or can be represented analytically (see, however, Section 3.6).

3.5. Noniterative Integral Equations Method

This method was first introduced by Sams and Kouri.[145-148] It was then developed by Henry et al.[55,56] who used it as the basis of a general computer program package NIEM for solving the collision of electrons by complex atoms and ions. They also showed how the nonlocal exchange potentials that arise in this application could be treated noniteratively.

Again the method is based on expansion (3), which leads to the coupled integro-differential equations (27). The integral solution of these equations can be written

$$F_{ij}(r) = \Delta_i \delta_{ij} G_i^{(1)}(k_i r) + G_i^{(2)}(k_i r) \int_0^r G_i^{(1)}(k_i r') S_{ij}(r') \, dr'$$

$$- G_i^{(1)}(k_i r) \int_0^r G_i^{(2)}(k_i r') S_{ij}(r') \, dr'$$

$$+ G_i^{(1)}(k_i r) \int_0^\infty G_i^{(2)}(k_i r') S_{ij}(r') \, dr', \tag{104}$$

where

$$\Delta_i = \begin{cases} 1, & i = 1, \ldots, n_a \\ 0, & i = n_a + 1, \ldots, n \end{cases} \tag{105}$$

and $G_i^{(1)}(k_i r)$ and $G_i^{(2)}(k_i r)$ are the regular and irregular solutions of the differential equation

$$\left[\frac{d^2}{dr^2} - \frac{l_i(l_i + 1)}{r^2} + \frac{2(Z - N)}{r} + k_i^2 \right] G_i^{(p)}(k_i r) = 0, \tag{106}$$

subject to the asymptotic boundary conditions which, for a neutral target, are

$$G_i^{(1)}(k_i r) \underset{r \to \infty}{\sim} \begin{cases} k_i^{-1/2} \sin(k_i r - \tfrac{1}{2} l_i \pi), & i = 1, \ldots, n_a \\ (2|k_i|)^{-1/2} e^{|k_i| r}, & i = n_a + 1, \ldots, n \end{cases} \tag{107}$$

and

$$G_i^{(2)}(k_i r) \underset{r \to \infty}{\sim} \begin{cases} -k_i^{-1/2} \cos(k_i r - \tfrac{1}{2} l_i \pi), & i = 1, \ldots, n_a \\ -(2|k_i|)^{-1/2} e^{-|k_i| r}, & i = n_a + 1, \ldots, n. \end{cases} \tag{108}$$

For ions these functions must be replaced by the equivalent regular and irregular Coulomb wave functions. Finally, in Eq. (104), $S_{ij}(r)$ are the remaining source terms in Eq. (27) which include the direct and nonlocal exchange potentials, and, as in Eqs. (8), (9), and (12), we have introduced a second index j to denote the linearly independent solutions. It follows from the definitions of $G_i^{(1)}(k_i r)$ and $G_i^{(2)}(k_i r)$ that Eq. (104) satisfies the boundary condition (12).

Another integral solution of Eq. (27) can be defined by omitting the normalization term in Eq. (104) giving

$$\psi_{ij}(r) = G_i^{(1)}(k_ir)H_{ij}^{(2)}(r) - G_i^{(2)}(k_ir)H_{ij}^{(1)}(r), \qquad (109)$$

where

$$H_{ij}^{(2)}(r) = \delta_{ij} - \int_0^r G_i^{(2)}(k_ir')S_{ij}(r')\,dr',$$

$$H_{ij}^{(1)}(r) = \int_0^r G_i^{(1)}(k_ir')S_{ij}(r')\,dr'. \qquad (110)$$

However, since the solutions of Eq. (27) are unique, the solutions defined by Eqs. (104) and (109) must be related by a linear transformation. Furthermore, the solution vectors in Eq. (109) can be combined so that the physical solutions have decaying waves in the closed channels. The K matrix is then defined by

$$K_{ij} = \sum_{l=1}^{n_a} H_{il}^{(1)}(\infty)\big(H^{(2)}(\infty)^{-1}\big)_{lj}, \qquad i, j = 1, \ldots, n_a \qquad (111)$$

The noniterative treatment of exchange in this method requires some comment. Let us confine our discussion to the nonlocal exchange operator defined by Eq. (68). We first write Eq. (109) in the form

$$\psi_{ij}(r) = \delta_{ij}G_i^{(1)}(k_ir) + \int_0^r G_i^{(2,1)}(r,r')\sum_l\left[V_{il}(r')\psi_{lj}(r')\right.$$

$$\left. + \int_0^\infty W_{il}(r',r'')\psi_{lj}(r'')\,dr''\right]dr', \qquad (112)$$

where

$$G_i^{(2,1)}(r,r') = G_i^{(2)}(k_ir)G_i^{(1)}(k_ir') - G_i^{(1)}(k_ir)G_i^{(2)}(k_ir'). \qquad (113)$$

Using Eqs. (68) and (69) we can then rewrite the nonlocal exchange operator in Eq. (112) in the form

$$\sum_{k=1}^{n_e} a_k\left[r'^{-\lambda_k-1}\int_0^{r'} P_{1_k}(r'')\psi_{l_kj}(r'')r''^{\lambda_k}\,dr'' - r'^{\lambda_k}\int_0^{r'} P_{1_k}(r'')\psi_{l_kj}r''^{-\lambda_k-1}\,dr''\right.$$

$$\left. + r'^{\lambda_k}\int_0^\infty P_{1_k}(r'')\psi_{l_kj}(r'')r''^{-\lambda_k-1}\,dr''\right]P_{2_k}(r'). \qquad (114)$$

The general solution of Eq. (112) is then given by

$$\psi_{ij} = \psi_{ij}^0 + \sum_{k=1}^{n_e} \tilde{\psi}_{ik} c_{kj}, \tag{115}$$

where ψ_{ij}^0 is the solution of Eq. (112), omitting all nonlocal exchange potentials, and the $\tilde{\psi}_{ik}$ are n_e linearly independent solutions including the nonlocal exchange potentials in a modified form. The difficulty in treating these exchange potentials noniteratively arises from the third term in Eq. (114) which involves an integral from 0 to ∞. The technique adopted involves setting all of these integrals equal to zero except one which is set equal to unity. This can be done in n_e independent ways giving the n_e independent solutions ψ_{ik}.

The final problem is to determine the so far unspecified coefficients c_{kj} in Eq. (115). These are determined by the self-consistent requirement that the final solution ψ_{ij} must indeed satisfy Eq. (112). Substituting Eq. (115) into Eq. (112) and using the equations satisfied by the ψ_{ij}^0 and $\tilde{\psi}_{ij}$ gives a set of linear algebraic equations to be solved.

The inclusion of correlation terms χ_i and orthogonality constraints can also be treated noniteratively by obtaining additional solutions of Eq. (112) and satisfying additional self-consistency conditions. However, the main difficulty with this method arises from the same cause as the difficulty experienced by the method described in Section 3.1.2, although not in such an acute form. That is, the method becomes inefficient for complex atoms and ions when the number of exchange terms n_e becomes large. Since the order of the matrices involved depends linearly on this number, the calculation of the $\tilde{\psi}_{ik}$ and the solution of the self-consistency equations eventually becomes very time consuming.

3.6. New Directions

In this section we briefly mention some new theoretical methods that have been proposed to extend the calculations to intermediate energies and new numerical methods for solving the collision problem at low energies.

We have seen that methods based on Eq. (3) can give, in principle, highly accurate results for nonrelativistic collisions of electrons by atoms and ions when all open channels can be included in the expansion. Some of these results will be presented in Section 3.7. On the other hand, at high energies, while an infinite number of channels are open, these are weakly coupled. In such situations, perturbation theory approaches based on the Born or eikonal series, perhaps supplemented by an exact treatment of the coupling between a few channels of interest, are then appropriate.[66,67,149,150] However, at intermediate energies ranging from an incident electron energy just below the first ionization threshold to an energy several times this threshold the situation is much more complicated, particularly when the charge on the ion is not large or is zero. There are then too many channels to include explicitly in Eq. (3), and the coupling of these channels is too strong to be treated by perturbation theory. Thus new theoretical methods must be developed to enable cross sections to be reliably calculated in this region.

Progress has been made in the treatment of this problem by expanding the total wave function in an L^2 integrable basis as reviewed by Reinhardt.[66] This corresponds to retaining only the second expansion in Eq. (3). The basic idea is to avoid the specification of channels and their corresponding asymptotic forms by using an expansion in terms of wave functions that vanish asymptotically. Then diagonalization of the Hamiltonian in this L^2 basis yields a discretization of the continuous spectrum from which scattering information can be extracted in certain circumstances. This extraction has been accomplished by analytical continuation of the T matrix from complex energies by Schlessinger, Schwartz, and Nuttall and coworkers,[151-154] and by Stieltjes imaging and moment T-matrix methods by Langhoff, Reinhardt, and Winick.[155-158] However, in principle, while these methods are quite general, in practice they have been limited to the elastic scattering of electrons and positrons on atomic hydrogen. Consequently, very much more work needs to be carried out to see if they can be developed as efficient methods to describe inelastic collisions from complex atoms and ions.

We have already mentioned in Section 2.1 that inclusion of pseudostates in Eq. (3) enables collisions at intermediate energies to be calculated.[70-73] However, a basic difficulty with this approach is that unphysical pseudostate thresholds and often pseudoresonances occur in the energy range of interest. The position of these can be changed by changing the pseudostate basis, but the accuracy of this approach at intermediate energies has so far been limited by these effects. An alternative approach considered by Burke, Berrington, and Sukumar[159] is to note that the L^2 correlation terms in the second expansion in Eq. (3) also give rise to pseudoresonances at intermediate energies. They argue that, in analogy with the optical model theory developed to describe nuclear reactions,[160-162] if the scattering amplitude is averaged over these pseudoresonances this allows approximately for loss of flux into channels that are not included explicitly in the first expansion of Eq. (3). Further work on both these approaches is required to see if accurate results can be obtained at intermediate energies.

Finally, we mention a very interesting approach by Poet,[163,164] who solved the model e–H scattering problem when both electrons are constrained to be in s-wave states. He obtained essentially exact results for elastic scattering at intermediate energies by solving the resultant partial differential equations in the (r_1, r_2) plane. However, there is still a question whether this approach can be extended to the full e–H scattering problem as well as considering the problem involving complex atoms and ions. Nevertheless, this type of approach may well have important implications for ionization as well as for excitation and it is under active investigation at present.[165]

New developments in numerical methods include the J-matrix method introduced by Heller and Yamani[166,167] and extended to multichannel electron scattering by complex atoms by Broad and Reinhardt.[168,169] They show that the zero-order Hamiltonian, obtained by omitting the interaction potential, reduces to a tridiagonal (Jacobi) matrix in a basis defined by a complete discrete set of nonorthogonal Laguerre functions. Although this matrix is infinite, it can be diagonalized analytically at any energy and thus the asymptotic form of the solution is accurately represented. The interaction potential, which is short range, can then be represented in a finite basis of Laguerre functions. This results in the need to diagonalize numerically a finite dimensional matrix which gives the S matrix at all energies. The method is analogous to expansion (3), with a different partition between the two expansions. So far, a general computer package has not been developed based on this approach, but if this is done it should be competitive with the programs based on the methods discussed in Sections 3.2–3.5.

The variational principle of Schwinger,[170] based on the Lippmann–Schwinger integral equation,[171] has been widely used for single-channel scattering.[172] Within the last few years McKoy, Rescigno, Watson, and Lucchese[173–176] have shown that it can be very effectively used to solve the multichannel problem which arises in electron–molecule scattering. Like the R-matrix method and the matrix variational method it allows the trial wave function to be represented by a discrete basis, and thus for electron–molecule collisions has the advantage, like these other methods, that standard molecule structure codes can be used with little modification. So far, the method in its multichannel form has had little application to the collision of electrons by complex atoms or ions, and no general computer package for this application exists. However, there is little doubt that this method would be very competitive with the methods described in Sections 3.2–3.5.

We conclude this section by mentioning some attempts to combine the advantages of the R-matrix method and the matrix variational method. We have pointed out that the functions $w_{0i}(r)$ and $w_{1i}(r)$ defined by Eq. (96) are often poorly adapted to represent the long-range multipole potentials that occur in electron–atom and electron–ion collisions. To overcome this problem, Oberoi and Nesbet[177] suggested redefining the function F_i in Eq. (95) to be numerical solutions of the coupled asymptotic equations beyond some radius $r > a$ (see Fig. 1) as in the R-matrix method. For $r \leqslant a$, these numerical functions are then continued in some more or less arbitrary way to go appropriately to zero at the origin. Nesbet[178] has recently suggested extending this concept by using these asymptotic solutions inside the radius $r = a$, at which exchange effects become negligible. He proposes introducing a second radius $r = b$ where $0 < b < a$ and using the numerical solution of the asymptotic equations for $r > b$. Of course in the range $b \leqslant r \leqslant a$ these solutions will not be exact solutions of the coupled integro-differential equations (27); however for a suitable choice of b the residual scattering effects will be quite small in this range. Thus, while the L^2 integrable correlation functions are still nonzero for $r \leqslant a$, their main role will be to represent the collision wave function for $r \leqslant b$. The advantage of this approach is that fewer correlation functions are required, but at the expense of making the bound–free matrix elements defined by Eq. (101) energy dependent.

3.7. Illustrative Results

In this section we consider some illustrative results selected from a vast number produced by the computer program packages described in the next section. Many of these results are stored in an atomic data bank set up at the SERC Daresbury Laboratory[179] and available on-line to U.K. scientists. The electron–ion results have been comprehensively reviewed by Seaton[180] and recently by Henry.[181]

Most of the calculations have been carried out on first- and second-row atoms and ions, although some isolated heavier systems such as Fe II, Cs, and Hg have been studied. Work using IMPACT has recently concentrated on forbidden transitions of second-row ions. Specifically, ions of the silicon $(3s^2 3p^2)$ and phosphorous $(3s^2 3p^3)$ isoelectronic sequences have been studied by Mendoza.[182] One example is Si-like S III, where the forbidden transitions

$$(3s^2 3p^2) \, {}^3P \rightarrow {}^1D, \quad {}^3P \rightarrow {}^1S, \quad \text{and} \quad {}^1D \rightarrow {}^1S \tag{116}$$

were studied using a calculation which included the 12 target states

$$(3s^2 3p^2) \, {}^3P, \, {}^1D, \, {}^1S,$$

$$(3s \, 3p^3) \, {}^5S^\circ, \, {}^3D^\circ, \, {}^3P^\circ, \, {}^1D^\circ, \, {}^1P^\circ, \, {}^3S^\circ, \qquad (3s^2 3p \, 3d) \, {}^3F^\circ, \, {}^3P^\circ, \, {}^3D^\circ. \tag{117}$$

A pseudo $4s$ orbital was introduced in order to obtain an accurate representation of the target wave functions, and the calculation was carried out on the NCAR CRAY-1 computer at Boulder.

In the case of the phosphorous sequence, IMPACT has been used to calculate the forbidden transitions in S II

$$(3s^2 3p^3)\,{}^4S^{\circ} \rightarrow {}^2D^{\circ}, \quad {}^4S^{\circ} \rightarrow {}^2P^{\circ}, \quad \text{and} \quad {}^2D^{\circ} \rightarrow {}^2P^{\circ}, \tag{118}$$

where the following six target states were included in the expansion

$$(3s^2 3p^3)\,{}^4S^{\circ},\,{}^2D^{\circ},\,{}^2P^{\circ},\,(3s\,3p^4)\,{}^4P^{e},\,{}^2D^{e},\,{}^2P^{e}. \tag{119}$$

Two pseudo-orbitals $4s$ and $4p$ were introduced to represent the target wave functions. Similar calculations were also carried out using RMATRX on the CRAY-1 at Daresbury and the results are presented in Fig. 2. The results differ markedly from previous distorted-wave calculations.

Calculations on the magnesium $(3s^2)$ and aluminum $(3s^2 3p)$ isoelectronic sequences are also under way using RMATRX on the CRAY-1 at Daresbury. The results of a recent 12-state calculation for the transition

$$(3s^2)\,{}^1S \rightarrow (3s\,3p)\,{}^3P^{\circ} \tag{120}$$

for magnesiumlike Si III are shown in Fig. 3.[183,184] Again the results are much larger than earlier distorted-wave calculations due mainly to resonant enhancement of the cross sections. Such enhancements at low energies are more important for second-row ions than for first-row ions and are particularly marked for forbidden transitions. The resonances are caused by the temporary capture of the incident electron by the Coulomb field of the ion in an excited state, and their importance increases as the strength of the coupling between low-lying states of the target increases. Recent results by Dufton and Kingston,[185] including the six lowest target states in S IV, confirm the importance of resonances for second-row ions.

In concluding this section we can summarize the present situation by remarking that, with program packages such as IMPACT and RMATRX and computers with the power of the CRAY-1, low-energy collision strengths accurate to about 10% can be obtained for first- and second-row atoms and ions. In the case of second-row systems this often involves solving about 20 coupled integro-differential equations. The extension to heavier systems such as Fe II, where preliminary work is now under way involving up to about 40 coupled equations, is close to the limit of the present programs and computers. However, it is expected that, even using present numerical methods, the development of better programming techniques and faster computers will change this situation very significantly

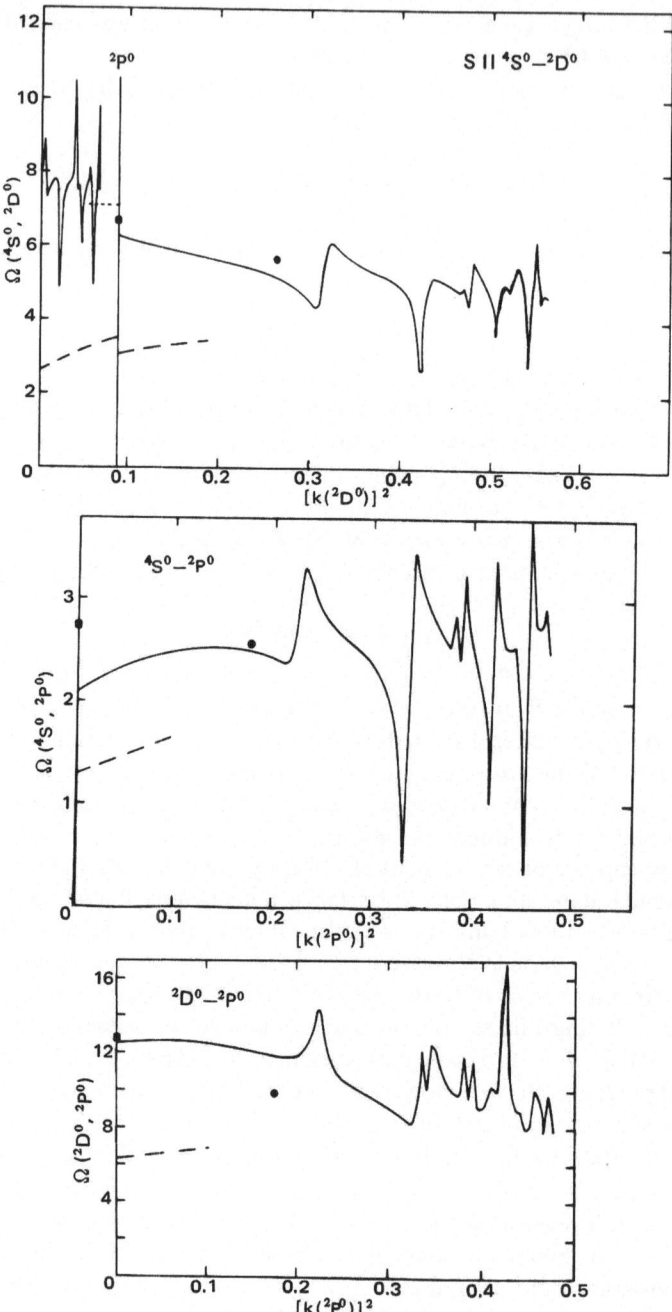

FIGURE 2. Electron collision strengths for the forbidden transitions in S II calculated by Mendoza,[182] compared with earlier three-state calculations (dots) and distorted-wave calculations (chain curve).

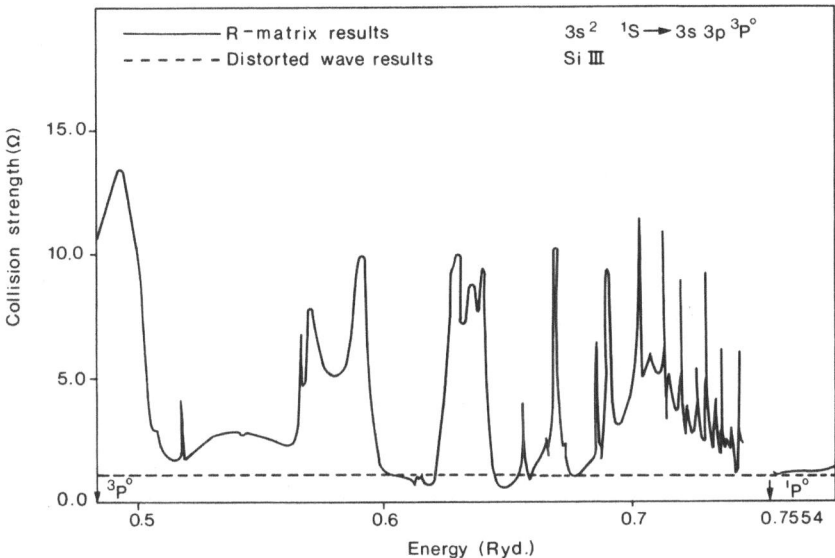

FIGURE 3. Electron collision strength for the $(3s^2)\,{}^1S \to (3s\,3p)\,{}^3P^{\circ}$ transitions in Si III calculated by Baluja et al.[183,184] compared with earlier distorted-wave calculations (dashed curve). (From Baluja et al.[183], Fig. 1.)

within the next 2–3 years. It is clear that these improvements will be necessary so that inelastic electron collisions with heavy atoms and ions where relativistic effects are important can be studied in detail.

4. Computer Program Packages

This survey is confined to a few major computer codes for calculating atomic structure properties and low-energy electron collision processes with atoms and ions in a partial-wave approach. Packages of general purpose programs[186] have been developed since the late 1960s at Queen's University, Belfast (with the RMATRX package, see Fig. 4), and at University College, London (around the program IMPACT, see Fig. 5). The intention of such general purpose codes can be summed up with Seaton's words in 1967: we should get away from "one man—one cross section." Other major centers of activity evolved at Louisiana State University in Baton Rouge (with the NIEM package[56]) and at the IBM Research Laboratory in San José, California (with the MATRIX VARIATIONAL code[58]). More recently, the Oxford group embarked on a fully relativistic package for structure (MCDF[64], multiconfiguration Dirac–Fock) and electron collisions. It has proved most fruitful that each center started from different

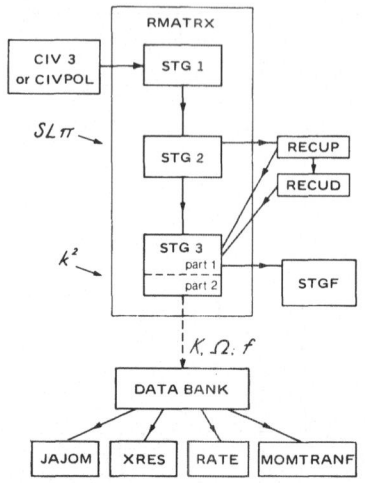

FIGURE 4. The RMATRX package and associated programs. The Atomic DATA BANK (see also Section 3), containing mainly electron collision data but also structure data such as oscillator strengths f, can be accessed by various programs for further processing; XRES, which computes resonance parameters, and RATE, a program that calculates rate coefficients assuming a Maxwell distribution of the colliding electrons are not described here.

approaches–often employing quite different methods—thus giving confidence when the results were converging for similar problems.

The amount of input required and of output produced by a general purpose program tend to be somewhat complementary, and the balance is usually determined by the applications one has in mind and to some extent by the methods employed. Attempts are under way at University College, London,[187] and at the Daresbury Laboratory to standardize input, each package converting it to its own format in a preprocessor step so that codes serving the same purpose can be used interchangeably. Such an option is particularly desirable since one code may be more efficient for one problem, another code more economical for some other problem or some section of it. The preprocessor codes make use of a portable NAMELIST facility written by Day.[188] General purpose programs will aim at minimal input, inserting default values for unspecified optional input. Ideally, a set of configurations C_i more or less suffices as input when computing the energy positions and associated radiative transition probabilities along an isoelectronic sequence. In practice, one may prefer to deal with a single ion at a time, possibly saving expensive intermediate data on some RESTART file—especially for the lower numbers of a sequence since correlation effects require a more elaborate expansion in terms of configurations. For the more highly ionized species one could do with even less input; rather than specifying configurations, the complexes involved in the states of interest would do! The concept of a complex and its significance in atomic structure owes a great deal to Layzer's work,[189,190] which was so much in Seaton's mind in those pioneering days of general purpose programs for atomic physics. It may also have appealed to him for a more formal reason in that the nuclear charge is treated as a continuous variable, with obvious parallels to a basic idea in quantum defect theory which is reviewed in the chapter by Moores and Saraph.[191] Scaling laws have since been a powerful tool in structure and collision work, while Layzer's nonrelativistic screening theory has been extended to the relativistic case.[192]

Printed output of general purpose programs easily grows prohibitively bulky for more complex problems and is therefore kept in check by various parameters in the input stream. In the context of standardization another class of output is more important: data for further processing in various subsequent programs, which are in card image form or more usually now on magnetic disc files. As early as in the late 1960s a standardized format for K-matrix elements was proposed for use by SIMMEG[193] and later extended for optional purposes in the program JAJOM.[100] Quantum defect theory is playing a vital role in quickly and economically obtaining collisional data for ions, e.g., through programs RANAL and JAJOM as indicated in Figs. 4 and 5, over a wide energy range from sparse results computed by elaborate codes such as RMATRX or IMPACT. Two more types of standardized formats may be mentioned: first, for term coupling coefficients, also used by JAJOM, and second, for radial orbital data as required by IMPACT,[48] which is an alternative for tabulated numerical input to the input format for STOs defined in Eq. (6). A special application of such numerical orbital data arises when successively building up an ion from some more highly ionized stage[182]; one may then feed back data obtained in a frozen-core approach as described for IMPACT by Seaton and Wilson.[194]

As vast amounts of collision data accumulate, recent work by Berrington, Elder, and Kingston[179] has concentrated on automatically storing K matrices, collision strengths; and rate coefficients in a DATA BANK, already mentioned in Section 3.7. See also Fig. 4.

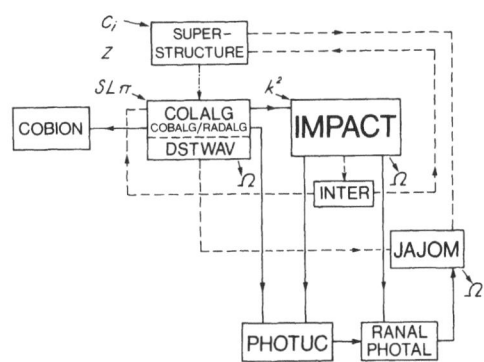

FIGURE 5. Program IMPACT and the major programs of its associated package. Outside the scope of this review are the photoionization branch RADALG/PHOTUC and the COBALG/COBION branch, which deals with electron impact ionization.

A final point concerns processibility. By a fortunate coincidence, Day's MACRO-PROCESSOR[195] became available just when Seaton's IMPACT code had completed its first job, $e + C$ III, and the size of a number of arrays had to be increased throughout some three boxes of cards to perform the next task. It was a few days before Christmas 1970, before the third or fourth debugging run: replacing the first six size parameters on all marked DIMENSION cards by symbolic macroprocessible names brought success within hours. All the major UCL codes have since been rewritten in preprocessible form, and conversion of other large codes is under way. Processibility extends to machine-dependent code or special options: e.g., activation of double precision replacements of single precision statements, or of an alternative code for use on vectorizing computers, is controlled by key words specified in the preprocessing step. Programs published up to date in the Computer Physics Communications program library, however, have not used the full MACRO-PROCESSOR. Instead, program-specific preprocessors were written in ordinary FORTRAN and employ subroutine STRING[196] and an auxiliary code, which is also incorporated in the MACRO-PROCESSOR.

Finally, we note that many of the program packages described below have been published in Computer Physics Communications and are available from the associated program library in Belfast.

4.1. Structure Packages

MCH77 is the recent robust version[197] of Charlotte Froese-Fischer's popular multiconfiguration Hartree–Fock program. It determines self-consistent field solutions for bound states in a nonrelativistic approach. A modified version[198] of subroutine OUTPUT provides orbitals $P_{nl}(r)$ as mentioned in Section 4.

CIV3 is one of the major codes which computes atomic structure and radiative transition properties in LS coupling[51] and in intermediate coupling using the Breit–Pauli Hamiltonian[54] [see Eqs. (41) and (42)]. Angular integrals are performed using Racah algebra, and radial functions $P_{nl}(r)$ are constructed as sums of STOs given by Eq. (6) so that radial integrals can be performed analytically. Common to general purpose programs, the parameters in the trial functions are evaluated variationally in an *ab initio* calculation. Among the special features is an adaptation that allows electric dipole and magnetic quadrupole hyperfine-structure parameters and energy corrections[199] to be evaluated. A modified version of CIV3, called CIV-POL, allows polarized pseudo-states to be calculated[85,86] (see section 2.2).

SUPERSTRUCTURE[47] computes atomic energies and radiative data in LS coupling and in intermediate coupling using the Breit–Pauli

Hamiltonian as in CIV3. It applies Slater techniques[200] to expand the angular terms in the usual multiconfiguration approach. Historically, Godfredsen's paper,[201] along with the advent of fast computers with short integer word length to store quantum numbers, provided the incentive to formulate such a fully automatic code.[43] Unless the user supplies radial orbitals $P_{nl}(r)$ in tabulated form the program internally calculates radial functions in a scaled statistical model potential[43,84]; the scaling parameters are determined variationally. Among subsequent intermediate coupling extensions the first feature[202,203] incorporated originates in the work of Blume and Watson,[204] who showed that mutual spin–orbit effects of a closed electronic shell on a valence electron behave like an effective screening on the spin–orbit coupling parameter. The latest feature that has been incorporated involves corrections to radiative operators that contribute to the transition amplitude with terms of equal order in α^2 as the Breit–Pauli components in the wave function Φ; it has been described recently[205] and applied to magnetic dipole transitions of O II and subsequently to the N I and P I sequences.[206]

The COWAN-ZEALOT suite of programs[207] originates in work by Cowan at Los Alamos, New Mexico, is based on Slater–Condon theory and Racah methods, and includes Cowan's[208] statistical approximation for exchange to the Hartree–Fock method. It computes atomic structure and radiative transition properties, either *ab initio* or semiempirically. Spin–orbit effects are included and optionally approximate relativistic corrections to orbitals.[209]

MCDF[64] sets up the relativistic Dirac–Fock equations and solves them within the framework of a multiconfiguration approximation, giving both atomic energy and radiative transition results for atoms and ions. The package is robust and easy to control; at various stages information is saved, which allows the program to restart for a similar problem.

4.2. Collision Packages

RMATRX is a general purpose package for calculating collision processes of electrons with a variety of targets, applying methods described in Section 3.3. The package[52,53] calculates collision strengths and K matrices for collisions in LS coupling with atoms and ions in three stages as shown in Fig. 4: STG1/STG2/STG3. Their role may be gauged from the input information in the figure, and a brief summary on their interaction can be found in a recent paper.[186] It is typical for these general purpose codes to be organized in such a manner that expensive data are saved at suitable stages; by exploiting such a facility in electron collision codes one may save computer time in runs for subsequent collision energies k_i^2.

Substantial parts of the code have been formulated for use on vectorizing machines. The program has been applied to numerous collision problems and it has a facility for solving the corresponding bound-state problem where all channels are closed. One recent variant of the package is of special interest to electron–atom collision processes: Scott and Taylor[210] have modified and extended RMATRX to deal with electron collisions in intermediate coupling, in a Breit–Pauli approach (see additional programs RECUP and RECUD in Fig. 4); the target atom or ion is also expanded in a Breit–Pauli approximation. Other properties calculated by RMATRX, such as polarizabilities and photoionization cross sections, are outside the scope of this review. So is program STGF, also shown in Fig. 4, which deals with free–free processes.

IMPACT[48] is the general purpose program for reducing the coupled integro-differential equations (27) to a system of linear algebraic (LA) equations as described in Section 3.2. In the bound-state case where all channels are closed, the LA equations (82) result in an eigenvalue problem. Since solutions $F = P_{nl}(r)$ from such frozen-core calculations have numerous applications, program INTER, shown in Fig. 5, provides standard tabulated output by interpolating from the sparse grid used in IMPACT. Among the additional features of IMPACT is the optional inclusion of polarization of the target core and also of dielectronic terms. An alternative code can be activated for running IMPACT on a vector processing machine. In addition, a CRAY-specific version has been created, using assembler code for a few central routines and CRAY utility routines. With this code CPU time is reduced by an order of magnitude compared with the standard version of IMPACT.

NIEM is a package based on the noniterative integral equations method, as outlined in Section 3.5, for solving the electron + target problem in LS coupling. In its present form[56] the package is organized in the stages POTC1/NIE2/ASYM3/GRN2; the names indicating which role these stages play.

DSTWAV computes electron–atom and electron–ion collision strengths and K matrices in LS coupling in a distorted-wave approach, approximating the collisional partial waves $F_i(r)$ by solutions in a suitable scaled statistical model potential. The coefficients a_i in expansion (3) are determined fully variationally,[44,45] and methods by Burgess[211] for Coulomb phases and amplitudes are employed in the evaluation of the elastic scattering phase δ^{DW} and of integrals required for I^t in Eq. (29). For evaluating long-range contributions to these integrals a code developed by Belling[212] is used. As indicated in Fig. 5, DSTWAV is derived from a basic program COLALG, which computes the partial-wave algebra required as a first stage by IMPACT. COLALG employs methods and codes developed earlier for SUPERSTRUCTURE; the collisonal partial waves are added to multiconfiguration target states using vector coupling formulas. VCCLSF, an alternative program to COLALG, has been written by Vo Ky Lan and Le Dourneuf.[213]

JAJOM[100] computes collision strengths Ω for transitions between fine-structure levels using K-matrix data in LS coupling according to Eq. (46) and optionally term coupling coefficients to make allowance for Breit–Pauli effects in the target.[99] A special version, called JAJOMII and described in Section 5.2.2 of the chapter by Moores and Saraph,[191] computes electron ion collision strengths Ω that contain resonances due to closed channels, starting with open-channel K-matrix input and RANAL as an intermediate step as indicated in Fig. 5. A detailed description of program RANAL is given by Moores and Saraph in Section 5.2.2. In the interfaced mode, program JAJOM can be used with input from the DATA BANK as shown in Fig. 4.

MOMTRANF[214] is an angular distribution code for electron–atom or electron–ion scattering in LS coupling, using as input T or K matrices.

ACKNOWLEDGMENTS

We wish to take this opportunity to thank Mike Seaton for his help, guidance, and friendship over many years. We also wish to thank our colleagues at University College, London, Queen's University, Belfast, and Daresbury Laboratory, Cheshire, for their continuing collaboration which has made this work possible.

References

1. M. J. SEATON, *Phil. Trans. R. Soc. London Ser. A* **245**, 469–499 (1953).
2. P. M. MORSE and W. P. ALLIS, *Phys. Rev.* **44**, 269–276 (1933).
3. D. R. BATES and H. S. W. MASSEY, *Proc. R. Soc. Lond. Ser. A* **177**, 329–339 (1941).
4. D. R. BATES and H. S. W. MASSEY, *Phil. Trans. R. Soc. Lond. Ser. A* **239**, 269–304 (1943).
5. D. R. BATES and H. S. W. MASSEY, *Proc. R. Soc. Lond. Ser. A* **192**, 1–16 (1947).
6. D. R. BATES and M. J. SEATON, *Mon. Not. R. Astron. Soc.* **109**, 698–704 (1949).
7. M. J. SEATON, *Mon. Not. R. Astron. Soc.* **110**, 247–255 (1950).
8. M. J. SEATON, *Proc. R. Soc. Lond. Ser. A* **208**, 408–418 (1951).
9. M. J. SEATON, *Proc. R. Soc. Lond. Ser. A* **208**, 418–430 (1951).
10. M. J. SEATON, *Proc. R. Soc. Lond. Ser. A* **218**, 400–416 (1953).
11. M. J. SEATON, in *The Airglow and the Aurora*, E. B. Armstrong and A. Dalgarno, Eds., Pergamon, London, 1956, pp. 289–301.
12. D. R. BATES, "Forbidden Atomic Lines in Auroral Spectra," Chapter 10 in this book.
13. D. R. FLOWER, "Applications to Gaseous Nebulae," Chapter 9 in this book.
14. P. G. BURKE and M. J. SEATON, in *Methods in Computational Physics*, Vol. 10, B. Alder, S. Fernbach, and M. Rotenberg, Eds., Academic, New York, 1971, pp. 1–80.

15. R. MARRIOTT, *Proc. Phys. Soc.* **72**, 121–129 (1958).
16. I. C. PERCIVAL, private communication to R. Marriott.
17. D. R. HARTREE, *The Calculation of Atomic Structures*, Wiley, New York, 1957.
18. I. C. PERCIVAL and, M. J. SEATON, *Proc. Camb. Phil. Soc.* **53**, 654–662 (1957).
19. P. G. BURKE, V. M. BURKE, I. C. PERCIVAL, and R. McCARROLL, *Proc. Phys. Soc.* **80**, 413–421 (1962).
20. V. M. BURKE and R. McCARROLL, *Proc. Phys. Soc.* **80**, 422–431 (1962).
21. T. L. JOHN, *Proc. Phys. Soc.* **76**, 532–538 (1960).
22. P. G. BURKE and H. M. SCHEY, *Phys. Rev.* **126**, 147–162 (1962).
23. K. SMITH, *Phys. Rev.* **120**, 845–847 (1960).
24. P. G. BURKE and K. SMITH, *Rev. Mod. Phys.* **34**, 458–502 (1962).
25. P. G. BURKE, H. M. SCHEY, and K. SMITH, *Phys. Rev.* **129**, 1258–1274 (1963).
26. P. G. BURKE, D. D. McVICAR, and K. SMITH, *Proc. Phys. Soc.* **83**, 397–407 (1964).
27. P. G. BURKE and D. D. McVICAR, *Proc. Phys. Soc.* **86**, 989–1006 (1965).
28. R. DAMBURG and M. GAILITIS, *Proc. Phys. Soc.* **82**, 1068–1070 (1963).
29. K. OMIDVAR, *Phys. Rev.* **133**, A970–A985 (1964).
30. R. P. McEACHRAN and P. A. FRASER, *Proc. Phys. Soc.* **82**, 1038–1045 (1963).
31. R. P. McEACHRAN, P. A. FRASER, J. B. G. WALLACE, and C. E. TULL, *Proc. Phys. Soc.* **86**, 473–489 (1965).
32. M. J. SEATON, Boulder Lecture Notes 1961, unpublished.
33. K. SMITH, R. J. W. HENRY, and P. G. BURKE, *Phys. Rev.* **147**, 21–28 (1966).
34. K. SMITH, R. J. W. HENRY, and P. G. BURKE, *Phys. Rev.* **157**, 51–68 (1967).
35. R. J. W. HENRY, P. G. BURKE, and A.-L. SINFAILAM, *Phys. Rev.* **178**, 218–225 (1969).
36. H. E. SARAPH, M. J. SEATON, and J. SHEMMING, *Proc. Phys. Soc.* **89**, 27–34 (1966).
37. S. J. CZYZAK, T. K. KRUEGER, P. DE A. P. MARTINS, H. E. SARAPH, and M. J. SEATON, *Mon. Not. R. Astron. Soc.* **148**, 361–365 (1970).
38. R. K. NESBET, *Phys. Rev.* **156**, 99–102 (1967).
39. K. SMITH and L. A. MORGAN, *Phys. Rev.* **165**, 110–122 (1968).
40. P. G. BURKE and A. HIBBERT, *Abstracts of VIth ICPEAC*, MIT Press, Cambridge, Mass., 1969, pp. 367–369.
41. P. G. BURKE, in *Computational Physics*, a Conference Digest, IoP, London, 1970, pp. 9–18.
42. M. J. SEATON, in *Computational Physics*, a Conference Digest, IoP, London, 1970, pp. 19–24.
43. W. EISSNER and H. NUSSBAUMER, *J. Phys. B* **2**, 1028–1043 (1969).
44. W. EISSNER and M. J. SEATON, *J. Phys. B* **5**, 2187–2198 (1972).
45. W. EISSNER, in *The Physics of Electronic and Atomic Collisions, Invited Papers and Progress Reports at the VIIth ICPEAC*, T. R. Govers and F. J. de Heer, Eds., North Holland, Amsterdam, 1972, pp. 460–479.
46. M. J. SEATON, *J. Phys. B* **7**, 1817–1840 (1974).
47. W. EISSNER, M. JONES, and H. NUSSBAUMER, *Comput. Phys. Commun.* **8**, 270–306 (1974).
48. M. A. CREES, M. J. SEATON, and P. M. H. WILSON, *Comput. Phys. Commun.* **15**, 23–83 (1978).
49. P. G. BURKE, A. HIBBERT, and W. D. ROBB, *J. Phys. B* **4**, 153–161 (1971).
50. P. G. BURKE and W. D. ROBB, *Adv. At. Mol. Phys.* **11**, 143–214 (1975).
51. A. HIBBERT, *Comput. Phys. Commun.* **9**, 141–172 (1975).
52. K. A. BERRINGTON, P. G. BURKE, J. J. CHANG, A. T. CHIVERS, W. D. ROBB, and K. T. TAYLOR, *Comput. Phys. Commun.* **8**, 149–198 (1974).
53. K. A. BERRINGTON, P. G. BURKE, M. LE DOURNEUF, W. D. ROBB, K. T. TAYLOR, and VO KY LAN, *Comput. Phys. Commun.* **14**, 367–412 (1978).
54. R. GLASS and A. HIBBERT, *Comput. Phys. Commun.* **16**, 19–34 (1978).
55. E. R. SMITH and R. J. W. HENRY, *Phys. Rev. A* **7**, 1585–1590 (1973); *Phys. Rev. A* **8**, 572–575 (1973).
56. R. J. W. HENRY, S. P. ROUNTREE, and E. R. SMITH, *Comput. Phys. Commun.* **23**, 233–273 (1981).

57. R. K. NESBET, *Comput. Phys. Commun.* **6**, 275–287 (1974).
58. R. K. NESBET, *Comput. Phys. Commun.* **17**, 163–169 (1979)
59. R. K. NESBET, *Variational Methods in Electron Atom Scattering Theory*, Plenum, New York, 1980.
60. N. S. SCOTT, and P. G. BURKE, *Comput. Phys. Commun.* **26**, 419–421 (1982).
61. R. K. NESBET, private communication.
62. M. JONES, *Phil. Trans. R. Soc. Lond. A* **277** 587–622 (1975).
63. N. S. SCOTT and P. G. BURKE, *J. Phys. B* **13**, 4299–4314 (1980).
64. I. P. GRANT, B. J. MCKENZIE, P. H. NORRINGTON, D. F. MAYERS, and N. C. PYPER *Comput. Phys. Commun.* **21**, 207–231 (1980).
65. P. H. NORRINGTON and I. P. GRANT, *J. Phys. B* **14**, L261–L267 (1981).
66. W. P. REINHARDT, *Comput. Phys. Commun.* **17**, 1–21 (1979).
67. B. H. BRANSDEN and M. R. C. MCDOWELL, *Phys. Rep.* **30C**, 207–303 (1977).
68. D. W. NORCROSS, "Asymptotic Solutions," Chapter 2 in this book.
69. R. DAMBURG and E. KARULE, *Proc. Phys. Soc.* **90**, 637–640 (1967).
70. P. G. BURKE and T. G. WEBB, *J. Phys. B* **3**, L131–L134 (1970).
71. P. G. BURKE and J. F. B. MITCHELL, *J. Phys. B* **6**, 320–328 (1973).
72. J. CALLAWAY and J. W. WOOTEN, *Phys. Rev. A* **9**, 1924–1931 (1974).
73. J. CALLAWAY and J. W. WOOTEN, *Phys. Rev. A* **11**, 1118–1120 (1975).
74. P. G. BURKE and A. J. TAULOR, *Proc. Phys. Soc.* **88**, 549–562 (1966).
75. A. J. TAYLOR and P. G. BURKE, *Proc. Phys. Soc.* **92**, 336–344 (1967).
76. A. W. WEISS, *Phys. Rev.* **162**, 71–80 (1967); A. W. WEISS, *Nucl. Instrum. Meth.* **90**, 121–131 (1970).
77. H. NUSSBAUMER, *Astrophys. J.* **166**, 411–422 (1971).
78. A. HIBBERT, *Rep. Prog. Phys.* **38**, 1217–1338 (1975).
79. L. CASTILLEJO, I. C. PERCIVAL, and M. J. SEATON, *Proc. R. Soc. Lond. Ser. A* **254**, 259–272 (1960).
80. Y. HAHN, T. F. O'MALLEY, and L. SPRUCH, *Phys. Rev.* **134**, B397–B404 (1964).
81. Y. HAHN, T. F. O'MALLEY, and L. SPRUCH, *Phys. Rev.* **134**, B911–B919 (1964).
82. M. GAILITIS, *Sov. Phys. JETP* **20**, 107–111 (1965).
83. W. A. MCKINLEY and J. H. MACEK, *Phys. Lett.* **10**, 210–212 (1964).
84. P. GOMBÁS, *Handb. Phys.* **36**, 109–231 (1956).
85. P. G. BURKE and J. F. B. MITCHELL, *J. Phys. B* **7**, 655–673 (1974).
86. VO KY LAN, M. LE DOURNEUF, and P. G. BURKE, *J. Phys. B* **9**, 1065–1078 (1961).
87. L. HULTHÉN, *K. Fysiogr. Sälsk. Lund. Förh.* **14**, No. 21, (1944).
88. W. KOHN, *Phys. Rev.* **74**, 1763–1772 (1948).
89. H. E. SARAPH, M. J. SEATON, and J. SHEMMINGS, *Phil. Trans. R. Soc. Lond.* **264**, 77–105 (1969).
90. J. M. BLATT and L. C. BIEDENHARN, *Rev. Mod. Phys.* **24**, 258–272 (1952).
91. A. M. LANE and R. G. THOMAS, *Rev. Mod. Phys.* **30**, 257–353 (1958).
92. M. H. HEBB and D. H. MENZEL, *Astrophys. J.* **92**, 408–423 (1940).
93. L. T. SINFAILAM, *Aust. J. Phys.* **33**, 261–281 (1980).
94. L. T. SINFAILAM and W. E. BAYLIS, *J. Phys. B* **14**, 559–571 (1981).
95. G. D. CARSE and D. W. WALKER, *J. Phys. B* **6**, 2529–2544 (1973).
96. D. W. WALKER, *J. Phys. B* **7**, 97–116 (1974).
97. J. J. CHANG, *J. Phys. B* **8**, 2327–2335 (1975).
98. J. J. CHANG, *J. Phys. B* **10**, 3335–3339 (1977).
99. H. E. SARAPH, *Comput. Phys. Commun.* **3**, 256–268 (1972).
100. H. E. SARAPH, *Comput. Phys. Commun.* **15**, 247–258 (1978).
101. M. H. ROSS and G. L. SHAW, *Ann. Phys. (N.Y.)* **13**, 147–186 (1961).
102. P. G. BURKE and J. F. B. MITCHELL, *J. Phys. B* **7**, 214–228 (1974).
103. H. A. BETHE and E. E. SALPETER, *Quantum Mechanics of One- and Two-Electron Atoms*, Springer-Verlag, Berlin, 1972.
104. A. I. AKHIEZER and V. B. BERESTETSKY, *Quantum Electrodynamics*, Interscience, New York, 1965.
105. G. RACAH, *Phys. Rev.* **61**, 537–539 (1942).

106. D. M. Brink and G. B. Satchler, *Angular Momentum*, Clarendon, Oxford, 1968.
107. N. F. Mott and H. S. W. Massey, *The Theory of Atomic Collisions*, Oxford University Press, London, 1965.
108. H. N. Browne and E. Bauer, *Phys. Rev. Lett.* **16**, 495–498 (1966).
109. I. P. Grant, *Adv. Phys.* **19**, 747–811 (1970).
110. I. P. Grant, *Comput. Phys. Commun.* **17**, 149–161 (1979).
111. B. J. McKenzie, I. P. Grant, and P. H. Norrington, *Comput. Phys. Commun.* **21**, 233–246 (1980).
112. K. Smith, R. P. McEachran, and P. A. Fraser, *Phys. Rev.* **125**, 553–558 (1962).
113. H. E. Saraph and M. J. Seaton, *Proc. Phys. Soc.* **80**, 1057–1066 (1962).
114. H. E. Saraph, *Proceedings of the IIIrd ICPEAC*, M. R. C. McDowell, Ed., North Holland, Amsterdam, 1964, pp. 359–363.
115. V. B. Sheorey, *Proc. Phys. Soc.* **92**, 531–538 (1967).
116. H. H. Robertson, *Proc. Camb. Phil. Soc.* **52**, 538–545 (1956).
117. D. G. Thompson, *Proc. R. Soc. Lond. Ser. A* **294**, 160–174 (1966).
118. M. A. Crees and D. L. Moores, *J. Phys. B* **8**, L195–L199 (1975).
119. B. I. Schneider and L. A. Collins, *J. Phys. B* **14**, L101–L106 (1981).
120. B. I. Schneider and L. A. Collins, *Phys. Rev. A* **24**, 1264–1266 (1981).
121. E. P. Wigner, *Phys. Rev.* **70**, 15–33 (1946).
122. E. P. Wigner, *Phys. Rev.* **70**, 606–618 (1946).
123. E. P. Wigner and L. Eisenbud, *Phys. Rev.* **72**, 29–41 (1947).
124. A. M. Lane and D. Robson, *Phys. Rev.* **151**, 774–787 (1966).
125. D. Robson and A. M. Lane, *Phys. Rev.* **161**, 982–993 (1967).
126. C. Mahaux and H. A. Weidenmüller, *Phys. Rev.* **170**, 847–856 (1968).
127. C. Bloch, *Nucl. Phys.* **4**, 503–528 (1957).
128. P. J. A. Buttle, *Phys. Rev.* **160**, 719–729 (1967).
129. U. Fano and C. M. Lee, *Phys. Rev. Lett.* **31**, 1573–1576 (1973).
130. C. M. Lee, *Phys. Rev. A* **10**, 584–600 (1974).
131. M. Danos and W. Greiner, *Phys. Rev.* **146**, 708–712 (1966).
132. B. I. Schneider, *Chem. Phys. Lett.* **31**, 237–241 (1975).
133. B. I. Schneider, *Phys. Rev. A* **11**, 1957–1962 (1975).
134. B. I. Schneider and P. J. Hay, *J. Phys. B* **9**, L165–L167 (1976).
135. B. I. Schneider, in *Electronic and Atomic Collisions*, G. Watel, Ed., North Holland, Amsterdam, 1977, pp. 257–269.
136. B. I. Schneider, M. Le Dourneuf, and P. G. Burke, *J. Phys. B* **12**, L365–L369 (1977).
137. B. I. Schneider, M. Le Dourneuf, and Vo Ky Lan, *Phys. Rev. Lett.* **43**, 1926–1929 (1977).
138. P. G. Burke, I. Mackey, and I. Shimamura, *J. Phys. B* **10**, 2497–2512 (1977).
139. B. D. Buckley, P. G. Burke, and Vo Ky Lan, *Comput. Phys. Commun.* **17**, 175–179 (1979).
140. C. J. Noble, P. G. Burke, and S. Salvini, to be published in *J. Phys. B* (1982).
141. H. S. W. Massey and B. L. Moiseiwitsch, *Proc. R. Soc. Lond. Ser. A* **205**, 483–496 (1951).
142. C. Schwartz, *Phys. Rev.* **124**, 1468–1471 (1961).
143. C. Schwartz, *Ann. Phys. (N.Y.)* **16**, 36–50 (1961).
144. R. K. Nesbet and R. S. Oberoi, *Phys. Rev. A* **6**, 1855–1862 (1972).
145. W. N. Sams and D. J. Kouri, *J. Chem. Phys.* **51**, 4809–4814 (1969).
146. W. N. Sams and D. J. Kouri, *J. Chem. Phys.* **51**, 4815–4819 (1969).
147. W. N. Sams and D. J. Kouri, *J. Chem. Phys.* **52**, 4144–4150 (1970).
148. W. N. Sams and D. J. Kouri, *J. Chem. Phys.* **53**, 496–501 (1970).
149. F. W. Byron, Jr. and C. J. Joachain, *Rep. Prog. Phys.* **34**, 233–324 (1977).
150. A. E. Kingston and H. R. J. Walters, *J. Phys. B* **13**, 4633–4662 (1980).
151. L. Schlessinger and C. Schwartz, *Phys. Rev. Lett.* **16**, 1173–1174 (1966).
152. L. Schlessinger, *Phys. Rev.* **171**, 1523–1527 (1968).
153. F. A. McDonald and J. Nuttall, *Phys. Rev. Lett.* **23**, 361–363 (1969).

154. G. Doolen, G. McCartor, F. A. McDonald, and J. Nuttall, *Phys. Rev. A* **4**, 108–111 (1971).
155. P. W. Langhoff, *Chem. Phys. Lett.* **22**, 60–64 (1973).
156. P. W. Langhoff and W. P. Reinhardt, *Chem. Phys. Lett.* **24**, 495–506 (1974).
157. J. R. Winick and W. P. Reinhardt, *Phys. Rev. A* **18**, 910–924 (1978).
158. J. R. Winick and W. P. Reinhardt, *Phys. Rev. A* **18**, 925–934 (1978).
159. P. G. Burke, K. A. Berrington, and C. V. Sukumar, *J. Phys. B* **14**, 289–305 (1981).
160. F. L. Friedman and V. F. Weisskopf, in *Niels Bohr and the Development of Physics*, W. Pauli, L. Rosenfeld, and V. F. Weisskopf, Eds., Pergamon, London, 1955, pp. 134–162.
161. H. Feshbach, *Ann. Phys. (N.Y.)* **5**, 537–590 (1958).
162. G. E. Brown, *Rev. Mod. Phys.* **31**, 893–919 (1959).
163. R. Poet, *J. Phys. B* **11**, 3081–3094 (1978).
164. R. Poet, *J. Phys. B* **13**, 2995–3008 (1980).
165. P. L. Altick, P. G. Burke, and K. T. Taylor, private communication.
166. E. J. Heller and H. A. Yamani, *Phys. Rev. A* **9**, 1201–1208 (1974).
167. E. J. Heller and H. A. Yamani, *Phys. Rev. A* **9**, 1209–1214 (1974).
168. J. T. Broad and J. Reinhardt, *J. Phys. B* **9**, 1491–1502 (1976).
169. J. T. Broad and J. Reinhardt, *Phys. Rev. A* **14**, 2159–2173 (1976).
170. J. Schwinger, *Phys. Rev.* **72**, 742 (1947).
171. B. A. Lippmann and J. Schwinger, *Phys. Rev.* **79**, 469–480 (1950).
172. B. L. Moiseiwitsch, *Variational Principles*, Wiley–Interscience, New York, 1966, pp. 256–261.
173. D. K. Watson and V. McKoy, *Phys. Rev. A* **20**, 1474–1483 (1979).
174. D. K. Watson, R. R. Lucchese, V. McKoy, and T. N. Rescigno, *Phys. Rev. A* **21**, 738–744 (1980).
175. R. R. Lucchese, D. K. Watson, and V. McKoy, *Phys. Rev. A* **22**, 421–426 (1980).
176. D. K. Watson, T. N. Rescigno, and V. McKoy, *J. Phys. B* **14**, 1875–1882 (1981).
177. R. S. Oberoi and R. K. Nesbet, *J. Comput. Phys.* **12**, 526–533 (1973).
178. R. K. Nesbet, Proceedings of Daresbury Laboratory Study Weekend, W. Eisner, ed. (1982).
179. K. A. Berrington, M. Elder, and A. E. Kingston, Queen's University–SERC Daresbury Atomic Data Bank.
180. M. J. Seaton, *Adv. At. Mol. Phys.* **11**, 83–142 (1975).
181. R. J. W. Henry, *Phys. Rep.* **68**, 1–91 (1981).
182. C. Mendoza, submitted to *Mon. Not. R. Astr. Soc.*; C. Mendoza, *Mon. Not. R. Astr. Soc.* **198**, 127–139 (1982).
183. K. L. Baluja, P. G. Burke, and A. E. Kingston, *J. Phys. B* **13**, L543–L545 (1980).
184. K. L. Baluja, P. G. Burke, and A. E. Kingston, *J. Phys. B* **14**, 1333–1340 (1981).
185. P. L. Dufton and A. E. Kingston, *J. Phys. B* **13**, 4277–4284 (1980).
186. K. A. Berrington and A. Crees, *Comput. Phys. Commun.* **17**, 181–205 (1979).
187. M. A. Crees, private communication.
188. A. C. Day, *Comput. Phys. Commun.* **22**, 403–410 (1981).
189. D. Layzer, *Ann. Phys. (N.Y.)* **8**, 271–296 (1959).
190. D. Layzer, Z. Horák, M. N. Lewis, and D. P. Thompson, *Ann. Phys. (N.Y.)* **29**, 101–124 (1964).
191. D. L. Moores and H. E. Saraph, "Applications of Quantum Defect Theory," Chapter 6 in this book.
192. D. Layzer and J. Bahcall, *Ann. Phys. (N.Y.)* **17**, 177–204 (1962).
193. H. E. Saraph, *Comput. Phys. Commun.* **1**, 232–240 (1970).
194. M. J. Seaton and P. M. H. Wilson, *J. Phys. B* **5**, L1–L3 (1972).
195. A. C. Day, A Macro-Processor for FORTRAN, Technical Report No. 2, University College, London, Computer Centre (1971).
196. A. C. Day, *Comput. Phys. Commun.* **13**, 101–105 (1977).
197. C. Froese-Fischer, *Comput. Phys. Commun.* **14**, 145–153 (1978).

198. H. E. SARAPH, private communication.
199. R. GLASS and A. HIBBERT, *Comput. Phys. Commun.* **11**, 125–140 (1976).
200. E. H. CONDON and G. H. SHORTLEY, *The Theory of Atomic Spectra*, Cambridge University Press, London, 1951.
201. E. GODFREDSEN, *Astrophys. J.* **145**, 308–332 (1966).
202. M. JONES, *J. Phys. B* **3**, 1571–1592 (1970).
203. M. JONES, *J. Phys. B* **4**, 1422–1439 (1971).
204. M. BLUME and R. E. WATSON, *Proc. R. Soc. Lond. Ser. A* **270**, 127–143 (1962).
205. W. EISSNER and C. J. ZEIPPEN, *J. Phys. B* **14**, 2125–2135 (1981).
206. C. MENDOZA and C. J. ZEIPPEN, private communication.
207. G. E. BROMAGE, The Cowan–Zealot Suite of Computer Programs for Atomic Structure, Science Research Council Report AL-R3, Abingdon, Oxon. (1978).
208. R. D. COWAN, *Phys. Rev.* **163**, 54–61 (1967).
209. R. D. COWAN and D. C. GRIFFIN, *J. Opt. Soc. Am.* **66**, 1010–1014 (1976).
210. N. S. SCOTT and K. T. TAYLOR, *Comput. Phys. Commun.* **25**, 347–387 (1982).
211. A. BURGESS, *Proc. Phys. Soc.* **81**, 442–452 (1963).
212. J. A. BELLING, *J. Phys. B* **1**, 136–138 (1968).
213. VO KY LAN and M. LE DOURNEUF, private communication.
214. S. A. SALVINI, *Comput. Phys. Commun.* **27**, 25–37 (1982).

2

NUMERICAL METHODS FOR ASYMPTOTIC SOLUTIONS OF SCATTERING EQUATIONS

D. W. Norcross

1. Introduction

The focus of our discussion is illustrated by Fig. 1, in which the radial extent of the scattering problem is divided into several regions a la Inokuti.[1] *Inner Regia* is the domain of hardy and resourceful individuals who are trained from birth to cope with such scourges as correlation and exchange. *Outer Regia*, in contrast, is a gentler clime of local potentials that are more easily controlled by its residents. These two kingdoms are so different that the rough folk of *Inner Regia* and the gentle folk of *Outer Regia* are forbidden to interact except at their common border.

Outer Regia is further subdivided into the duchies of *Intermedia* and *Asymptopia*. The latter is a rather flat and monotonous land, the inner border of which is defined as that point from which the tourist can peruse the entire domain. The precise location of the border however, depends on the power of the tourist's vision. Knowledge of *Intermedia*, on the other hand, requires thorough solution of the scattering equations over the entire domain, despite the simplicity of the potential.

The honoree of this volume is one of the very few individuals who has traveled extensively in all three regions. His contributions to the development of general techniques for solving the complex equations describing electron–atom (ion) scattering in *Inner Regia*, and their significance for astrophysics, are described elsewhere in this volume. Here I wish to review his travels in *Outer Regia*, and perhaps to suggest future directions based on his original explorations.

D. W. NORCROSS ● Joint Institute for Laboratory Astrophysics, University of Colorado and National Bureau of Standards, Boulder, Colorado 80309.

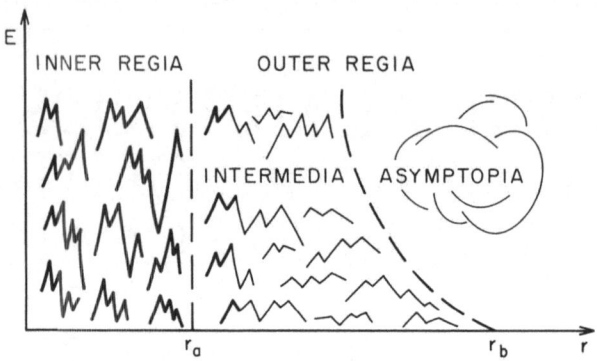

FIGURE 1. Map of the three general regions of coordinate space of concern to us here. The geographical features become more gentle as r is increased, but less rapidly for energies close to zero.

The fundamental equation at issue can be written

$$\left[\frac{d^2}{dr^2} - \frac{l(l+1)}{r^2} + \frac{2z}{r} + \mathbf{V}(r) + k^2\right]\mathbf{H}(r) = 0, \tag{1}$$

where

$$\mathbf{V}(r) = \sum_{n=2} \mathbf{C}^{(n)} r^{-n} \tag{2}$$

for $r \geqslant r_a$, and $\mathbf{C}^{(n)}$ is a matrix (here and after indicated by bold notation) of known constants. Scattering channels are open or closed (hereafter subscripts o and c) as k^2 is > 0 or < 0, respectively, and will be assumed ordered by decreasing values of k^2. A column of $\mathbf{H}(r)$ corresponds to a particular initial channel.

Our problem starts at the border of *Inner Regia*, where its residents present us with a set of solutions $\mathbf{\Psi}(r)$ that satisfy regular boundary conditions at the origin and that are presumably composed of columns having good linear independence. We, in exchange, must provide a set of solutions $\mathbf{I}(r)$ and $\mathbf{J}(r)$ of Eq. (1) at r_b having correct asymptotic forms and find a way to match them to $\mathbf{\Psi}(r)$. We have only two options: either to propagate $\mathbf{I}(r)$ and $\mathbf{J}(r)$ from r_b to r_a, or to propagate $\mathbf{\Psi}(r)$ from r_a to r_b. In either case the matching equation (for two points, or function and derivative),

$$\mathbf{\Psi}(r_m) = \mathbf{I}(r_m)\mathbf{A} - \mathbf{J}(r_m)\mathbf{B} \tag{3}$$

at the point r_m yields the solution over all space and the reactance matrix $\mathbf{R}$, defined by

$$\mathbf{R} = \mathbf{B}\mathbf{A}^{-1}. \tag{4}$$

In Section 2 we consider the asymptotic forms of solutions of Eq. (1) in some detail with particular attention to their uniqueness. Techniques for propagating solutions across the domain of *Intermedia* will be discussed in Section 3. Here our major concern is the problem of obtaining stable, linearly independent solutions, and a new technique, based on one of Seaton's original ideas, will be introduced. The development of tools for solving Eq. (1) on the border of *Asymptopia* will be reviewed in Section 4, where some improvements to existing techniques will be suggested.

2. Specification of Asymptotic Forms

It will be useful to have established at the outset the notation for solutions (the Green's functions) of the reduced equation

$$\left[\frac{d^2}{dr^2} - \frac{l(l+1)}{r^2} + \frac{2z}{r} + k^2 \right] \mathcal{K}(r) = 0, \tag{5}$$

where linearly independent solutions for all r that have unit Wronskian can be expressed, for $z = 0$, in terms of Ricatti–Bessel functions, and for $z \neq 0$, in terms of Coulomb and Whittaker functions.[2] We display here only their asymptotic forms, which for $k^2 > 0$ are

$$\mathcal{F}_o(r) \underset{r\to\infty}{\sim} k^{-1/2}\sin\left[kr - \tfrac{1}{2}l\pi + \frac{z}{k}\ln(2kr) + \sigma \right] \tag{6}$$

and

$$\mathcal{G}_o(r) \underset{r\to\infty}{\sim} -k^{-1/2}\cos\left[kr - \tfrac{1}{2}l\pi + \frac{z}{k}\ln(2kr) + \sigma \right], \tag{7}$$

and for $k^2 < 0$ (in which case we take $k = |k|$) are

$$\mathcal{F}_c(r) \underset{r\to\infty}{\sim} (2k)^{-1/2}(2kr)^{-z/k}e^{kr} \tag{8}$$

and

$$\mathcal{G}_c(r) \underset{r\to\infty}{\sim} -(2k)^{-1/2}(2kr)^{z/k}e^{-kr}, \tag{9}$$

where $\sigma = \arg\Gamma[(1 + 1 - iz)/k]$.

Solutions of Eq. (1) with asymptotic forms

$$F(r) \underset{r\to\infty}{\sim} \begin{bmatrix} \mathcal{F}_o(r) & O(r^{-2})\mathcal{G}_c(r) \\ O(r^{-2}) & \mathcal{G}_c(r) \end{bmatrix} \tag{10}$$

and

$$G(r) \underset{r\to\infty}{\sim} \begin{bmatrix} \mathcal{G}_o(r) & O(r^{-2})\mathcal{G}_c(r) \\ O(r^{-2}) & \mathcal{G}_c(r) \end{bmatrix} \tag{11}$$

are appropriate for the solution of Eq. (3). The identity of the closed-channel columns of Eqs. (10) and (11) need not concern us, as can be seen by casting Eq. (3) into the form

$$\Psi(r_m)A^{-1} = I(r_m) - J(r_m)R. \tag{12}$$

Only the open-channel columns of Eq. (12) are physically significant, and the matching equation actually involves the determination of only the four unknowns $(A^{-1})_{oo}$, $(A^{-1})_{co}$, R_{oo}, and R_{co}. The closed-channel columns of $I(r)$ are thus of no direct interest, but the reason for the particular choice in Eq. (10) will become apparent. The primary goal of the exercise is, of course, R_{oo}, with which cross sections are specified.

Functions adequate for the solution of Eq. (12) are not uniquely specified, however, by the asymptotic forms of Eqs. (10) and (11), since we could also use functions $I(r)$ and $J(r)$ that are related to them by a constant transformation, for example,

$$I(r) = F(r)X, \tag{13}$$

where

$$X = \begin{pmatrix} 1 & 0 \\ X_{co} & X_{cc} \end{pmatrix}. \tag{14}$$

That the submatrices X_{co} and X_{cc} need not be specified further can be seen by noting first that they do not alter the open-channel columns of $I(r)$, and second that R_{oo} is not changed in the product XR.

This ambiguity, or lack of uniqueness, is in fact a disguised blessing. An analogous ambiguity has long been exploited to suppress "creeping linear dependence"[3-6] and to maintain numerical significance[7] in solutions of integral equations being propagated in r. It has also been used in

another guise to guarantee linear independence in Seaton's[8] linear algebraic technique for the solution of scattering equations. In these works
advantage is taken of the fact that the result for **R** would be unchanged
where $\mathbf{\Psi}(r_m)$ to be replaced in Eq. (12) by

$$\mathbf{\Phi}(r_m) = \mathbf{\Psi}(r_m)\mathbf{Y}, \tag{15}$$

where **Y** is any constant matrix with nonzero determinant.

We will see in Section 3 how the lack of uniqueness displayed in Eqs.
(13) and (14) has been exploited[9] to suppress the growth of linear
dependence in solutions being propagated from r_b to r_a. It will then be
shown how the key idea in this very powerful device (for which Seaton's
co-author disclaims any credit) can be used to exploit the lack of uniqueness displayed in Eq. (15) in the development of a technique for propagating $\mathbf{\Psi}(r)$ from r_a to r_b.

3.   Travels in Intermedia

In this section we presuppose the existence of solutions of Eq. (1) at r_b
with asymptotic forms given by Eqs. (10) and (11). Techniques for solving
the differential equations of scattering theory are legion (the recent review
by Thomas et al.[10] nicely compares many). Our problem is not how to
propagate solutions between r_a and r_b, but how to do so most efficiently
without losing either linear independence or numerical significance. One
major obstacle is exponential growth from closed-channel components
swamping the oscillatory open-channel components.[11] Another problem is
loss of numerical significance in particular open channels that might occur
even in the absence of closed channels.[7]

We restrict our attention to techniques that permit the wave function
to be recovered in *Intermedia* (a minor limitation), but do impose one
major condition—they must permit the specification of the solution at one
point in terms of the solution at an adjacent point and some function of the
potential. Many techniques other than the two considered here satisfy these
criteria.

The second requirement is the essential feature of the Fox–Goodwin
method[12] for the solution of two-point boundary-level problems. As we
shall see, it permits these techniques to be developed into methods for the
most effective exploitation of the lack of uniqueness displayed in Eqs.
(13)–(15). The first technique to be discussed is now in general use[13] in
the powerful and general computer packages based on the *R*-matrix
method of Burke et al.[14] and the linear algebraic method of Seaton,[8] but

it can be greedy in computer time. The second has yet to be tested, but I believe it has the potential for both greater speed and accuracy.

3.1. Numerical Integration of the Differential Equations

Here we consider the propagation of the solutions $\mathbf{I}(r)$ and $\mathbf{J}(r)$ defined by Eqs. (10), (11), (13), and (14) from r_b to r_a. The three-point Numerov formula,

$$\boldsymbol{\alpha}(t)\mathbf{H}(t-1) - 2\boldsymbol{\beta}(t)\mathbf{H}(t) + \boldsymbol{\gamma}(t)\mathbf{H}(t+1) = 0, \tag{16}$$

relates the solution $\mathbf{H}(t+1)$ at r_{t+1} to solutions at the two previous points r_{t-1} and r_t. The matrices $\boldsymbol{\alpha}(t)$, $\boldsymbol{\beta}(t)$, and $\boldsymbol{\gamma}(t)$ are defined in terms of

$$\mathbf{W}(r) = -\frac{l(l+1)}{r^2} + \frac{2z}{r} + \mathbf{V}(r) + k^2 \tag{17}$$

by

$$\boldsymbol{\alpha}(t) = 1 + \frac{h^2}{12}\mathbf{W}(t-1), \tag{18a}$$

$$\boldsymbol{\beta}(t) = 1 - \frac{5h^2}{12}\mathbf{W}(t), \tag{18b}$$

and

$$\boldsymbol{\gamma}(t) = 1 + \frac{h^2}{12}\mathbf{W}(t+1), \tag{18c}$$

where $h = r_{t+1} - r_t = r_t - r_{t-1}$.

It can be shown that the solution at r_{t+1} is related to that at r_t by

$$\mathbf{D}(t)\mathbf{H}(t) = \mathbf{H}(t+1), \tag{19}$$

where $\mathbf{D}(t)$ and $\mathbf{D}(t-1)$ are related, using Eqs. (16) and (19), by

$$\mathbf{D}(t-1) = \left[2\boldsymbol{\beta}(t) - \boldsymbol{\gamma}(t)\mathbf{D}(t)\right]^{-1}\boldsymbol{\alpha}(t). \tag{20}$$

Thus the solution at r_b is formally connected to that at r_a by the global propagator $\mathbf{D}$ defined by

$$\mathbf{D} = \prod_{t=a}^{b-1} \mathbf{D}(t). \tag{21}$$

Construction of $\mathbf{D}$ can be initiated by using solutions $\mathbf{H}(r)$ at r_b and r_{b+1}, and Eq. (19), to define

$$\mathbf{D}(b) = \mathbf{H}(b+1)\big[\mathbf{H}(b)\big]^{-1}, \tag{22}$$

and the progressive application of Eq. (20). The relation connecting the solutions at r_a and r_b is then

$$\mathbf{D}\mathbf{H}(a) = \mathbf{H}(b)\mathbf{X}. \tag{23}$$

Having now four equations in six unknowns, we are free to choose the closed-channel rows of $\mathbf{H}(a)$ to suit our convenience, leaving the previously arbitrary elements of $\mathbf{X}$ to be determined.

We can ensure good linear independence of the solution at r_a by the choice

$$\mathbf{H}(a) = \begin{pmatrix} \mathbf{H}_{oo}(a) & \mathbf{H}_{oc}(a) \\ 0 & 1 \end{pmatrix}, \tag{24}$$

and proceed to solve Eq. (23) for $\mathbf{H}_{oo}(a)$ and $\mathbf{H}_{oc}(a)$ by eliminating the unknowns $\mathbf{X}_{co}$ and $\mathbf{X}_{cc}$, their noble purpose having been served. The results are[9]

$$\mathbf{H}_{oo}(a) = \mathbf{P}\big[\mathbf{H}_{oo}(b) - \mathbf{Q}\mathbf{H}_{co}(b)\big] \tag{25a}$$

and

$$\mathbf{H}_{oc}(a) = \mathbf{P}\big[\mathbf{Q}\mathbf{D}_{cc} - \mathfrak{D}_{oc}\big], \tag{25b}$$

where

$$\mathbf{P} = \big[\mathbf{D}_{oo} - \mathbf{Q}\mathfrak{D}_{co}\big]^{-1} \tag{26a}$$

and

$$\mathbf{Q} = \mathbf{H}_{oc}(b)\big[\mathbf{H}_{cc}(b)\big]^{-1}. \tag{26b}$$

The asymptotic forms specified for the closed-channel columns of Eqs. (10) and (11) ensure the suppression of exponential growth in the open-channel columns of both $\mathbf{I}(r)$ and $\mathbf{J}(r)$. Having solved Eqs. (25a) and (25b) for both $\mathbf{I}(a)$ and $\mathbf{J}(a)$, the solutions throughout *Intermedia* are easily recovered if we have remembered to save the two-point propagator matrices $\mathbf{D}(t)$.

Curiosity as to why the global propagator **D** is designed to operate in the *opposite* direction from that in which we wish to travel should be satisfied by the form of Eq. (23). Had we constructed the global propagator in the other "obvious," direction, it would require inversion before use in Eq. (23). If the technique outlined above is valuable, it is precisely because **D**, however constructed, has unpleasant properties with respect to inversion.

This device, as implemented in my original[15] computer program ASYM and its later revisions,[16–18] facilitates the calculation of reliable asymptotic solutions even when r_b is quite large, as near threshold or in the presence of long-range potentials and strongly closed channels. It is impossible to assess the magnitude of its impact, since it is buried within a computer code within other computer codes, but it has undoubtedly been helpful.

Its major limitation[19] is a consequence of the facts that **D** must be constructed twice—for both **I**(r) and **J**(r), that calculation of **D**(t) involves a matrix inversion at every point r_t, that all channels must be included (in spite of the fact that r_b may be large for only one, or at most a few, channels), and that use of the Numerov algorithm requires a relatively small value of h (as compared with, e.g., an integral equations technique). It may thus be rather time consuming.

Another potential limitation is the fact that **I**(r) and **J**(r) are propagated independently. There is thus no control over the development of poor linear independence between the open–open parts of the solutions, even in the absence of closed channels.

3.2. Noniterative Integration of the Phase-Amplitude Equations

We now consider the alternative of propagating the solutions $\Psi(r)$ from r_a to r_b. The approach used is based on variable phase theory,[20,21] as first applied to multichannel scattering by Matese and Henry.[22]

The solution $\Psi(r)$ is first expressed as a linear combination of the Green's functions $\mathcal{K}(r)$ as

$$\Psi(r) = \mathcal{F}(r)\mathbf{A}(r) - \mathcal{G}(r)\mathbf{B}(r) \tag{27}$$

[the functions $\mathbf{A}(r)$ and $\mathbf{B}(r)$ become, in the limit $r \to \infty$, the constant matrices in Eq. (3)]. With the subsidiary condition

$$\Psi'(r) = \mathcal{F}'(r)\mathbf{A}(r) - \mathcal{G}'(r)\mathbf{B}(r), \tag{28}$$

the properties of the Green's functions permit us to form the differential equations

$$\mathbf{A}'(r) = \mathcal{G}(r)\mathbf{V}(r)\mathbf{\Psi}(r) = \mathbf{Z}^{(11)}(r)\mathbf{A}(r) + \mathbf{Z}^{(12)}(r)\mathbf{B}(r), \qquad (29a)$$

and

$$\mathbf{B}'(r) = \mathcal{F}(r)\mathbf{V}(r)\mathbf{\Psi}(r) = \mathbf{Z}^{(21)}(r)\mathbf{A}(r) + \mathbf{Z}^{(22)}(r)\mathbf{B}(r). \qquad (29b)$$

These can be combined to form a simple first-order nonlinear differential equation for $\mathbf{R}(r)$, the initial value of which is easily determined from Eqs. (27), (28), and (4). This equation has been used to treat the asymptotic region,[23] but this approach has been found[13] to be no more efficient than the method outlined in Section 3.1. In any event, it does not satisfy our requirement that the wave function be easily recoverable throughout *Intermedia*.

Instead we can transform Eqs. (29a) and (29b) into integral equations. If these are presumed integrated from r_t to r_{t-1} with a simple second-order quadrature (e.g., a trapezoidal with weights w_t) and used to evaluate $\mathbf{\Psi}(t - 1)$ by substitution into Eq. (27), the result is

$$\mathbf{\Psi}(t - 1) = \mathcal{F}(t - 1)\left[\mathbf{D}^{(11)}(t)\mathbf{A}(t) + \mathbf{D}^{(12)}(t)\mathbf{B}(t)\right]$$

$$- \mathcal{G}(t - 1)\left[\mathbf{D}^{(21)}(t)\mathbf{A}(t) + \mathbf{D}^{(22)}(t)\mathbf{B}(t)\right], \qquad (30)$$

where

$$\mathbf{D}^{(ij)}(t) = \mathbf{1}\delta_{ij} - \mathbf{Z}^{(ij)}(t)\omega_t . \qquad (31)$$

It is important to note that Eq. (30) does not involve an iterative approximation to the integration of Eqs. (29a) and (29b). The terms involving $\mathbf{A}(t - 1)$ and $\mathbf{B}(t - 1)$ simply cancel when the result is substituted into Eq. (27). In this regard, Eq. (30) has the same property that is exploited in noniterative treatments[7,24,25] of the integral equations for $\mathbf{\Psi}(r)$.

Now using Eq. (30), we can construct the global propagator for this approach as

$$\mathbf{D} = \prod_{t=b}^{a+1} \begin{bmatrix} \mathbf{D}^{(11)}(t) & \mathbf{D}^{(12)}(t) \\ \mathbf{D}^{(21)}(t) & \mathbf{D}^{(22)}(t) \end{bmatrix}, \qquad (32)$$

which is completely composed of known functions. Then the result analogous to Eq. (23) is found, using Eqs. (15), (27), and (32), to be

$$\mathbf{A}(a)\mathbf{Y} = \mathbf{D}^{(11)}\mathbf{A}(b)\mathbf{D}^{(12)}\mathbf{B}(b) \qquad (33a)$$

and

$$\mathbf{B}(a)\mathbf{Y} = \mathbf{D}^{(21)}\mathbf{A}(b) + \mathbf{D}^{(22)}\mathbf{B}(b). \tag{33b}$$

We have then, considering the matrices as usually partitioned, eight equations in twelve unknowns. Four may therefore be specified almost at will, and we find it convenient to choose

$$\mathbf{A}(b) = \begin{pmatrix} \mathbf{A}_{oo}(b) & \mathbf{A}_{oc}(b) \\ 0 & 1 \end{pmatrix} \tag{34a}$$

and

$$\mathbf{B}(b) = \begin{pmatrix} 1 & 0 \\ \mathbf{B}_{co}(b) & \mathbf{B}_{cc}(b) \end{pmatrix}. \tag{34b}$$

Upon eliminating $\mathbf{Y}$ between Eqs. (33a) and (33b) we obtain

$$\mathbf{P}\mathbf{A}(b) = \mathbf{Q}\mathbf{B}(b), \tag{35}$$

where

$$\mathbf{P} = \mathbf{B}(a)\big[\mathbf{A}(a)\big]^{-1}\mathbf{D}^{(11)} - \mathbf{D}^{(21)} \tag{36a}$$

and

$$\mathbf{Q} = \mathbf{D}^{(22)} - \mathbf{B}(a)\big[\mathbf{A}(a)\big]^{-1}\mathbf{D}^{(12)}. \tag{36b}$$

The elements of Eqs. (34a) and (34b) are then found to be

$$\mathbf{A}_{oo}(b) = \big(\mathbf{P}_{oo} - \mathbf{Q}_{oc}\mathbf{Q}_{cc}^{-1}\mathbf{P}_{co}\big)^{-1}\big(\mathbf{Q}_{oo} - \mathbf{Q}_{oc}\mathbf{Q}_{cc}^{-1}\mathbf{Q}_{co}\big), \tag{37a}$$

$$\mathbf{A}_{oc}(b) = \big(\mathbf{P}_{oo} - \mathbf{Q}_{oc}\mathbf{Q}_{cc}^{-1}\mathbf{P}_{co}\big)^{-1}\big(\mathbf{Q}_{oc}\mathbf{Q}_{cc}^{-1}\mathbf{P}_{cc} - \mathbf{P}_{oc}\big), \tag{37b}$$

$$\mathbf{B}_{co}(b) = \mathbf{Q}_{cc}^{-1}\big[\mathbf{P}_{co}\mathbf{A}_{oo}(b) - \mathbf{Q}_{co}\big], \tag{37c}$$

and

$$\mathbf{B}_{cc}(b) = \mathbf{Q}_{cc}^{-1}\big[\mathbf{P}_{co}\mathbf{A}_{oc}(b) + \mathbf{P}_{cc}\big]. \tag{37d}$$

Again, having obtained $\mathbf{A}(b)$ and $\mathbf{B}(b)$, it is a simple matter to reconstruct the solution over the entire domain of *Intermedia* using the stored matrices $\mathbf{D}^{(ij)}(t)$ and Eq. (30).

Despite the implication from Eqs. (33a) and (33b) that $\mathbf{A}(a)$ and $\mathbf{B}(a)$

must be uniquely specified, it is clear from Eqs. (36a) and (36b) that only $B(a)[A(a)]^{-1}$ is required. Thus the technique can be used even if only the logarithmic derivative of $\Psi(r)$ at r_a is known, since[22]

$$\mathbf{B}(a)\left[\mathbf{A}(a)\right]^{-1} = \left[\mathcal{F}' - \mathcal{F}\Psi'\Psi^{-1}\right]\left[\mathcal{G}' - \mathcal{G}\Psi'\Psi^{-1}\right]^{-1}\bigg|_{r_a}. \tag{38}$$

It is interesting to note the similarity in form of Eqs. (25a) and (25b) and Eqs. (37a) and (37b). The expressions for X_{co} and X_{cc} are also similar in form to Eqs. (37c) and (37d).

The forms adopted in Eqs. (34a) and (34b) are but one of a continuum of possible choices. One extreme would be to force upper- and lower-triangular form for $A(b)$ and $B(b)$, respectively. The other would be to adopt the unit matrix for $A(b)$, leaving all elements of $B(b)$ to be determined. The former would ensure the greatest degree of linear independence, the latter the greatest degree of numerical significance. The choice adopted in Eqs. (34a) and (34b) should be adequate for most practical purposes.

While Eqs. (36a) and (36b) and (37a) and (37d) are a little more complex than their counterpart Eqs. (25a), (25b), and (26a), they offer several possible advantages. Only one global propagator is required, and its construction does not involve a matrix inversion at every step. Thus computation times should increase as the square rather than the cube of the number of channels. This technique also permits us to enforce good linear independence between the two components $A(r)$ and $B(r)$, as compared with the earlier separate treatment of $I(r)$ and $J(r)$.

The mesh used may also be more coarse (and variable), in spite of the crudeness of the quadrature, since we are propagating the phase amplitudes, not the wave functions themselves, the former being much more slowly varying in r. That it may be significantly faster in comparison with the Numerov approach is suggested by the results of the tests carried out by Thomas et al.[10] By analogy with Seaton's linear algebraic method,[8] the fact that Eqs. (25a), (25b), and (26a), and Eqs. (36a), (36b), and (37a) and (37d), effectively transform a set of coupled second-order differential equations into sets of linear algebraic equations may permit the use of an even coarser mesh than in conventional applications of Numerov or integral equations techniques.

Application of this approach requires efficient routines for generating the Green's functions, and these are available.[26-28] The existence of minor errors in the program of Rountree et al.[27] has been noted,[23] and these have been corrected.[29]

Of course, the value of r_b may be taken as large as necessary for subsequent application of the techniques to be discussed in Section 4. It may even be more efficient simply to compute $R(b) = B(b)[A(b)]^{-1}$ at several widely spaced values of r_b and terminate the solution when $\mathbf{R}_{oo}(b)$ has converged to some specified accuracy.

It should also be noted that this approach could be used to stabilize solutions in an integral equations treatment of the scattering equations in *Inner Regia*. Earlier work[3-7,24,30] involved such devices as upper-triangularization or partial unitarization every few mesh points. This is not only more time consuming, but as White and Hayes[5] point out "how often reorthogonalization is necessary is not clear, nor was any good testing procedure found." Use of the present technique should turn the generation of solutions with good linear independence over a large range of r into a mindless exercise—once coded thence forgotten. In this regard it would share a lovely property of the stabilization technique in Seaton's linear algebraic method.[8]

4. At the Border of Asymptopia

Techniques that are routinely employed for obtaining the functions $F(r)$ and $G(r)$ can be classified as either iterative or as involving asymptotic expansions. It is probably safe to say that the latter will generally yield a smaller value of r_b than iterative methods for a specified accuracy and expenditure of computer time, but for nearly degenerate channels or near-threshold scattering iterative techniques may be more efficient.

Which of several available techniques to use involves, in addition, the issue of the rather delicate distribution of labor between *Intermedia* and *Asymptopia*. Significant additional expense associated with achieving a smaller value of r_b may not be warranted, for example, if the same result could be achieved less expensively by the use of rapid techniques for propagating between r_b and r_a. Given the possibility that the new method discussed in Section 3.2 may be of some value, it behooves us to pay some renewed attention to iterative techniques. We will therefore devote more of our attention to the iterative techniques in this section, after briefly discussing asymptotic expansions.

4.1. Asymptotic Expansions

These techniques have been in steady use[13] for both atoms and ions since their development[31,32] early in the computer era. To illustrate the essential idea of this approach, we postulate a solution of Eq. (1) of the form

$$\mathbf{H}(r) = \left[\mathbf{A}(r) + i\mathbf{B}(r)\right]\left[f(r) + ig(r)\right]. \tag{39}$$

Since both real and imaginary parts of $\mathbf{H}(r)$ must be solutions, they are linearly independent if $f(r)$ and $g(r)$ have nonzero Wronskian, and the desired solutions are

$$\mathbf{F}(r) = \mathbf{A}(r)f(r) - \mathbf{B}(r)g(r) \tag{40a}$$

and

$$\mathbf{G}(r) = \mathbf{A}(r)g(r) + \mathbf{B}(r)f(r). \tag{40b}$$

If the derivatives of $f(r)$ and $g(r)$ can be expressed as products of $f(r)$ and $g(r)$, and polynomials in r, we can obtain recursion relations for $\mathbf{A}^{(n)}$ and $\mathbf{B}^{(n)}$ by expressing $\mathbf{A}(r)$ and $\mathbf{B}(r)$ as power series in r, for example,

$$\mathbf{A}(r) = \sum_{n=0}^{\infty} \mathbf{A}^{(n)}r^{-n}, \tag{41}$$

substituting either $\mathbf{F}(r)$ or $\mathbf{G}(r)$ into Eq. (1), and equating to zero the coefficients of each power of r in the two series multiplying $f(r)$ and $g(r)$. The value of r_b is defined as that value for which the term at the minimum of the series contributes less than some specified amount, but this is a very ill-defined problem for multichannel scattering. This is one of the pervasive (and little mentioned) difficulties with asymptotic expansions.

In the original work[31,32] the necessary conditions were satisfied by the choices

$$f_o(r) = \lim_{r \to \infty} \mathscr{F}(r) \tag{42a}$$

and

$$g_o(r) = \lim_{r \to \infty} \mathscr{G}(r), \tag{42b}$$

where the forms given in Eqs. (6)–(9) are implied. With these choices r_b diverges at threshold as k^{-1} for $z = 0$, and as k^{-3} for $z \neq 0$. Since the coefficients $\mathbf{A}^{(n)}$ and $\mathbf{B}^{(n)}$ must also incorporate the Coulomb and angular momentum terms in Eq. (1), r_b also increases with z and l; and because the coefficients contain terms proportional to $(k_i^2 - k_j^2)^{-1}$ the value of r_b can also become large for nearly degenerate channels.

It has recently been shown[33] that several of these latter problems may be nicely resolved by developing expansions based on the Green's functions themselves, rather than on their asymptotic forms. The Coulomb Green's functions, however, do not satisfy the required relationship between the functions and their derivatives. Gailitis showed that the approach based on Eq. (39) could be retained by first introducing asymptotic expansions of the Green's functions, for example,

$$\mathcal{F}(r) = f_0(r) \sum_{m=0}^{\infty} A_0^{(m)} r^{-m}, \tag{43}$$

where $f_0(r)$ is defined by Eq. (42a). This leads to recursion relations combining the coefficients in Eq. (43) and those obtained using the original forms of Eqs. (42a) and (42b). Gailitis also suggested two other approaches based on alternatives to Eq. (39).

Because the contributions from the Coulomb and angular momentum terms are incorporated in the Green's functions, less is required of the expansion coefficients and r_b may in consequence be smaller. The order of the divergence of r_b near threshold for ions is also reduced to k^{-1}, but difficulties with nearly degenerate channels remain.

An attractive feature of these techniques is the possibility of specifically choosing the Green's functions to suit a particular problem. Gailitis suggests, for example, that the exact solutions developed by Seaton[34] for the problem of degenerate states coupled by a dipole interaction (e.g., hydrogenic systems or polar molecules) could be used. One can also imagine using the Green's functions based on the well-known Mathieu functions[35] for systems with strong dipole polarizabilities.

There have been as yet no exhaustive tests of Gailitis' innovation, but it seems reasonable to suspect that it will result in a generally lower value of r_b for a modest increase in computational time. It is not obvious, however, that this will result in any real overall efficiency.

4.2. Iterative Techniques

Seaton has had an interest in iterative techniques from the beginning of his career, in particular, those involving the WBK approximation.[36] He and his students and collaborators[37–39] developed several pioneering applications to the single-channel Coulomb problem. Another of his students was one of the first to attempt an iterative solution of the integral equations of multichannel scattering theory in the asymptotic region.[40] Elements of these two ideas were combined in the multichannel iterated WBK (IWBK) method developed with another of his students.[41]

Another technique, also based on integral equations, was introduced at about the same time.[24,42] The latter and the IWBK method are still incorporated in the current generation of scattering codes[13,29] and are summarized in this section. A very simple expedient is suggested for improving the utility of the iterated IWBK technique, which (as coded[15]) has rightly been criticized[17] because of its excessive use of time. This is achieved by effectively *increasing* the minimum value of r_b for which it will provide a specified accuracy. It is also shown that the matricant approach[24,42] can be generalized to include the treatment of closed as well as open channels, thereby *reducing* the useful value of r_b.

4.2.1. *The Iterated WBK (IWBK) Method.* The WBK approximation was adopted for the functions $f(r)$ and $g(r)$, which take the form

$$f(r) = \chi^{-1/2}(r)\sin \phi(r) \tag{44a}$$

and

$$g(r) = -\chi^{-1/2}(r)\cos \phi(r), \tag{44b}$$

where

$$\chi(r) = \left(k^2 + \frac{2z}{r}\right)^{1/2} \tag{45}$$

and

$$\phi(r) = \chi r + \frac{z}{k}\left[1 - \ln\left(\frac{z}{k} + kr + \chi r\right)\right] + \sigma - \frac{1}{2}l\pi. \tag{46}$$

Integral equations were then developed for $\mathbf{A}(r)$ and $\mathbf{B}(r)$, which can be iterated N times if $2N - 1$ derivatives of χ^{-1} and $2N - 2$ derivatives of χ, $\mathbf{V}$, and the subsidiary function

$$q(r) = r^{-3}\chi^{-2}\left(\frac{5}{4}r^{-1}\chi^{-2} - 1\right) \tag{47}$$

are initially specified. This suggests that computational time will increase rapidly as the number of iterations increases, particularly for a multichannel problem.

In the original code[15] six iterations were taken, with the object of obtaining a value of r_b less than, or comparable to, that produced using the asymptotic expansion, but at the price of larger computational times. This was not a serious problem until the advent of extremely fast programs[13] for treating *Inner Regia*, which were also capable of handling a much larger number of channels. It became a tail-wagging-dog problem. With a much faster technique for propagating between r_a and r_b, the utility of the IWBK method could be restored by the simple expedient of *reducing* the number of iterations to, say, two. A rough estimate of the reduction in computer time is a factor of 100, thus rendering the IWBK method comparable in time to the asymptotic expansion. Consequently, the value of k^2 for which the IWBK method yields the lower value of r_b will be lower than previously, perhaps $\sim z^2/100$, compared with $\sim z^2/15$.

4.2.2. *The Generalized Matricant.* The equations of this technique are developed along the lines discussed in Section 3.2, and we therefore presume that we have solutions at r_b produced by propagation from r_a. If r_b is sufficiently large, then $\mathbf{A}(b)$ and $\mathbf{B}(b)$ will have very nearly reached their asymptotic values. This suggests the use of the first iteration of the integrals of Eqs. (29a) and (29b), that is, Eqs. (33a) and (33b) become

$$\mathbf{A}(b)\mathbf{Y} = \mathbf{D}^{(11)}\mathbf{A}(\infty) + \mathbf{D}^{(12)}\mathbf{B}(\infty) \tag{48a}$$

and

$$\mathbf{B}(b)\mathbf{Y} = \mathbf{D}^{(21)}\mathbf{A}(\infty) + \mathbf{D}^{(22)}\mathbf{B}(\infty), \tag{48b}$$

where now

$$\mathbf{D}^{(11)} = 1 - \int_b^\infty \mathcal{G}(r)\mathbf{V}(r)\mathcal{F}(r)\,dr, \tag{49a}$$

$$\mathbf{D}^{(12)} = \int_b^\infty \mathcal{G}(r)\mathbf{V}(r)\mathcal{G}(r)\,dr, \tag{49b}$$

$$\mathbf{D}^{(21)} = -\int_b^\infty \mathcal{F}(r)\mathbf{V}(r)\mathcal{F}(r)\,dr, \tag{49c}$$

and

$$\mathbf{D}^{(22)} = 1 + \int_b^\infty \mathcal{F}(r)\mathbf{V}(r)\mathcal{G}(r)\,dr. \tag{49d}$$

These equations may also be viewed[42] as just the first iteration on the Born series expansion for the solutions of the integral equations. It may be noted that we have again developed the global propagator to operate in the opposite direction from Knirk's original suggestion and common usage.[24,29]

We now need to know only the result for $\mathbf{R}_{oo}(\infty)$, given by

$$\mathbf{R}_{oo}(\infty) = \left\{ \mathbf{B}(\infty)\left[\mathbf{A}(\infty)\right]^{-1} \right\}_{oo}. \tag{50}$$

The integrals in Eqs. (49a)–(49d) involving $\mathcal{F}_c(r)$ disappear from the equation for $\mathbf{R}_{oo}(\infty)$ by finally imposing the boundary condition that $\mathbf{B}_{co}(\infty)$

$= 0$. The result is

$$\mathbf{R}_{oo}(\infty) = \left[\mathbf{R}_{oo}(b)\mathbf{D}_{oo}^{(12)} + \mathbf{R}_{oc}(b)\mathbf{D}_{co}^{(12)} - \mathbf{D}_{oo}^{(22)} \right]^{-1}$$
$$\times \left[\mathbf{D}_{oo}^{(21)} - \mathbf{R}_{oo}(b)\mathbf{D}_{oo}^{(11)} - \mathbf{R}_{oc}(b)\mathbf{D}_{co}^{(11)} \right], \qquad (51)$$

where

$$\mathbf{R}(b) = \mathbf{B}(b)\left[\mathbf{A}(b) \right]^{-1}. \qquad (52)$$

If Eqs. (37a)–(37d) are used in Eq. (52), we then have a direct connection between $\mathbf{R}_{oo}(\infty)$ and $\mathbf{R}(a)$. It should be noted, however, that use of Eq. (35), that is, $\mathbf{R}(b) = \mathbf{Q}^{-1}\mathbf{P}$, would be counterproductive.

The integrals for the elements of $\mathbf{D}_{co}^{(11)}$ and $\mathbf{D}_{co}^{(12)}$ that introduce closed-channel contributions are easily obtained, but calculation of the elements of $\mathbf{D}_{oo}^{(ij)}$ is not trivial. Two techniques have been devised,[30,43] but they both have a major weakness. The first of these[30] encounters difficulties for z/k large for any one channel, since it is based on asymptotic expansions for the Coulomb functions.[38] The second[43] is untroubled by thresholds, but breaks down for nearly degenerate channels because it is based on asymptotic expansions for the integral. Marrying the two techniques in a single package would leave near threshold for nearly degenerate channels as the sole remaining difficulty. In this case one could apply the IWBK method instead of the matricant. Seaton is involved[44] in an effort to develop a new technique to circumvent these difficulties.

The difficulty of taking more than one iteration is a disadvantage of the matricant technique, but the fact that the full Green's functions are to be used should partially outweigh this. There is also nothing to preclude using Green's functions that incorporate more of the total potential field. One of its nicest features is that, given a symmetric $\mathbf{R}_{oo}(b)$, any lack of symmetry in the resulting $\mathbf{R}_{oo}(\infty)$ is a direct measure of the errors in the approximation. In the event of poor symmetry, it would be a simple matter to add more terms to the product Eq. (32), and reevaluate Eqs. (37a), (37d,) and (50). All of this could be done without the active intervention of the user of a computer program based on these ideas.

5. Concluding Remarks

It is not possible in this short space to adequately summarize all the methods that have been, or might be, used to treat the asymptotic problem. For references to several earlier approaches, the reader is referred to the review by Burke and Seaton.[45] Methods based on the R-matrix propagator technique[46] are also being used.[47] Its major advantages are inherent stability (closed channels introduce none of the difficulties discussed above) and the fact that the global propagator is energy independent. Its main disadvantage is that both diagonalization of $V(r)$ and matrix inversion are required at every mesh point.

With the advent of larger and faster hardware and software for carrying out calculations in *Inner Regia*, we may do well to "rediscover" some older approaches for treating *Outer Regia*, and perhaps modify them to suit the current era.

ACKNOWLEDGMENT

To think in terms of matrices, and partitioned ones at that, is but one of very many valuable lessons learned from Professor Seaton. I am pleased to dedicate any useful new ideas contained in this chapter to him, as they are based on his work and due to his inspiration.

Note Added in Proof

A program based on the phase-amplitude equations outlined in Section 3.2 has been written.[48] However, it employs numerical integration of the first-order differential Eqs. (29a) and (29b), rather than the integral equations approach, and a technique for stabilizing the solutions that is more akin to earlier reorthogonalization methods mentioned in that section.

The method outlined in Section 3.2 will fail for electron–molecule scattering in *Inner Regia*, even in the absence of asymptotically closed channels, if strong mixing of large and small partial waves occurs at the nuclear singularities. In the present case $\mathcal{D}$,[21] and hence P_{oo}, can become singular, for the same reason that prompted the development of the original reorthogonalization methods. It may be possible to surmount this difficulty by regarding the offending group of high angular momentum channels as closed for the purpose of evaluating Eqs. (37a) to (37d), which is consistent with the fact that only the remaining subset of low angular momentum channels are of interest asymptotically.

References

1. M. INOKUTI, *Comments At. Mol. Phys.* **10**, 99–106 (1981).
2. M. ABRAMOWITZ and I. A. STEGUN, *Handbook of Mathematical Functions*, U.S. Department of Commerce, National Bureau of Standards Applied Mathematics Series 55 (1965).

3. R. G. GORDON, *J. Chem. Phys.* **51**, 14–25 (1969).
4. D. SECREST, in *Methods in Computational Physics*, Vol. 10, B. Alder, S. Fernbach, and M. Rotenburg, Eds., pp. 243–286, Academic, New York, 1971.
5. R. W. WHITE and E. F. HAYES, *J. Chem. Phys.* **57**, 2985–2993 (1972).
6. W. EASTES and D. SECREST, *J. Chem. Phys.* **56**, 640–649 (1972).
7. M. A. MORRISON, N. F. LANE, and L. A. COLLINS, *Phys. Rev. A* **15**, 2186–2201 (1977).
8. M. J. SEATON, *J. Phys. B* **7**, 1817–1840 (1974).
9. D. W. NORCROSS and M. J. SEATON, *J. Phys. B* **6**, 614–621 (1973).
10. L. D. THOMAS, M. H. ALEXANDER, B. R. JOHNSON, W. A. LESTER, JR., J. C. LIGHT, K. D. MCLENITHAN, G. A. PARKER, M. J. REDMON, T. G. SCHMALZ, D. SECREST, and R. B. WALKER, *J. Comput. Phys.* **41**, 407–426 (1981).
11. D. W. NORCROSS, *J. Phys. B* **4**, 1458–1475 (1971).
12. L. FOX, *The Numerical Solution of Two-Point Boundary-Value Problems in Ordinary Differential Equations*, Oxford University Press, London, 1957.
13. K. BERRINGTON and M. CREES, *Comput. Phys. Commun.* **17**, 181–205 (1979).
14. P. G. BURKE, A. HIBBERT, and W. D. ROBB, *J. Phys. B* **4**, 153–166 (1971).
15. D. W. NORCROSS, *Comput. Phys. Commun.* **1**, 88–96 (1969).
16. A. T. CHIVERS, *Comput. Phys. Commun.* **5**, 416–429 (1973).
17. M. A. CREES, *Comput. Phys. Commun.* **19**, 103–137 (1980).
18. M. A. CREES, *Comput. Phys. Commun.* **23**, 181–198 (1981).
19. D. W. NORCROSS, *Comput. Phys. Commun.* **6**, 257–264 (1973).
20. F. CALOGERO, *Variable Phase Approach to Potential Scattering*, Academic, New York, 1967.
21. C. ZEMACH, *Nuovo Cimento* **33**, 939–947 (1964).
22. J. J. MATESE and R. J. W. HENRY, *Phys. Rev. A* **5**, 222–226 (1972).
23. M. LEDOURNEUF and VO KY LAN, *J. Phys. B.* **10**, L35–42 (1977).
24. E. R. SMITH and R. J. W. HENRY, *Phys. Rev. A* **7**, 1585–1590 (1973).
25. W. N. SAMS and D. J. KOURI, *J. Chem. Phys.* **51**, 4309–4314 (1969).
26. A. R. BARNETT, D. H. FENG, J. W. STEED, and L. J. B. GOLDFARB, *Comput. Phys. Commun.* **8**, 377–395 (1974).
27. S. P. ROUNTREE, T. BURNETT, and R. J. W. HENRY, *Comput. Phys. Commun.* **11**, 27–35 (1976).
28. M. J. SEATON, *Comput. Phys. Commun.* **25**, 87–95 (1982).
29. R. J. W. HENRY, S. P. ROUNTREE, and E. R. SMITH, *Comput. Phys. Commun.* **23**, 233–273 (1981).
30. E. R. SMITH, *J. Comput. Phys.* **18**, 201–223 (1975).
31. P. G. BURKE and H. M. SCHEY, *Phys. Rev.* **126**, 147–162 (1962).
32. P. G. BURKE, D. D. MCVICAR, and K. SMITH, *Proc. Phys. Soc.* **83**, 397–407 (1964).
33. M. GAILITIS, *J. Phys. B* **9**, 843–854 (1976).
34. M. J. SEATON, *Proc. Phys. Soc.* **77**, 174–183 (1961).
35. S. WANTANABE and C. H. GREEN, *Phys. Rev. A* **22**, 158–169 (1980).
36. D. R. BATES and M. J. SEATON, *Mon. Not. R. Astron. Soc.* **109**, 698–704 (1949).
37. M. J. SEATON and G. PEACH, *Proc. Phys. Soc.* **79**, 1296–1297 (1962).
38. A. BURGESS, *Proc. Phys. Soc.* **81**, 442–452 (1963).
39. P. DE A. MARTINS, *J. Phys. B* **1**, 154–162 (1968).
40. F. H. M. FAISAL, *J. Phys. B* **1**, 181–194 (1968).
41. D. W. NORCROSS and M. J. SEATON, *J. Phys. B* **2**, 731–739 (1969).
42. D. L. KNIRK, *J. Chem. Phys.* **57**, 4782–4788 (1972).
43. J. BELLING, *J. Phys. B* **1**, 136–138 (1968).
44. N. C. SIL, M. A. CREES, and M. J. SEATON, *J. Phys. B.*, in press.
45. P. G. BURKE and M. J. SEATON, in *Methods in Computational Physics*, Vol. 10, B. Alder, S. Fernbach, and M. Rotenburg, Eds., pp. 1–80, Academic, New York, 1971.
46. J. C. LIGHT and R. B. WALKER, *J. Chem. Phys.* **65**, 4272–4282 (1976).
47. L. A. COLLINS and B. I. SCHNEIDER, *Phys. Rev. A* **24**, 2387–2401 (1981); K. L. BALUJA, P. G. BURKE, and L. A. MORGAN, *Comput. Phys. Commun.*, in press.
48. J. P. CROSKERY, N. S. SCOTT, K. L. BELL, and K. A. BERRINGTON, *Comput. Phys. Commun.*, in press.

3

COLLISIONS BETWEEN CHARGED PARTICLES AND HIGHLY EXCITED ATOMS

Ian Percival

1. Introduction

Atoms in states of low excitation have radii of order $a_0 = 1$ bohr, ionization energies of order 1 rydberg (Ry), and charged particle cross sections of order πa_0^2. Optically allowed radiative transitions from these states are faster than the corresponding transitions produced by collisions in most astrophysical environments.

Highly excited atoms have one electron in a state of high principal quantum number n. They are relatively very large, with radii close to $n^2 a_0$, and very delicate, with binding energies close to Ry/n^2. In a plasma, neighboring charged particles are always colliding with them and producing changes of state with cross sections of order $n^4 \pi a_0^2$ or even $n^5 \pi a_0^2$. Highly excited atoms cannot remain in that state very long unless the plasma densities are low.

However, in the outer atmospheres of the stars, in the gaseous nebulae, and particularly in interstellar space, the densities are so low that these atoms survive well. Electrons recombine with ions to form atoms or ions in highly excited states, which are therefore important for ionization balance, and radio-recombination lines have been observed with frequencies corresponding to transitions between states with n as high as 300.

The rates of spontaneous radiative transitions from states of level n decrease as $n^{-4.5}$, which is roughly the same proportion as the corresponding *increase* in cross sections, so for sufficiently high n, whose value depends on the transition and on the environment, the radiative and collisional processes are of comparable importance. The populations of these levels determine the ionization balance and their observation provides

IAN PERCIVAL ● Department of Applied Mathematics, Queen Mary College, University of London, Mile End Road, London E1 4NS, England.

information about density and temperature, provided that the cross sections or rate coefficients for the important transitions are known.

Collisions producing changes Δn in principal quantum number and collisions producing changes Δl in angular momentum quantum number l are both important. The fastest processes with the largest rates in each category are the most important, and that usually means that processes with smaller $|\Delta n|$ or $|\Delta l|$ are more important than those with large $|\Delta n|$ or $|\Delta l|$.

The largest collisional rates of all are those for the optically allowed transitions produced by collisions with charged particles, usually electrons or protons. For an atom in a state of level n, the cross sections are often very much larger than the geometrical cross section of the atom $n^4 \pi a_0^2$, whereas the wavelength of the incident charged particle is very much smaller than the atom. In that case, the incident particle may be considered to move on a classical path with impact parameter b, and most of the cross section comes from values of b significantly greater than the radius of the atom, so that a dipole approximation can be used for the interaction that produces the transition.

Seaton[1] went further and used first-order perturbation theory with a low b cutoff based on conservation of probability. He applied this impact parameter (IP) approximation to electron and proton collisions and to $|\Delta n| = 1$ and $|\Delta l| = 1$ transitions. The results have a very simple form, and within their range of validity provide rates that are perfectly adequate for many astrophysical and other applications. The IP method is discussed in Section 2.

Although the transitions with $|\Delta n| > 1$ and $|\Delta l| > 1$ have cross sections that are smaller, they are not negligible. To obtain them, to improve the precision of the theory, and to widen its range of validity, we have to depart from the simple picture presented by the IP approximation. But if we try to use the standard methods that are of such value for low n, and described elsewhere in this book, we run into all kinds of problems. The number of coupled states is so large that the solution of close-coupling equations takes too long even for $n = 6$ or 7, which is very low for many astrophysical applications. The Born approximation is satisfactory for sufficiently high incident particle energies, but it can be shown that it breaks down for $|\Delta n| = 1$ and incident electrons of energy E less than 1 Ry, so it is not very useful in interstellar space and most of the stellar atmospheres.

Furthermore the number of states is so large that even the tabulation of cross sections is a problem. It is preferable to obtain formulas.

Because quantum numbers are large it is tempting to use classical mechanics, but it must be used with care, particularly when the changes in the quantum numbers are small. In an earlier review, Percival and

Richards[2] treated in some detail the application of classical and correspondence principle methods to n-changing transitions for hydrogenlike systems in which the states of the level n are treated as completely degenerate and equally populated. This chapter will mainly be concerned with l-changing collisions, with the emphasis on the use of the sudden approximation for small impact parameters.

2. The Impact Parameter (IP) Method

Seaton[1] used this method to obtain cross sections for optically allowed transitions produced by incident charged particles. It consists of a combination of approximations that provides simple analytical forms and is therefore particularly suited to collisions with highly excited states.

The first approximation is to assume that the incident particle moves on a classical path. The essential idea was due to Bohr, it was developed by von Weizsäcker and Williams, and applied extensively to nuclear problems.[3] If the impact parameter for a collision is b, and the probability for the $i \to j$ transition with this impact parameter is $P_{i \to j}(b)$, then the cross section for the transition is

$$\sigma(i \to j) = 2\pi \int_0^\infty P_{i \to j}(b) b \, db. \tag{1}$$

The probability is given by the absolute value of the square of the transition amplitude:

$$P_{i \to j}(b) = |S_{ji}(b)|^2. \tag{2}$$

For most atomic applications the deviation of the classical path is not great, and it can be assumed that it is a straight line.

The next approximation is an adaptation of an earlier method that used partial waves.[4]

Two different procedures are used for different ranges of the incident particle energy E. For the first procedure the lower limit of the integral is replaced by a value b_0, such that the cross section agrees with the Bethe value at high energies. Because b_0 is of atomic dimensions, a mean radius of the atom is sometimes used instead. The Bethe cross section requires only the dipole matrix element and is therefore an approximation to the Born cross section, but, in practice, it is often smaller than Born at the lower energies and thus in better agreement with the exact cross section. This applies only to optically allowed transitions. This procedure deals with the higher E, where weak-coupling theory is valid, but for lower E, where the coupling can be strong, it includes contributions from impact parameters b in which the weak coupling, or first-order perturbation theory transition probability $P^1_{i \to j}(b)$, is greater than unity, which clearly violates the conservation of probability.

So for the low E strong-coupling region, a second procedure is used, where the cross section is approximated by

$$\sigma(i \to j) = 2\pi \int_0^{b_1} \tfrac{1}{2} b \, db + 2\pi \int_{b_1}^{\infty} P^1_{i \to j}(b) b \, db$$

$$= \tfrac{1}{2} \pi b_1^2 + 2\pi \int_{b_1}^{\infty} P^1_{i \to j}(b) b \, db \qquad (3)$$

and b_1 is chosen so that

$$P^1_{i \to j}(b_1) = \tfrac{1}{2}, \qquad (4)$$

making the probability a continuous function of impact parameter. The justification for this approximation is that for small b, where the coupling is large, rapid oscillations due to interference effects average out to a probability of $\tfrac{1}{2}$ for the initial and final state. This is reasonable where coupling to any other state than i and j is negligible, but for highly excited initial and final states this is not always true, and other possibilities are considered in later sections.

In practice, for a given incident particle energy E, the smaller of the two cross sections is chosen.

The third approximation is to use the dipole term in the expansion of interaction potential. For optically allowed transitions this is very satisfactory when most of the cross section comes from impact parameters b larger than the atomic radius a, and that includes most cases of astrophysical interest. Of course, this approximation cannot be used for optically forbidden transitions, since it then gives zero.

The first-order transition amplitude is

$$S^1_{ji}(b) = -\frac{i}{\hbar} \int_{-\infty}^{\infty} \langle j | v_b(t) | i \rangle \exp i\omega t \, dt \qquad (5)$$

where for incident particles of charge $Z_1 e$ and impact parameter b, the interaction potential is

$$V_b(t) = -Z_1 e^2 / |\mathbf{R}_b(t) - \mathbf{r}|. \qquad (6)$$

The rectilinear path of the incident particle is

$$\mathbf{R}_b(t) = b\hat{\mathbf{x}} + Vt\hat{\mathbf{z}}, \tag{7}$$

where V is its velocity. The position of the excited atomic electron is given by $\mathbf{r}$.

For large cross sections the dominant contributions to the transition amplitude come from $|\mathbf{r}| \ll |\mathbf{R}|$, so we can use the dipole approximation and neglect the parts of the atomic wave function outside the path of the incident particle. This gives the first-order dipole transition probability as

$$P_{i \to j}^{1d}(b) = \frac{4}{3} \frac{Z_1^2 e^4}{\hbar^2} \frac{|\mathbf{r}_{ji}|^2}{(bV)^2} \beta^2 \left[K_0^2(\beta) + K_1^2(\beta) \right], \tag{8}$$

where

$$\beta = \frac{b}{b_{\mathrm{ad}}} = \frac{b\omega}{V} = \frac{b|\Delta E|}{\hbar V} \tag{9}$$

and

$$\mathbf{r}_{ji} = \langle j|\mathbf{r}|i\rangle. \tag{10}$$

In Eq. (9)

$$\Delta E = E_j - E_i \tag{11}$$

is the energy difference between the initial and final states, $\omega = |\Delta E|/\hbar$ is the corresponding angular frequency and

$$b_{\mathrm{ad}} = V/\omega \tag{12}$$

is the adiabatic impact parameter. For $b > b_{\mathrm{ad}}$, i.e., $\beta > 1$, the effective collision time is longer than the oscillation time characteristic of the $i \to j$ transition, so the collision is adiabatic; otherwise the collision is said to be sudden. The functions $K_0(\beta)$ and $K_1(\beta)$ are modified Bessel functions that fall off at least exponentially for $\beta > 1$, so the first-order transition probability rapidly becomes negligible as the impact parameter increases into the adiabatic region.

For transitions between degenerate levels, the probability is summed over final states and averaged over initial states.

In all cases the IP cross section is obtained by integrating the probability over the impact parameter using one of the above procedures. In the strong-coupling region the cross section is

$$\sigma(i \to j) = \frac{1}{2} \pi b_1^2 + \frac{8\pi}{3} \frac{Z_1^2 e^4}{(\hbar V)^2} |\mathbf{r}_{ji}|^2 \int_{b_1/b_{ad}}^{\infty} \left[K_0^2(\beta) + K_1^2(\beta) \right] \beta \, d\beta, \quad (13)$$

where b_1 is chosen so that $P_{i \to j}^1(b_1) = \frac{1}{2}$. But the cutoff at b_1 can only be justified if the second term of Eq. (13) is large compared to the first, and this happens when $b_1 \ll b_{ad}$.

The integral is then well approximated by[1]

$$\int_{b_1/b_{ad}}^{\infty} \left[K_0^2(\beta) + K_1^2(\beta) \right] \beta \, d\beta = \ln\left(\frac{1.12 b_{ad}}{b_1} \right), \qquad b_1 \ll b_{ad} \quad (14)$$

and the transition probability at b by

$$P_{i \to j}^{1d}(b) = \frac{4}{3} \frac{Z_1^2 e^4}{\hbar^2} \frac{|\mathbf{r}_{ji}|^2}{(bV)^2}, \qquad b_1 < b \ll b_{ad}, \quad (15)$$

So in the region where the strong-coupling form of the IP method is valid we have

$$b_1^2 = \frac{\frac{8}{3} Z_1^2 e^4 |\mathbf{r}_{ji}|^2}{(\hbar V)^2} \quad (16)$$

and

$$\sigma^{IP}(i \to j) = \pi b_1^2 \left[\frac{1}{2} + \ln(1.12 b_{ad}/b_1) \right] \qquad \text{(strong coupling form).} \quad (17)$$

In applications, it is important to maintain the detailed balance relations between forward and reverse transitions. For large n and small $|\Delta n|$ and $|\Delta l|$ transitions the exact method is unimportant, but the first applications of the IP method were to transitions with fairly low n. For these transitions averaged values were taken for V and b, according to a symmetrizing method based on comparison with full quantum treatments.

The method was first applied by Seaton to transitions between states of hydrogen and helium with a maximum n of 6. He extended it to $n \to n \pm 1$ transitions of hydrogenlike systems with arbitrarily high n for use in calculations on recombination spectra.[5] Improved results were obtained by Saraph.[6]

The IP method had to be modified to deal with transitions between degenerate l states, and these modifications are discussed in the appropriate section.

3. The Sudden Approximation

The IP method assumes that the probability of the transition from the initial to the final levels is $\frac{1}{2}$ over most of the strong-coupling region. This is adequate where i and j are the only strongly coupled levels but breaks down as more and more states become closely coupled together. This is just the condition under which the classical theory with the simple density-of-states correspondence principle can be used,[2] but there is an intermediate region where neither theory is adequate.

For most collisions of astrophysical interest much of the strong-coupling region can be treated adequately using the sudden approximation,[2,7] which is also valid for values of b where the coupling is weak, provided b is much less than b_{ad}.

The sudden transition amplitude is given by

$$S_{ji}^s = \left\langle j \left| \exp\left[-\frac{i}{\hbar} \int_{-\infty}^{\infty} dt\, V_b(t) \right] \right| i \right\rangle \tag{18}$$

and this can be used when the commutators between the interaction operators $V_b(t)$ in interaction representation at different times t are negligible. This holds when the interaction takes place in a time shorter than any of the natural periods of oscillation $2\pi/\omega_{ij} = 2\pi\hbar/|\Delta E_{ij}|$ of the significantly coupled states, i.e., $b < b_{ad}(\omega_{ij})$.

Where both sudden and dipole approximations are valid for a range of impact parameters between b' and b'', the contribution to the cross section can be written

$$2\pi \int_{b'}^{b''} \left| \left\langle j \left| \exp\frac{2iZ_1e^2x}{\hbar bV} \right| i \right\rangle \right|^2 b\, db \qquad \text{(sudden dipole)}, \tag{19}$$

where x is the coordinate of the bound electron. Richards[24] has pointed out that, on making the substitution

$$K = 2Z_1e^2/\hbar bV,$$

this becomes

$$\frac{8\pi Z_1e^2}{\hbar V} \int_{K'}^{K''} |\langle j|\exp iKx|i\rangle|^2 K^{-3}\, dK,$$

which has the same form as the Born approximation. But they are not the same and are valid for quite different regions of incident particle velocity V. Although both require V to be large, the sudden dipole approximation takes account of strong coupling and is valid at lower V than the Born approximation, which does not assume a dipole interaction or a classical path. However, the formal similarity between the approximations saves labor when both are needed.

On comparing the first-order and sudden approximations, Eqs. (5) and (18), it is evident that, where *both* are valid, the transition amplitude is given by

$$S_{ji}^{1s}(b) = \frac{-i}{\hbar} \int_{-\infty}^{\infty} \langle j|V_b(t)|i\rangle \, dt, \tag{20}$$

and where the dipole approximation is also valid, the transition probability is

$$P_{i\to j}^{1sd}(b) = \frac{4}{3} \frac{Z_1^2 e^4}{\hbar^2 b^2 V^2} |\mathbf{r}_{ji}|^2 = \frac{4}{3} \frac{Z_1^2 R}{\mathcal{E}} \frac{|\mathbf{r}_{ji}|^2}{b^2}, \tag{21}$$

which is identical to the first-order dipole approximation (15) when $b \ll b_{ad}$. $\mathcal{E} = \frac{1}{2} m_e V^2$ is the energy of an electron with the same energy as the incident charged particle.

In the spirit of the IP approximation, we can estimate the contribution $\sigma(> b', i \to j)$ from impact parameters greater than b' by using first-order perturbation theory *without* the sudden approximation. Using the modified Bessel functions, the probability decreases below the sudden value by an exponential factor beyond $b = b_{ad}$, and using the asymptotic forms we find that

$$\sigma(> b') = \frac{8\pi}{3} \frac{Z_1^2 R}{\mathcal{E}} |\mathbf{r}_{ji}|^2 \ln\left(\frac{1.12 b_{ad}}{b'}\right). \tag{22}$$

Therefore, the adiabatic effects at large impact parameters can be accounted for by assuming that the sudden approximation is valid up to $b = 1.12 b_{ad}$ and introducing a sharp cutoff at that value.

At low impact parameters it is possible to use the sudden approximation to improve upon the IP method, and this is illustrated for particular cases in the next section.

Methods based on the sudden approximation with an adiabatic cutoff are useful over a range of incident particle velocities V of astrophysical interest, but they are not to be used for all[2] V (see Fig. 1). For sufficiently high V the Born or asymptotic Born[2] approximation can be used, because the coupling between the initial level and *any* other level is then weak and there is no need to use the sudden approximation. For a given V, the

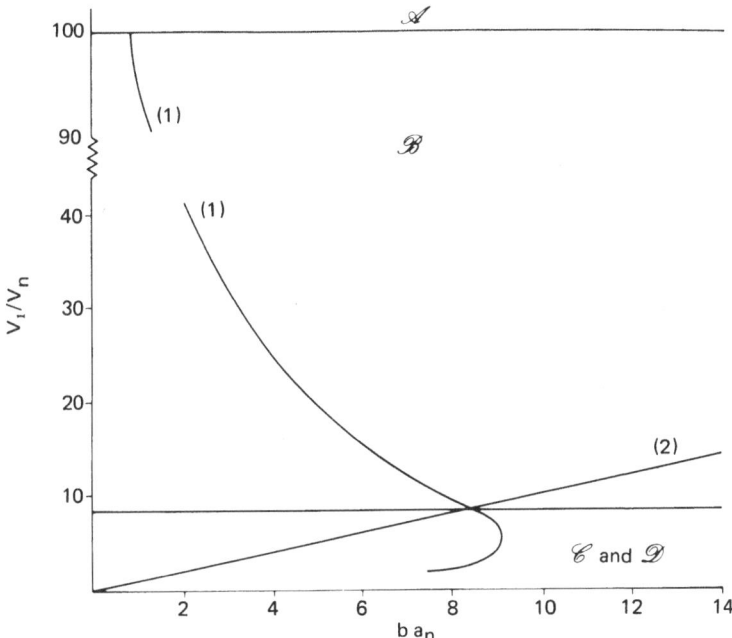

FIGURE 1. Regions $\mathscr{A}$, $\mathscr{B}$, $\mathscr{C}$, and $\mathscr{D}$ for $n \to n \pm 1$ transitions. Curve (1) is the strong-coupling impact parameter ($\approx b_1$) and curve (2) is the adiabatic impact parameter b_{ad}. The boundary between $\mathscr{B}$ and $\mathscr{C}$ is shifted for l-changing collisions.

impact parameter b_1 below which strong coupling is important is obtained by summing (15) or (21) over final states, equating to unity, and neglecting factors of order unity. Remembering that the mean atomic radius a is given approximately by

$$a^2 \approx \sum_j |\mathbf{r}_{ji}|^2, \tag{23}$$

we have

$$b_1 = Z_1 e^2 a / \hbar V \quad \text{(strong boundary)}. \tag{24}$$

This is similar to the boundary for the IP approximation, but a different use will be made of it. For all cases of interest the initial and final mean atomic radii are approximately equal.

It can be shown that a first-order theory is seriously in error when the strong boundary penetrates into the atom, so for the Born approximation to be valid we must have

$$V > |Z_1|e^2/\hbar \qquad \text{(Born, region } \mathscr{C}). \qquad (25)$$

If

$$\mathscr{E} = \tfrac{1}{2}m_e V^2 \qquad (26)$$

is the energy of an electron with the same velocity as the incident charged particle, this gives

$$\mathscr{E} > |Z_1|\text{Ry} \qquad \text{(Born, region } \mathscr{C}) \qquad (27)$$

Note the independence of this result from the target, ionic charge, the level $n_i l_i$, and the process.

Below this value, a strong-coupling method is required, for example, one based on the sudden approximation. But there is a lower V limit to the applicability of these methods. They begin to break down when V becomes so low that the strong coupling is important at $b = b_{ad}$, that is, when $b_1 = b_{ad}$. So the range of application of the sudden dipole with adiabatic cutoff is

$$|Z_1|\text{Ry} > \mathscr{E} > |Z_1 Z_3| \frac{\text{Ry}}{n} \frac{|\Delta E_{ij}|}{\Delta E_n} \qquad \text{(sudden dipole, region } \mathscr{B})$$

$$\approx |Z_1 Z_3|\text{Ry} \, \delta_l/n. \qquad (28)$$

Here ΔE_{ij} is the energy difference between initial and final levels,

$$\Delta E_n = 2(Z_3^2/n^3)\text{Ry} \qquad (29)$$

is the energy difference between neighboring n levels of the same l, and Z_3 is the charge on the ionic core of the target system. The second form of the inequality in Eq. (28) is given in terms of the quantum defect of an initial state. Note that the lower bound of region $\mathscr{B}$ does depend on the target, the level $n_i l_i$, and the process, and that the lower bound is smaller for smaller transfers of energy, such as occur for l-changing transitions with the same n.

4. Transitions between Levels with Quantum Defects

Lasers have allowed a large number of different groups to carry out a range of experiments on atoms excited to nl levels with low l and $n = 8$ to $n = 70$ or more. Usually these atoms are not hydrogenic, and, unlike the astrophysical situation, few states are excited. Nevertheless, some of these experiments have astrophysical significance because they allow a comparison between experiment and theory that checks the validity of the assumptions underlying each.

Schiavone et al.[8] reported absolute cross sections for l-changing transitions between 30 and 300 eV electrons and highly excited rare-gas atoms, but the initial states were broadly distributed in n and l because they were excited by electron impact. Delpech et al.[9] reported rate coefficients for electron-induced transitions with $n = 8$–17 in helium in an afterglow at 300–2200 K, but for these values the astrophysical collision rates for $n \to n'$ do not compete effectively with the radiative rates and no more than a bound was obtained on rates for l-changing processes. Foltz et al.[10] have obtained preliminary results on l-changing collisions in sodium.

More detailed results have been reported by MacAdam et al.[11] for the process

$$X^+ + \mathrm{Na}(nd) \to \mathrm{Na}(n, l \geqslant 3) + X^+, \tag{30}$$

where $X = $ He, Ne, or Ar at 400–2000 eV and n is between 21 and 28. Most of the results were for incident Ne^+. The reduced velocity V/v_n varied between 0.6 and 1.7, where v_n is the rms velocity of the excited electron. Now in the inequality (28), $Z_1 = Z_3 = 1$, and $V/v_n = 1$ when $\mathscr{E} = \mathrm{Ry}/n^2$, so the lower limit of region $\mathscr{B}$ for process (30) is given by

$$V/v_n = (n\delta_d)^{1/2} \quad \text{(lower limit of } \mathscr{B}), \tag{31}$$

where

$$\delta_s = 1.35, \quad \delta_p = 0.85, \quad \delta_d = 0.014 \tag{32}$$

are the quantum defects of high Na levels. The experiment was therefore carried out in region $\mathscr{B}$, near the lower limit of the validity of the sudden dipole method. This is an important region for many astrophysical processes.

Percival and Richards[12] used this theory to investigate cross sections for l-changing collisions of the type

$$nl \to n'l' \tag{33}$$

and MacAdam et al. found agreement, as shown in Fig. 2, between this theory and experiment for the dependence of their signal R_y on n at fixed $\mathscr{E}$, which was

$$R = Cn^\beta, \quad \beta \approx 5,$$

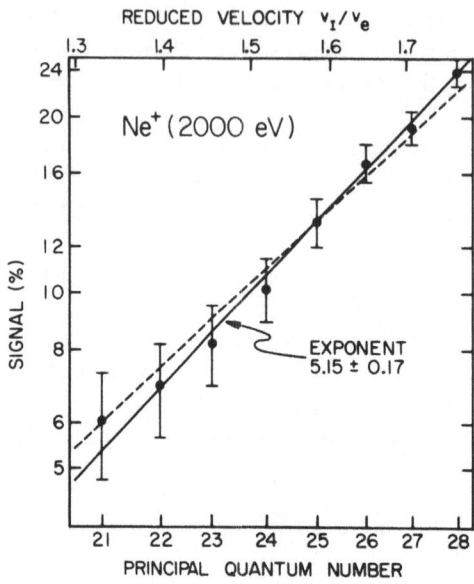

FIGURE 2. Exponent β for experiments of MacAdam *et al.*[11]

but not for the dependence of cross section on reduced velocity V/v_n at low values of the latter.

This disagreement can be traced to the difference between processes (30) and (33). No theory was presented for process (30), so by default the comparison had to be made for different processes. It so happens that the theory for processes of the type (30), summed over final levels, can be carried out by a sudden dipole method that is actually easier and more accurate than for the individual processes (33), and the difference between the summed cross section and the dominant individual process can be significant, particularly for low $\mathscr{E}$.

We discuss the merits of some of the approximations and then present the theory for

$$X^+ + Na(nl) \rightarrow Na(\text{all } n^1 l^1 \neq nl) + X^+, \qquad l \geqslant 2 \qquad (34)$$

which is particularly simple to calculate by the methods given in reference 12. This is not exactly the same as (30), but it was actually measured by MacAdam *et al.* as a drop in the signal for Na(nd).

Some discussion of the meaning of *dipole approximation* is necessary. When the cross sections are very large compared to the geometric cross section, impact parameters $b \gg a$ must dominate, and for these the dipole approximation is perfectly adequate. However, this does *not* mean that nondipole transitions can be neglected. They will be neglected if a first-order theory such as the Bethe (Born dipole) approximation is used, but the sudden dipole approximation (19) takes account of strong coupling, and so includes contributions from nondipole transitions, which result from a succession of dipole transitions during a *single* collision. Such single-collision multiple processes must be carefully distinguished from the multiple-collision processes that so often plague the experimenter.

It is single-collision multiple processes of the type

$$d \to f \to g \to \ldots, \qquad d \to f \to d \to f \to g \to \ldots,$$
$$d \to p \to d \to f \to g \to \ldots, \qquad \text{etc.} \tag{35}$$

that are responsible for the differences between the cross sections for the processes (30), (33), and (34).

Now consider the theory for process (34) at energy $\mathscr{E}$ in range $\mathscr{B}$. The contribution to the cross section from $b' > b_1$ is given by summing Eq. (22) over final states:

$$\sigma(> b', i \to \Sigma' j) = \frac{8\pi}{3} \frac{Z_1^2 \mathrm{Ry}}{\mathscr{E}} \sum_j{}' |\mathbf{r}_{ji}|^2 \ln\left(\frac{1.12 V}{b' \omega_{ji}}\right), \tag{36}$$

where $\sum'$ denotes a sum over all states except those of the initial level. It is independent of the initial m quantum number. Since the logarithmic dependence is slow, little error is made if we replace the ω_{ji} by its values for the dominant terms. This is clearcut for d states, but may be a little problematical for s and possibly p states.

Using the sum rules given by Bethe and Salpeter,[13] we have

$$\sum_{n'm'} |\langle n'\, l+1\, m'|\mathbf{r}|n\, l\, m\rangle|^2 = \frac{5}{2} \frac{a^2(l+1)}{(2l+1)} \left[1 - \frac{3l(l+1)-1}{n^{*2}} \right] \tag{37a}$$

$$\sum_{n'm'} |\langle n'\, l-1\, m'|\mathbf{r}|n\, l\, m\rangle|^2 = \frac{5}{2} \frac{a^2 l}{(2l+1)} \left[1 - \frac{3l(l+1)-1}{n^{*2}} \right], \tag{37b}$$

where we have adapted the hydrogenic formula by taking $a = n^{*2} a_0$ as the radius of the atom and

$$n^* = n - \delta_l, \qquad E_n = -\mathrm{Ry}/n^{*2}. \tag{38}$$

For initial d levels and higher initial l, the dominant contributions always come from transitions with constant n, so if we write

$$E_{nl+1} - E_{nl} = \hbar\omega^+ \quad \text{and} \quad E_{nl} - E_{nl-1} = \hbar\omega^- , \tag{39}$$

we have from Eqs. (21) and (22)

$$P^{1sd}(b, nl \to \Sigma' n'l + 1) = \frac{5}{3} \frac{Z_1^2 Ry}{\mathcal{E}} \frac{a^2}{b^2} \frac{(l+1)}{(2l+1)} \left[1 - \frac{3l(l+1) - 1}{n^{*2}} \right], \tag{40a}$$

$$P^{1sd}(b, nl \to \Sigma' n'l - 1) = \frac{5}{3} \frac{Z_1^2 Ry}{\mathcal{E}} \frac{a^2}{b^2} \frac{l}{(2l+1)} \left[1 - \frac{3l(l+1) - 1}{n^{*2}} \right], \tag{40b}$$

$$\sigma^{1d}(> b', nl \to \Sigma' n'l') = \pi a^2 \frac{10}{3} \frac{Z_1^2 Ry}{\mathcal{E}} \left[1 - \frac{3l(l+1) - 1}{n^{*2}} \right]$$

$$\times \left[\frac{l+1}{2l+1} \ln\left(\frac{1.12 V}{b\omega^+} \right) + \frac{l}{2l+1} \ln\left(\frac{1.12 V}{b\omega^-} \right) \right]. \tag{41}$$

Here πa^2 is the geometric cross section.

Using correspondence principle methods,[2] it can be shown that the sudden dipole probability has the form[12]

$$P^{sd}(b, nl \to n'l') = 1 - \left[\frac{\sin X}{X} J_0(X) \right]^2, \quad X = b_1/b \tag{42}$$

where b_1 can be obtained from the classical orbit corresponding to an s state. A better estimate for b_1 is obtained by expanding in powers of X and comparing the first significant term with (40), giving

$$b_1^2 = \frac{a^2 2 Z_1^2 Ry}{\mathcal{E}} \left[1 - \frac{3l(l+1) - 1}{n^{*2}} \right]. \tag{43}$$

To obtain cross sections for small b, we have to integrate (42) to large X, where it decreases very rapidly. A reasonable approximation to the cross section integral up to impact parameter b' gives

$$\sigma^{sd}(< b') = \pi b_1^2 \left[0.620 + \tfrac{5}{6} \ln\left(b'^2/b_1^2 \right) \right], \quad b' > b_1. \tag{44}$$

This can be combined with Eq. (41) to give the dipole cross section for region $\mathscr{B}$.

$$\sigma^{d}(nl \to \textstyle\sum' n'l')$$

$$= \pi b_1^2 \left\{ 0.812 + \frac{5}{3} \left[\frac{l+1}{2l+1} \ln \left(\frac{V}{b_1\omega^+} \right) \right. \right.$$

$$\left. \left. + \frac{l}{2l+1} \ln \left(\frac{V}{b_1\omega^-} \right) \right] \right\}, \qquad \text{region } \mathscr{B}. \quad (45)$$

For sodium D states the second logarithmic term is negative in the energy region of the experiments, which means that this is in region $\mathscr{C}$ for these processes, but in any case these transitions are certainly negligible compared to $d \to f$ transitions, except at significantly higher energies. Neglecting this term we get

$$\sigma^{d}(nd \to \textstyle\sum' n'l') = \pi b_1^2 \left\{ 0.812 + \ln \left[\frac{(V/v_n)^2}{\sqrt{2}\, n\delta} \right] \right\}$$

$$= \pi a^2 2n^2 \left(\frac{v_n}{V} \right)^2 \left[1.41 + \ln \left(\frac{V}{v_n} \right)^2 \right], \qquad n = 28$$

$$(46)$$

In Fig. 3 this result is compared with the experiments. Although the

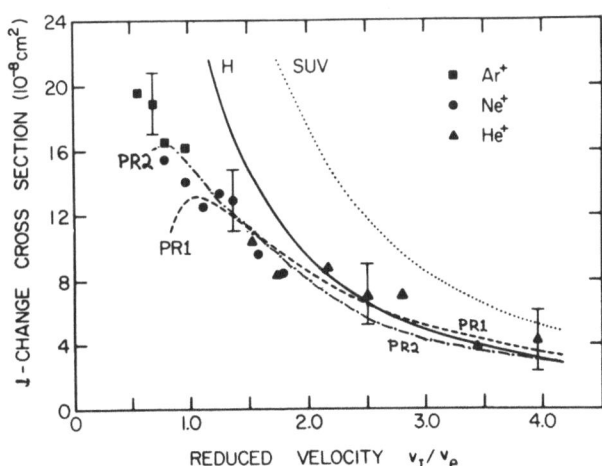

FIGURE 3. PR1, sudden dipole theory for $nd \to nf$ transitions (33), and PR2 for $nl \to \sum' n'l'$ transitions (34) compared to the experiments.[11] H and SUV are other theories discussed in the paper by MacAdam et al.[11]

calculation is carried out near the boundary of region $\mathscr{B}$, the agreement is excellent. The difference between the two theoretical curves PR is due to strong coupling producing $f \to g$ transitions *during* the collisions. At V/v_n = 0.5 the cross section (46) is drastically reduced. This is because $b_{\mathrm{ad}} < b_1$ and the assumptions of region $\mathscr{B}$ are not valid. Experience with similar calculations suggests that there should be a broad maximum in the cross section in the neighborhood of the $\mathscr{B}-\mathscr{C}$ boundary given by (31), but detailed calculations have not yet been carried out.

There is a clear resemblance of the sudden dipole method to the IP method. There are two main differences. One is that the sudden approximation can be used to make a precise estimate of b_1, which plays a slightly different role in the two methods. The other is that the form of the cross section for b below b_1 is given by the sudden approximation and is different for different processes. For process (34) there are large numbers of final states coupled together, and the probability for low impact parameter approaches unity for $b < b_1$.

It should be noted that Monte Carlo trajectory integration methods are not appropriate to the calculation of cross sections in region $\mathscr{B}$ of the incident particle energy. A significant proportion of the cross section comes from impact parameters for which the classical energy transfer is less than the quantal energy level splitting. To deal with this, Heisenberg's form of the correspondence principle, or some equivalent theory, must be used; but this is difficult to apply to a numerical trajectory integration without making well justified simplifying assumptions about the nature of the classical motion that reduce the theory to the strong-coupling correspondence principle.[2] In region $\mathscr{B}$ the application of this principle results in a theory that is very similar to that described here.

Consider individual $nl \to n'l'$ transitions of type (33). These are more difficult to obtain than sums over final states. For example, consider an $nd \to nf$ transition. The final nf level is much closer to the higher nearly degenerate nl level—including the states of the ng level—than it is to the initial nd level. For a collision with a given b and V, the upward transitions from the nf state are always sudden whenever the $nd \to nf$ transition is sudden, and the number of n levels is very large, of order n^2, so the losses into this "sea" of nearly degenerate n levels *during* the collision must be taken into account when calculating the $np \to nd$ cross section.

Let $a = n_a l_a$ be the initial level and let c represent the sea of $n_a l$ states with $l > l_a$. Let $b = n_b l_b$ be the final level and d the sea of all $n_b l$ states with $l > l_b$. We restrict our consideration to those cases in which $n_a \neq n_b$, so c and d are different. This case and the case with $n_a = n_b$ are considered by Percival and Richards.[12]

The important transitions that may take place during the course of a single collision are

$$
\begin{array}{lll}
\text{level} & b \to d & \text{degenerate sea} \\
& \updownarrow \\
\text{level} & a \to c & \text{degenerate sea}
\end{array}
\tag{47}
$$

The reverse transitions from d to b and from c to a are negligible statistically because of the large number of states in the degenerate seas, whereas transitions between c and d, although probable, are irrelevant.

To calculate all the transitions that take place during the course of a single collision precisely would require an elaborate time-dependent close-coupling theory, which would take an exorbitant amount of computer time. In order to avoid this, approximations must be made. The principal approximation is to assume that during a collision, while the incident particle is moving along its rectilinear path, the relative populations of a, b, c, and d change with time t according to the ordinary laws of probability, instead of the quantal laws of combinations of amplitudes. The procedure is to calculate the transition probabilities for the processes indicated by the arrows in Eq. (47), using quantum mechanics but neglecting the other transitions of the diagram. These probabilities are then combined together by making reasonable assumptions about the time dependence of the processes, and following the evolution of the populations as a function of t.

For example, the transition probability $P^0(a \to b)$ might be calculated as in the IP approximation, but this does not give the probability $P(a \to b)$ because the initial $n_a l_a$ states can be depleted *before* the $a \to b$ transition takes place by the $a \to c$ process, which is usually just as fast or faster. Similarly, the b states can be depleted by the $b \to d$ process *after* the $a \to b$ transition. Thus the $a \to b$ transition is not very probable for any impact parameter, because where $P^0(a \to b)$ approaches unity, $P(a \to c)$ and $P(b \to d)$ are just as large or larger. The greatest value of $P(a \to b)$ is attained as the result of a delicate balance between the processes. Because $P(a \to b)$ is never very close to unity, it is possible to neglect the $b \to a$ transition in the probability analysis. It is *not* neglected, however, in the calculation of $P^0(a \to b)$.

In practice, it is easier to calculate the probabilities $P(a \to a)$ and $P(b \to b)$, that is, the atom remains at level a or b when it starts there, than it is to calculate the transitions to the seas. The probability analysis then gives us

$$
P(a \to b) = P^0(a \to b) \left[\frac{P(a \to a) - P(b \to b)}{\ln P(a \to a) - \ln P(b \to b)} \right].
\tag{48}
$$

By the definition of the resonant seas, the transitions to them are *always* sudden when the transition $a \to b$ is sudden, but when $a \to b$ is not sudden it is adiabatic, and $P^0(a \to b)$ is then small. So it makes sense to use the sudden approximation to evaluate $P(a \to a)$ and $P(b \to b)$ in Eq. (48), for it is then a good approximation when $P^0(a \to b)$ is large, and makes no significant contribution when $P^0(a \to b)$ is small.

Clearly $P(a \to a)$ and $P(b \to b)$ are similar when both states are coupled to degenerate seas and both are in the sudden region. In that case, formula (48) simplifies to

$$P(a \to b) = P^0(a \to b)P(a \to a). \qquad (49)$$

This has the interpretation that *during* the collision the $a \to b$ transition may be lost if the atom undergoes a transition to one of the c states before the $a \to b$ transition has a chance to take place or if it makes a $b \to d$ transition after the $a \to b$ transition takes place. Either of these processes depleted $P(a \to b)$ and thus the $a \to b$ cross section.

A similar argument applies if $n_a = n_b$, except that there is then only one degenerate sea. It is this depletion process that is responsible for the difference between the two PR theoretical results of Fig. 3.

In this section we have concentrated on region $\mathcal{B}$ of the incident particle energies, because this is of particular importance for laboratory experiment and is relevant to astrophysics. For l-changing transitions, as for the n-changing transitions,[2] there are four distinct regions—$\mathcal{A}$, $\mathcal{B}$, $\mathcal{C}$, $\mathcal{D}$—where different kinds of approximations are appropriate, and more detailed calculations are necessary to obtain the precise form of the cross sections in the neighborhoods of the boundaries between the regions.

5. Transitions within the Degenerate Sea

States with the same n and different l are almost completely degenerate for hydrogen and hydrogenic ions. For other atoms and ions most of the l states are nearly degenerate for sufficiently high n also. The radio-recombination lines emitted in interstellar space from high n non-hydrogenics are not distinguishable by virtue of the quantum defects of the low l states, but because of shifts due to differences in the masses of the nuclei. Otherwise they cannot be distinguished observationally from the hydrogenics.

Pengelly[14] solved the capture cascade equations for a hydrogenic system in the limit of low densities, and the agreement with observations in planetary nebulae was less than satisfactory. He concluded that it was necessary to take account of l-changing collisions.

Subsequently Pengelly and Seaton[15] calculated the cross sections for the $nl \to nl \pm 1$ collisions using a modified form of the impact parameter method. The usual form gives an infinite cross section for dipole transitions between completely degenerate states. This is clearly unrealistic. They took into account two phenomena that reduce the cross section to a finite value. One is that the atom may radiate *during* a collision. The other is that a charged particle in a plasma that is more distant than the Debye radius D is effectively shielded by oppositely charged particles, so Coulomb interactions at distances beyond D are rapidly damped to zero. They calculated the cross sections and rates on the basis that there are three possible cutoffs:

$$b_{ad} \approx \hbar V / \Delta E, \qquad b_D \approx D, \qquad b_{rad} \approx V T_{rad}, \tag{50}$$

where T_{rad} is the radiative lifetime. The smallest of these is the important one. Since the cutoff only appears logarithmically, any errors in the cutoff are unlikely to have a significant effect.

For the summed cross section

$$\sigma(nl) = \sigma(nl \to nl - 1) + \sigma(nl \to nl + 1) \tag{51}$$

they obtain, using the IP method,

$$\sigma(nl) = \pi b_1^2 \left[\tfrac{1}{2} + \ln(b_c/b_1) \right], \tag{52}$$

where b_c is the smallest cutoff (50) and

$$b_1^2 = 6a^2 \frac{Z_1^2}{Z_3^2} \frac{Ry}{\mathcal{E}} \left[1 - \frac{l(l+1) + 1}{n^2} \right]. \tag{53}$$

This differs from the sudden approximation result in the constant term $\tfrac{1}{2}$, which is usually relatively unimportant. For these cross sections the lower b cutoff is not so significant, because region $\mathcal{C}$ does not exist for transitions between completely degenerate states. Seaton[16] used these cross sections in a later study of recombination spectra.

However, strong coupling does have a significant effect, due to the possibility of large changes Δl, which have smaller cross sections than the $\Delta l = \pm 1$ transitions. However, a change Δl has an effect on the l distribution that is proportional to $(\Delta l)^2$, since the collisions produce a diffusion in *l*. These large Δl changes are taken into account explicitly in the following classical treatment of angular momentum diffusion[17] by Percival and Richards.

When the collisions are very distant, it is difficult to specify when one collision ends and another begins, and arbitrary assumptions lead to errors. We treat the effect of all collisions of charged particles as a time-dependent electric field $\mathcal{E}(t)$ in a similar fashion to Purcells'[18] treatment of individual fine-structure transitions. The electric field due to several charged particles is simply the sum of the fields due to the particles individually, and arbitrary numbers of collisions may proceed simultaneously. The fields due to individual electrons and ions are statistically independent within the Debye radius, beyond which their combined field tends to zero rapidly. In this treatment it is the Debye radius only that provides the effective cutoff making the transition rates finite.

An ion moving sufficiently slowly in the neighborhood of an atom produces a very strong Stark effect, which removes some of the degeneracy. In such circumstances, or in the presence of strong external electric fields, it might be more appropriate to use the Stark quantum members (n, n_e, m) rather than the angular momentum quantum numbers (n, l, m) to represent the atom. The different representation must lead to the same physical result, but it suggests a very different picture of the process taking place, which leads to a simpler analysis in different environments.

The theory is based on the classical theory of a hydrogen atom in crossed electric and magnetic fields—a theory described in the Born reference[19] and based on the researches of Bohr, Lenz, and Klein. The quantum mechanical Stark theory with the relation to the Fock $O(4)$ symmetry and the transformation from angular momentum to Stark representation is given in convenient form by Hughes.[20] The symmetry is also used in the quantum theory of a hydrogen atom in crossed electric and magnetic fields by Demkov,[21] for single-particle transitions between degenerate states caused by charged particles by Demkov et al.[22] and by Ostrovskii and Solov'ev.[23]

Transitions between degenerate states of hydrogenic atoms due to time-dependent electric fields are very special. For them it is possible to obtain correspondence identities, whereby the quantal results may be obtained exactly from the classical results. This is because the equations of motion for the important dynamical variables are linear, so that Heisenberg's equations for them are identical to the classical equations.

The classical picture is also helpful for visualizing what happens to the atom. In the absence of an electric field the electron with mass m_e and charge e moves in an elliptic orbit with its focus situated at the nucleus and with semimajor axis $a = n^2 a_0 / Z_3$. The eccentricity ε of the orbit of a given

(n, l) state is

$$\varepsilon^2 = 1 - (l + \tfrac{1}{2})^2/n^2 = 1 - I_l^2/I_n^2$$

$$I_l = (l + \tfrac{1}{2})\hbar, \qquad I_n = n\hbar, \tag{54}$$

where I_n and I_l are the classical actions.[2] The fundamental angular frequency of the electron is

$$\omega_n = v/a = (Z_3 e^2/m_e)^{1/2} a^{-3/2}. \tag{55}$$

By Bohr's correspondence principle, this classical frequency is approximately equal to the frequency for transitions between quantum states of neighboring principal quantum number, n, so that we have

$$\Delta E_n = E_{n+1} - E_n \approx \hbar \omega_n. \tag{56}$$

We consider only those electric fields $\mathcal{E}$ that vary adiabatically with respect to the fundamental frequency:

$$\left| \frac{d\mathcal{E}}{dt} \right| \ll \omega_n |\mathcal{E}|. \tag{57}$$

This puts a limit on the effective collision time T_{coll}:

$$\omega_n T_{\text{coll}} \gg 1. \tag{58}$$

For unperturbed degenerate states the energy splitting and the corresponding frequency are zero. There are *no* collisions that are adiabatic with respect to them in the sense that they can be adiabatic with respect to ω_n. This is the reason why the theory is essentially different from the theory of nondegenerate collisions.

In order to understand the behavior of the atom in the time-dependent field of passing charged particles it is first necessary to understand the motion in a static field. The application of a static electric field produces a change in the eccentricity and the plane of the elliptic orbit. Except for the very strongest fields, the length of the semimajor axis a remains unchanged. The plane of the orbit changes periodically with angular frequency ω_S, the Stark frequency, whereas the eccentricity changes periodically with double this frequency. The correspondence principle relates this Stark frequency to the splitting between Stark quantum levels

$$\Delta E_S = |E_{n_e+2} - E_{n_e}| \approx \hbar \omega_S, \tag{59}$$

where n_e is the "electric" quantum number which changes in steps of 2.

We only consider fields that are weak in the sense that

$$\omega_S \ll \omega_n, \qquad \Delta E_S \ll \Delta E_n. \tag{60}$$

In geometrical terms this means that the electron moves once around its orbit in a time short by comparison with any time characteristic of the changes in the elliptical orbit itself. The orbit may then be considered a dynamical entity in its own right, and its motion described in terms of the parameters defining it, with the exception of the semimajor axis a which always remains constant.

The orbit is characterized at any given time t by an angular momentum vector $\mathbf{L}(t)$, pointing in a direction perpendicular to the plane of the ellipse, and a Runge–Lenz vector $\mathbf{A}(t)$ pointing in the direction of its perihelion. The normalization of $\mathbf{A}(t)$ is not universally agreed upon. We can choose it to have the dimensions of angular momentum, which simplifies the subsequent analysis and also the relation with quantum theory. The value of $\mathbf{A}(t)$ is then given in terms of the position $\mathbf{r}(t)$, velocity $\mathbf{v}(t)$, and angular momentum of the electron:

$$\mathbf{A} = (-m_e/2E_n)^{1/2}(\mathbf{v} \times \mathbf{L} - Z_3 e^2 \mathbf{r}/r). \tag{61}$$

In a *fixed* electric field the component of $\mathbf{A}(t)$ in the field direction is given by $n_e \hbar$, n_e being the electric quantum number.

It is shown by Born[19] that the motion is particularly simple when expressed in terms of the sum and difference vectors

$$\mathbf{X} = \tfrac{1}{2}(\mathbf{L} - \mathbf{A}), \qquad \mathbf{Y} = \tfrac{1}{2}(\mathbf{L} + \mathbf{A}). \tag{62}$$

These vectors also arise in the quantal description of the motion in terms of the dynamical $O(4)$ symmetry.[20] For a static field the mean motion of the $\mathbf{X}$ and $\mathbf{Y}$ vectors is described by the equations

$$\frac{d\mathbf{Y}}{dt} = \omega_S \times \mathbf{Y}, \qquad \frac{d\mathbf{X}}{dt} = -\omega_S \times \mathbf{X}, \tag{63}$$

where the angular frequency vector is

$$\omega_S = \frac{3a}{2I_n} e\mathcal{E}. \tag{64}$$

Equation (63) remains valid even if $\mathcal{E}$ varies with time, provided the variation is adiabatic in the sense of Eq. (57).

It follows from the definitions in Eq. (62) and the differential equations of (63) that $\mathbf{X}$ and $\mathbf{Y}$ have the same constant magnitude,

$$|\mathbf{X}|^2 = |\mathbf{Y}|^2 = I_n^2/4 = n^2\hbar^2/4 \tag{65}$$

and that for constant field $\mathbf{\mathcal{E}}$ they trace out two right circular cones with axis parallel to $\mathbf{\mathcal{E}}$ and passing through the origin. The components of $\mathbf{X}$ and $\mathbf{Y}$, and thus of $\mathbf{A}$ and $\mathbf{L}$, are constant in the direction of $\mathbf{\mathcal{E}}$ and are linearly related to the classical electric and magnetic action variables I_e and I_m and the corresponding quantum numbers n_e and m as follows. If $\mathbf{\mathcal{E}}$ is used as the axis of quantization,

$$\mathbf{Y} \cdot \hat{\mathbf{\mathcal{E}}} = \tfrac{1}{2}(I_m + I_e) = \tfrac{1}{2}(m + n_e)\hbar$$

$$\mathbf{X} \cdot \hat{\mathbf{\mathcal{E}}} = \tfrac{1}{2}(I_m - I_e) = \tfrac{1}{2}(m - n_e)\hbar$$

$$\mathbf{A} \cdot \hat{\mathbf{\mathcal{E}}} = I_e = n_e\hbar = (n_2 - n_1)\hbar \tag{66}$$

$$\mathbf{L} \cdot \hat{\mathbf{\mathcal{E}}} = I_m = m\hbar.$$

Also, it follows from Eq. (62) and (54) that

$$|\mathbf{A}| = \varepsilon I_n . \tag{67}$$

The vectors $\mathbf{X}$ and $\mathbf{Y}$ each rotate with frequency ω_S, in opposite directions about a common axis in the direction of $\mathbf{\mathcal{E}}$.

For time-dependent fields the analysis used by Born to obtain Eq. (63) is invalid unless the time variation of the field is adiabatic with respect to the stark frequency ω_S. However, the equations are valid under the weaker condition that the field variation is adiabatic with respect to the fundamental frequency, $\omega_n \gg \omega_S$.

Plasmas exist in a very wide variety of physical conditions, leading to a wide variety of different kinds of collisions. Different effects occur for different kinds of collisions, and so different assumptions and approximations may have to be made. A broad classification of the collisions is provided here.

Consider the effect of many collisions of electrons and ions with a highly excited atom, and take into account the possibility of large numbers of simultaneous collisions. The time dependence of the trajectory is defined by

$$\mathbf{R}(t) = \mathbf{b} + \mathbf{V}t, \qquad \mathbf{b} \cdot \mathbf{V} = 0, \tag{68}$$

where $\mathbf{V}$ is the velocity and $\mathbf{b}$ the vector impact parameter. Since $b = |\mathbf{b}|$ is large by comparison with the size a of the atom, the electric field is given by the dipole approximation

$$\mathcal{E}(t) = Z_1 e\mathbf{R}/R^3, \qquad R < D$$
$$= 0, \qquad\qquad R > D \tag{69}$$

where

$$D = \left(\frac{kT}{4\pi Ne^2}\right)^{1/2} \tag{70}$$

is the Debye radius, as discussed in the introduction, $Z_1 e$ is the charge on the incident particle, and N is the density of electrons. $Z_1 = -1$ for electrons.

The collision time T_{coll} of Eq. (58) is given in terms of impact parameter and velocity by the equation

$$T_{coll} = b/V, \tag{71}$$

and the Stark rotation time by

$$T_S = 2\pi/\omega_S. \tag{72}$$

At the point of closest approach, a charged particle produces an electric field that causes a rotation of the ellipse, with frequency ω_S. The collision may be adiabatic with respect to the Stark rotation

$$\omega_S T_{coll} > 1 \qquad \text{(Stark adiabatic)} \tag{73}$$

in which case the ellipse rotates many times due to the collision; or it may be impulsive with respect to the Stark rotation,

$$\omega_S T_{coll} < 1 \qquad \text{(Stark sudden)} \tag{74}$$

in which case the ellipse rotates through a relatively small angle due to the collision. The second condition can also be written

$$b > b_c \qquad \text{where} \quad \frac{b_c}{a} = \frac{v_n}{V}\left|\frac{Z_1}{Z_3}\right| \tag{75}$$

Note that, unlike the usual case, the Stark sudden collisions occur for larger impact parameters. For most conditions these collisions are more important then the Stark adiabatic collisions, so we consider them in more detail. For them the **X** and **Y** vectors move through a small angle during the collision. Classical perturbation theory may be used to solve Eq. (63) for the time-dependent electric field $\mathcal{E}(t)$ given by (69), giving an angle of rotation for each vector, in opposite senses, about the **b** direction of

$$
\theta = \left| \int_{-T_D}^{T_D} dt\, \omega_S(t) \right|
$$

$$
= 3 \left| \frac{Z_1}{Z_3} \right| \frac{a v_n}{b V} \left(1 - \frac{b^2}{D^2} \right)^{1/2} \qquad \text{(Stark sudden)}.
$$

(76)

The effect of many collisions is to move each vector on the surface of a sphere in a series of small arcs, each arc being a rotation about a fixed axis. The directions of these axes are isotropically distributed. Thus the vectors perform rotational Brownian motion on a spherical surface. After many collisions the vectors are rotated through a large angle about an axis isotropically distributed in space. However, since the small rotations of **X** and **Y** are correlated, their motion cannot be treated independently.

We suppose that the external conditions are isotropic, in which case we only need the distribution function of $\mu = \cos\chi$, χ being the angle between **X** and **Y**. We have

$$
\varepsilon = \sin(\chi/2), \qquad \mu = 1 - 2\varepsilon^2,
$$

(77)

so that the distribution functions for l and μ are related by

$$
\rho(l) = \frac{2(2l+1)}{n^2} \rho(\mu), \qquad \mu = \frac{2(l+\frac{1}{2})^2}{n^2} - 1.
$$

(78)

The equation of motion of μ is, from Eq. (63),

$$
\frac{d\mu}{dt} = -2\sqrt{1-\mu^2}\, \omega_S \cdot \hat{n},
$$

(79)

where $\hat{n}$ is a unit vector in the direction of $\mathbf{X} \times \mathbf{Y}$. Then the mean square change in μ is

$$
\langle \Delta\mu^2 \rangle = \tfrac{4}{3}(1-\mu^2)\langle \theta^2 \rangle.
$$

(80)

The diffusion equation for ρ is

$$
\frac{\partial \rho}{\partial t} = \lambda_D \left[(1-\mu^2)\rho'' - 2\mu\rho' \right],
$$

(81)

where λ_D is the diffusion constant,

$$\lambda_D = \frac{2}{3} \frac{\langle \theta^2 \rangle}{\Delta t}. \tag{82}$$

The diffusion constant is obtained by averaging over a Boltzmann distribution of velocities and all relevant impact parameters

$$\lambda_D = \frac{2}{3} N \frac{2}{\pi^{1/2}(kT)^{3/2}} \int_0^\infty \exp\left(\frac{-E}{kT}\right) VE^{1/2} \, dE \int_{b_1}^D 2\pi\theta^2 b \, db, \tag{83}$$

where N is the density of ions at temperature T, and b_1 is the larger of the boundaries between the Stark sudden and Stark adiabatic region [Eq. (75)] and the sudden and adiabatic region [Eq. (58) and (71)],

$$b_1 = \max\left(\frac{V}{v_n} a, \ \frac{v_n}{v} \frac{Z_1}{Z_3} a\right). \tag{84}$$

Using Eq. (76) and assuming that b_1/D is negligible, we obtain

$$\lambda_D = A \left\{ e^{-x_c} \ln\left[3.14 \times 10^{11} \frac{Z_3^2}{n^3} \sqrt{\frac{m_i}{m_e}} \left(\frac{N}{cm^{-3}}\right)^{-1/2} \right] \right.$$

$$+ (1 - e^{-x_c}) \ln\left[1.98 \times 10^6 \left(\frac{T}{K}\right) \left(\frac{N}{cm^{-3}}\right)^{-1/2} \left(\frac{Z_3}{nZ_1}\right) \sqrt{\frac{m_e}{m_i}} \right]$$

$$\left. - E_1(x_c) - e^{-x_c} \ln x_c - 0.29 \right\}. \tag{85}$$

$$A = 1.03 \times 10^{-4} \left(\frac{N}{cm^{-3}}\right) \left(\frac{T}{K}\right)^{-1/2} \left(\frac{nZ_1}{Z_3}\right)^2 \left(\frac{m_i}{m_e}\right)^{1/2} s^{-1} \tag{86}$$

$$x_c = \frac{1.58 \times 10^5}{T/K} \frac{Z_1 Z_3}{n^2} \left(\frac{m_i}{m_e}\right). \tag{87}$$

In this expression m_i is the reduced mass of the colliding ion, T/K is the temperature in Kelvin, and $E_1(x)$ is the exponential integral approximated to within 2% for all x by the simple form

$$E_1(x) = e^{-x}\ln\left\{1 + \frac{\exp[-0.577/(1+2x)]}{x}\right\}. \tag{88}$$

For high temperatures $x_c \ll 1$ and the second term of the diffusion constant, Eq. (85), is insignificant and λ_D may be obtained by putting $b_1 = Va/v_n$ in integral (83). For low temperatures, $x_c \gg 1$ and only the second term is significant; λ_D may be obtained by putting $b_1 = v_n Z_1 a / V Z_3$.

The diffusion constant, λ_D, is proportional to the square root of the reduced mass and the square of the incident particle charge. If there are a number of species, the combined diffusion constant is the sum of the individual diffusion constants. In general, the diffusion constant for electrons may be ignored by comparison with that of the ions.

The time for an ensemble of atoms to reach the equilibrium distribution from Stark sudden collisions is approximately

$$\tau_{eq} \approx 1/\lambda_D \tag{89}$$

where λ_D is the constant of the diffusion equation whose value is given by Eq. (85)–(87).

The physical picture of collisions in the small impact parameter region is quite different from that in the Stark sudden region. In general, there are fewer collisions, but each collision produces a large change in the angle between X and Y. If τ_c were the mean time between collisions in this region and if each collision produced a uniform distribution in μ the process would be a Poisson sequence and the probability of the distribution *not* being uniform at some time $t > 0$ would be given by $\exp(-t/\tau_c)$. However, this region is not usually important.

This theory takes account of changes in l of arbitrary size, but all processes are included together, and comparison with individual processes is not immediate. Bohr found out that for the stoppage of charged particles classical estimates of diffusion are almost identical to quantal rates, *even* when the latter are dominated by small changes in quantum number, and classical estimates of individual processes are poor. This follows from his correspondence principles and almost certainly applies to the present problem also.

References

1. M. J. SEATON, *Proc. Phys. Soc.* **79**, 1105 (1962).
2. I. C. PERCIVAL and D. RICHARDS, *Adv. At. Mol. Phys.* **11**, 1 (1975).

3. C. F. von Weizsäcker, *Z. Phys.* **88**, 612 (1934); E. J. Williams, *Phys. Rev.* **45**, 729 (1934); K. Alder, A. Bohr, T. Huus, B. Mottelson, and A. Winther, *Rev. Mod. Phys.* **28**, 432 (1956).
4. M. J. Seaton, *Proc. Phys. Soc.* **68A**, 457 (1955).
5. M. J. Seaton, *Mon. Not. R. Astron. Soc.* **127**, 177 (1964).
6. H. E. Saraph, *Proc. Phys. Soc.* **83**, 763 (1964).
7. J. Calloway and E. Bauer, *Phys. Rev. A* **140**, 1072 (1965). See also reference 2.
8. J. A. Schiavone, S. M. Tarr, and R. S. Freund, *Phys. Rev. A* **20**, 71 (1979).
9. J. F. Delpech, J. Boulmer, and F. Devos, *Phys. Rev. Lett.* **39**, 1400 (1977); F. Devos, J. Boulmer, and J. F. Delpech, *J. Phys. (Paris)* **40**, 215 (1979).
10. G. W. Foltz, E. J. Beiting, T. H. Jeys, K. A. Smith, F. B. Dunning, and R. F. Stebbings, *Bull. Am. Phys. Soc.* **24**, 1204 (1979) and *Phys. Rev. A* **25**, 187 (1982).
11. K. B. MacAdam, R. Rolfes, and D. A. Crosby, *Phys. Rev. A* **24**, 1286 (1981); K. B. MacAdam, D. A. Crosby, and R. Rolfes, *Phys. Rev. Lett.* **44**, 980 (1980).
12. I. C. Percival and D. Richards, *J. Phys. B* **10**, 1497 (1977).
13. H. A. Bethe and E. E. Salpeter, *Quantum Mechanics of One- and Two-Electron Atoms*, Plenum, New York 1977, p. 257.
14. R. M. Pengelly, *Mon. Not. R. Astron. Soc.* **127**, 145 (1964).
15. R. M. Pengelly and M. J. Seaton, *Mon. Not. R. Astron. Soc.* **127**, 165 (1964).
16. M. J. Seaton, *Mon. Not. R. Astron. Soc.* **127**, 177 (1964).
17. I. C. Percival and D. Richards, *J. Phys. B* **12**, 2051 (1979).
18. E. M. Purcell, *Astron. J.* **116**, 457 (1952).
19. M. Born, *Mechanics of the Atom*, Bell, London, 1960.
20. J. W. B. Hughes, *Proc. Phys. Soc.* **91**, 810 (1967).
21. Yu N. Demkov, B. S. Monozon, and V. N. Ostrovskii, *Sov. Phys.–JETP* **30**, 775 (1970).
22. Yu N. Demkov, V. N. Ostrovskii, and E. A. Solov'ev, *Sov. Phys.–JETP* **39**, 57 (1974).
23. V. N. Ostrovskii and E. A. Solov'ev, *Sov. Phys.–JETP* **39**, 779 (1974).
24. D. Richards, *J. Phys. B: Atom. Molec. Phys.* **6**, 823 (1973).

4

PROTON IMPACT EXCITATION OF POSITIVE IONS

A. DALGARNO

1. Introduction

In hot hydrogenic plasmas with an impurity content of heavier elements, the emission spectrum and the distribution of the impurity ionization stages are largely determined by electron impact processes. There are, however, particular instances where transitions are caused preferentially by proton impacts. In addition to modifying the plasma characteristics, such transitions have a special utility as diagnostic probes of the proton component of the plasma.

When the energy ΔE of a transition is large compared to the kinetic energy E of the proton, the phase $\Delta E t / \hbar$ undergoes severe cancellation when integrated over the collision path and the proton impact excitation cross section is small. When ΔE is small compared to E, little cancellation occurs and the longer interaction time of the proton with the target leads to a proton impact cross section that may exceed considerably the electron impact cross section. Seaton[1] pointed out that, when ΔE is much less than the mean thermal energy kT, the rate coefficient is controlled by collisions in the energy range where the cross section has passed through its maximum and is decreasing approximately as the inverse square of the particle velocity, v^{-2}. The rate coefficient for proton impact excitation of neutral targets is accordingly a factor $(M_p/m)^{1/2}$ greater than that for electron impact excitation where M_p is the proton and m is the electron mass. Thus it is collisions with protons that are effective in redistributing angular momenta in high-lying Rydberg levels

$$H^+ + H(nl) \rightleftarrows H + H(nl'). \tag{1}$$

A. DALGARNO ● Harvard–Smithsonian Center for Astrophysics, Cambridge, Massachusetts 02138. This research was supported by the Division of Chemical Sciences of the U.S. Department of Energy.

Following the original discussion of

$$H^+ + H(2s) \rightleftarrows H^+ + H(2p) \qquad (2)$$

by Seaton,[1] Pengelly and Seaton[2] used semiclassical perturbation theory to calculate the cross sections for reactions (1). For high n, proton impacts bring about a statistical equilibrium of the l-level populations. Protons may be effective also in n-changing collisions,

$$H^+ + H(n) \rightleftarrows H^+ + H(n'), \qquad (3)$$

and cross sections for reactions (3) have been derived by Burgess and Summers.[3] For the excitation of positive ion targets more detailed analysis may be necessary because the Coulomb interaction diminishes the rate coefficients for proton impact while enhancing those for electron impact. For

$$H^+ + He^+ (nl) \rightleftarrows H^+ + H^+ (nl'), \qquad (4)$$

Pengelly and Seaton[2] demonstrated that the effects of the Coulomb interaction were small for large values of n at the temperatures of the order of 10^4 K, characteristic of astrophysical plasmas.

Seaton[4] also investigated the fine-structure excitation process

$$H^+ + Fe^{13+} \left(3p\,^2P_{1/2}\right) \rightarrow H^+ + Fe^{13+} \left(3p\,^2P_{3/2}\right). \qquad (5)$$

The excited fine-structure level may radiate with the emission of the coronal green line at 5304.3 Å. Seaton[4] showed that the proton excitation rate exceeds the electron excitation rate at temperatures above 1.3×10^6 K. This initial exploration of proton-induced fine-structure transitions in positive ions stimulated a considerable body of calculations—still in progress.

2. Excitation of Fine-Structure Transitions

If ρ denotes collectively the position vectors ρ_i of the target electrons measured with respect to the target nucleus and r denotes the vector joining the nucleus to the proton, the instantaneous electrostatic interaction is given in atomic units by

$$V(\mathbf{r},\boldsymbol{\rho}) = \frac{Z}{r} - \sum_{i=1}^{N} \frac{1}{|\mathbf{r} - \boldsymbol{\rho}_i|}, \qquad (6)$$

where Z is the nuclear charge of the positive ion and the summation is taken over all the N target electrons. Expression (6) may be expanded in the form

$$V(\mathbf{r},\boldsymbol{\rho}) = \sum_\lambda \frac{4\pi}{2\lambda+1} \sum_\mu \sum_i v_\lambda(r,\rho_i) Y_{\lambda\mu}(\hat{\mathbf{r}}) Y^*_{\lambda\mu}(\hat{\boldsymbol{\rho}}_i), \tag{7}$$

where $\hat{\mathbf{r}}$ and $\hat{\boldsymbol{\rho}}_i$ are unit vectors in the directions of $\mathbf{r}$ and $\boldsymbol{\rho}_i$, respectively, and the coefficients $v_\lambda(r,\rho_i)$ are given by

$$v_\lambda(g,\rho_i) = \frac{Z}{r}\delta_{\lambda 0} - \frac{1}{\rho_i}\left(\frac{r}{\rho_i}\right)^\lambda, \qquad r < \rho_i \tag{8}$$

$$= \frac{Z}{r}\delta_{\lambda 0} - \frac{1}{r}\left(\frac{\rho_i}{r}\right)^\lambda, \qquad r > \rho_i. \tag{9}$$

Seaton[4] assumed that the fine-structure transitions are driven by the interaction between the proton and the undistorted charge distribution of the target ion. Suppose the transition takes place from a state of total electronic angular momentum $\mathbf{J}$ to a state of total electronic angular momentum $\mathbf{J}'$, each composed of states of the same orbital angular momentum $\mathbf{L}$ and spin angular momentum $\mathbf{S}$ and suppose that all other electronic states are distant in energy and may be neglected. The coupling matrix elements take the form

$$\langle JM_J| V(\mathbf{r},\boldsymbol{\rho})|J'M_{J'}\rangle$$

$$= \sum_\lambda \frac{4\pi}{2\lambda+1}\sum_\mu Y_{\lambda\mu}(\hat{\mathbf{r}})\langle JM_J|\sum_i v_\lambda(r,\rho_i)Y^*_{\lambda\mu}(\hat{\boldsymbol{\rho}}_i)|J'M_{J'}\rangle. \tag{10}$$

Because the parity is unchanged, the dipole terms $\lambda = 1$ vanish. In the systems of most interest, J changes by 1 or 2. Then, only the terms with $\lambda = 0$ and $\lambda = 2$ survive, and (10) may be reduced to

$$\langle JM_J|V(\mathbf{r},\boldsymbol{\rho})|J'M_{J'}\rangle = \langle JM_J|\sum_i v_0(r,\rho_i)|J'M_{J'}\rangle\delta_{JJ'}\delta_{M_J M_{J'}}$$

$$+ \frac{4\pi}{5}\sum_\mu Y_{2\mu}(\hat{\mathbf{r}})\langle JM_J|\sum_i v_2(r,\rho_i) Y^*_{2\mu}(\hat{\boldsymbol{\rho}}_i)|J'M_{J'}\rangle.$$

$$\tag{11}$$

For the 0–1 transition, the matrix elements vanish and the transition cannot occur in first order.

Formal expressions for

$$\langle JM_J | \sum_i v_2(r, \rho_i) Y^*_{2\mu}(\hat{\rho}_i) | J' M_{J'} \rangle$$

may be obtained by the standard procedures of atomic scattering theory.[5] The angular coefficients have been listed by Nikitin,[6] Masnou-Seeuws and McCarroll[7] (see also references 8 and 9), and Faucher.[10]

At large distances r, the interaction is dominated by the Coulomb repulsion ζ/r, where $\zeta = Z - N$ is the excess charge of the target ion. Because the Coulomb repulsion prevents the close approach of the proton and the target, the region in which $r < \rho$ is unimportant at low energies and the transition is controlled by the long-range interaction of the proton and the permanent quadrupole moment, q, of the positive ion. In his original study, Seaton[4] adopted the long-range form q/r^3 at all separations, as have subsequently many authors. At high energies it is important to use an interaction in which the singularity at the origin is suppressed. Failure to do so leads to a cross section that decreases at high energies as $E^{-1/2}$, whereas the correct behavior is E^{-1}. The consequences of modifications in the short-range form have been explored by Reid and Schwartz,[11] Faucher and Landman,[12] and Doyle et al.[9] The effects are serious only at energies where other excitation mechanisms are important. Thus for O^{4+} and Ne^{7+}, the rate coefficients for the $J = 0$–1 transition are in error by only 10% at 10^7 K when the asymptotic interaction is used.[9]

2.1. Semiclassical Theory

In solving the scattering problem, Seaton[4] employed a semiclassical approximation in which the nuclei are assumed to move classically under the action of the Coulomb repulsion ζ/r. The separation r is a function of time which depends parametrically on the impact parameter b. The cross section $\sigma(E)$ can be written

$$\sigma(E) = 2\pi \int_0^\infty P(b) b \, db, \tag{12}$$

where $P(b)$ is the probability that a transition occurs during motion along the classical orbit specified by the energy E and the impact parameter b. The transition occurs as the electronic state evolves under the influence of the time-dependent potential $V(\rho, t)$. The wave function $\Psi(\rho, t)$ satisfies the time-dependent Schrödinger equation

$$(H + V)\Psi(\rho, t) = \frac{i}{\hbar} \frac{\partial \Psi}{\partial t} (\rho, t), \tag{13}$$

where H is the target Hamiltonian and

$$H\Phi(JM_J|\rho) = E_J\Phi(JM_J|\rho). \tag{14}$$

In the impact parameter formalism, $\Psi(\rho, t)$ is expanded according to

$$\Psi(\rho, t) = \sum_{J, M_J} a_{JM_J}(t)\Phi(JM_J|\rho)\exp(-iE_Jt/\hbar) \tag{15}$$

and the coefficients satisfy the first-order equations

$$\dot{a}_{JM}(t) = \frac{i}{\hbar}\sum_{J', M_{J'}} a_{J'M_{J'}}(t)\langle JM_J|F|J'M_{J'}\rangle\exp\left[-\frac{i(E_{J'} - E_J)t}{\hbar}\right]. \tag{16}$$

If the initial state is specified by J, M_J and Eq. (16) is solved subject to the initial conditions

$$a_{J'M_{J'}}(t = 0) = \delta_{JJ'}\delta_{M_JM_{J'}} \tag{17}$$

$$P(JM_JJ'M_{J'}|b) = |a_{J'M_{J'}}(t = \infty)|^2. \tag{18}$$

Equation (12) then gives the cross section $\sigma(JM_J \rightarrow J'M_{J'})$, and the fine-structure cross section $\sigma(J \rightarrow J')$ is obtained by averaging over the initial M_J and summing over the final $M_{J'}$:

$$\sigma(J \rightarrow J') = \frac{1}{2J + 1}\sum_{M_J}\sum_{M_{J'}}\sigma(JM_J \rightarrow J'M_{J'}). \tag{19}$$

Seaton[4] estimated $P(b)$ by using a first-order perturbation solution[13] of Eq. (16) for $P(b) \leqslant 0.5$ and taking $P(b) = 0.5$ for b less than b', where b' is the largest value of b at which $P(b) = 0.5$. Seaton[4] demonstrated that the first-order treatment is adequate at low energies where the Coulomb repulsion prevents a close approach and ensures that $P(b)$ remains small at all values of b that contribute to the cross section, but that with increasing energy it is important to enforce unitarity.

Bely and Faucher[14] and Landman[15] followed Seaton[4] in using a semiclassical first-order method but with different procedures for preserving unitarity at intermediate energies. Bely and Faucher[14] presented results for several ions in $2p$, $2p^5$, $3p$, and $3p^5\,{}^2P_J$ levels. For the ${}^3P_0 - {}^3P_1$ transitions, the first-order theory yields a vanishing probability $P(b)$. Landman[15] investigated transitions of Fe^{12+} for which he adopted a statistical distribution among the accessible levels when the total transition probability $P(b)$ exceeded 0.5. Useful representations of the data of Bely and Faucher[13] were constructed by de Boer *et al.*[16]

Apart from additional complexity in the algebra of angular momentum coupling, the same scattering calculations can provide cross sections for transitions in which the changes in projection quantum number M_J are identified. The process may be important in the depolarization of the emitted radiation. Semiclassical first-order cross section data for changes in M_J have been computed by Sahal–Brechot[17] and Malinovsky.[18]

The coupled equations (16) may be solved exactly and this was done by numerical integration at a small number of energies by Bahcall and Wolf[19] for several ions in ^2P_J and ^3P_J states. Some additional calculations by Wolf were quoted by Chevalier and Lambert for Fe^{12+} [20] and Ca^{14+}.[21] In constructing analytical fits to their limited numerical data, Bahcall and Wolf[19] adopted an incorrect high-energy extrapolation that affected the high-temperature limits of their rate coefficients.[11] In subsequent calculations of cross sections from numerical solutions of the coupled impact parameter equations, results have been obtained over an extended range of energies[7,8,9,11,22–24] and the correct limiting form at high energies has been reproduced. Cross sections for transitions between magnetic sublevels of Fe^{12+} and Fe^{13+} have been given by Landman.[25] More approximate estimates of the rate coefficients have been made by Kastner[26] and Kastner and Bhatia[27] who combined the results of semiclassical first-order theory at low energies with empirical representations of the coupled intermediate energy data.

The semiclassical approximation with its assumption of a common trajectory in all the scattering channels must fail at energies near threshold. Various extensions of the approximation have been advanced and applied to fine-structure excitation induced by neutral particle collisions[28–30] but they are probably unnecessary for collisions between positively charged particles because the cross section becomes uninterestingly small as the energy decreases toward threshold.

2.2. Quantal Theory

A quantum mechanical description of the nuclear motion[5] was employed by Faucher[10] in an investigation of Fe^{12+}. The theory is a straightforward modification of electron–ion scattering theory and reduces to the integration of coupled sets of differential equations

$$\left[\frac{d^2}{dr^2} - \frac{l(l+1)}{r^2} - \frac{2M\zeta}{r} + Mk^2 \right] F_i(r) = 2M \sum_j V_{ij}(r) F_j(r), \quad (20)$$

where k is the channel wave number and $V_{ij}(r)$ are radial coupling matrix elements. The cross section is a sum

$$\sigma(J \to J') = \frac{\pi}{2M(2J+1)} \frac{1}{E} \sum_l (l+1)\Omega_l(J,J'), \qquad (21)$$

of collision strengths $\Omega_l(J,J')$ which may be obtained from the asymptotic solutions of Eq. (20). The quantal collision strengths and semiclassical transition probabilities are directly related. For $^3P_J-^3P_{J'}$ transitions,[12]

$$P(b) = \frac{1}{2(2J+1)} \Omega_l(J,J'), \qquad (22)$$

where

$$b = (2ME)^{-1/2}l. \qquad (23)$$

Further quantal calculations, some in conjunction with semiclassical results, have been carried out by Faucher and coworkers.[10,23]

Explicit comparison of full close-coupled quantal and semiclassical calculations[12] shows no differences at low energies outside the threshold region, but with increasing energy the semiclassical transition probabilities contain more pronounced oscillations at small impact parameters. The deviations occur for classical orbits in which the colliding particles approach to distances at which the proton penetrates the electronic charge distribution of the target ion. At such distances the representations adopted for the interaction of the proton and the ion are inadequate.

Fine-structure excitation can be regarded alternatively as a molecular process in which the transitions occur during the evolution of the electronic Σ and Π states of the quasimolecule formed by the approach of the proton and the positive ion. In place of the atomic representation $|JM_J\rangle$, the molecular representation $|\Lambda M_L\rangle$ is used[5,31] and the total wave function is expanded in the set of all the molecular states that dissociate to the proton and the target ion in its fine-structure states. The molecular approach was applied by Masnou-Seeuws and McCarroll[7] to the ^2P_J levels of Fe^{13+} and to the $np^2\,^3P_J$ levels of several ions.

An approximate version of the quantal formulation, the elastic approximation,[32] was developed using the molecular approach. The limitations of the elastic approximation have been analyzed by Reid and Rankin,[33] Harel et al.,[34] and Nikitin and Reznikov.[30] For the excitation of positive ions, it overestimates the cross section at all energies.[35]

The molecular approach offers the possibility of incorporating more accurate descriptions of the interaction potentials. Asymptotically, the interaction potentials $V_\Sigma(r)$ and $V_\Pi(r)$ of the Σ and Π electronic states vary according to

$$V_\Sigma(r) \sim \frac{\zeta}{r} - \frac{q}{r^3} \qquad (24)$$

and

$$V_\Pi(r) \sim \frac{\zeta}{r} + \frac{1}{2}\frac{q}{r^3}, \qquad (25)$$

forms that are equivalent to assuming that the interaction is determined by the long-range interaction of the proton and the permanent quadrupole moment q of the ion. Conventional molecular structure calculations can be used to improve the accuracy of V_Σ and V_Π. The arbitrariness at short distances can be avoided and with appropriately chosen molecular wave functions the attractive long-range polarization forces can be included. For the ion Fe^{13+}, preliminary calculations[35] suggest that the effects are small. The situation may be different for less highly ionized targets.

A more accurate representation of the interaction can be obtained in the atomic approach by adding further atomic eigenfunctions to the expansion of the total wave function. In a semiclassical study of the proton impact excitation of S^{3+},

$$S^{3+}\left({}^2P_{1/2}\right) + H^+ \rightarrow S^{3+}\left({}^2P_{3/2}\right) + H^+, \qquad (26)$$

Schwarz, Reid, and Dalgarno[36] employed a three-state expansion, consisting of the $3s^2 3p\,{}^2P_{1/2}$, $3s^2 3p\,{}^2P_{3/2}$ and $3s3p^2\,{}^2S_{1/2}$ states. The ${}^2S_{1/2}$ state is accessible by dipole transitions from the ${}^2P$ states and the fine-structure transition can occur by a virtual dipole–dipole sequence. At energies well below the ${}^2S_{1/2}$ excitation threshold, the retention of the ${}^2S_{1/2}$ state is equivalent to the inclusion of a long-range polarization force. Schwarz et al.[36] found that the fine-structure transition cross section obtained from the three-state expansion exceeded that obtained from the two-state expansion everywhere but by no more than 7%. They pointed out that for impact energies exceeding 200 keV, direct impact excitation of the excited ${}^2S_{1/2}$ level followed by radiative decay to the ${}^2P$ levels is a more rapid means of bringing about a change in the fine-structure level populations.

3. Excitation of Metastable Transitions

Landman[15] pointed out that the quadrupole interaction between the proton and the target ion also drives transitions between the ${}^3P$ and the metastable ^1D and ^1S states of Fe XIII when their configurations are described in the intermediate-coupling representation. The $^3P_0$ eigenfunction contains an admixture of the $^1S_0$ LS-coupled state, and the $^3P_2$ eigenfunction contains an admixture of the $^1D_2$ LS-coupled state. The quadrupole interaction also drives transitions between the $^1S_0$ and $^1D_2$ states. Landman presented the results of unitarized first-order and fully coupled semiclassical theories for[15,22] Fe^{12+} and for[24] Fe^{11+} and Kastner and Bhatia[27] have presented some semiempirical data for the ^4S–$^2D_{3/2}$ and ^4S–$^2D_{5/2}$ transitions of S^{9+}. The cross sections for the excitations of metastable levels are smaller in magnitude than those for the excitation of fine-structure levels.

A fully coupled quantal treatment of the proton impact excitation of the transition from $C^{2+}(2p^2\,{}^1S)$ to $C^{2+}(2p^2\,{}^1D)$ has been carried out by Bienstock, Heil, and Dalgarno.[37] For this case, in addition to the long-range quadrupole coupling, the excitation may occur by virtual transitions to the charge-transfer molecular state dissociating to $C^{3+}(2s\,{}^2S)$ and a ground-state hydrogen atom. Avoided crossings between the direct excitation and charge-transfer states occur in the $^1\Sigma^+$ and $^3\Sigma^+$ molecular symmetries.[38]

4. Charge-Transfer Ionization

In some circumstances, proton impacts may be competitive with electron impacts as an ionization mechanism. For those cases in which the charge transfer of multiply charged ions,

$$X^{n+} + H \rightarrow X^{(n-1)+} + H^+, \tag{27}$$

occurs rapidly and preferentially into the ground state of $X^{(n-1)+}$, the reverse process of charge-transfer ionization

$$H^+ + X^{(n-1)+} \rightarrow H + X^{n+} \tag{28}$$

may be an important source of X^{n+} ions at thermal energies. Theoretical calculations by Bates, Johnston, and Stewart[39] show that charge transfer of Mg^{2+} with H occurs mainly to the ground state of Mg^+. Their prediction has been verified experimentally by measurements of the forward and reverse reactions.[40,41] Charge transfer of Ti^{2+} is another system where the ground-state species determine the cross section and for which experimental data are available.[41]

Theoretical studies have been carried out for several systems of astrophysical interest.[42] Of particular significance to the ionization structure of the solar chromosphere are the charge-transfer ionization processes[43]

$$H^+ + Si^+ \rightarrow H + Si^{2+} \tag{29}$$

and

$$He^+ + Si^{2+} \rightarrow He + Si^{3+}. \tag{30}$$

Most of Seaton's research has involved electron impact phenomena. Yet it was his studies[1,2,4] that established the importance of proton impacts with the ionized constituents of astrophysical nebulae and laboratory gases and initiated a lengthy series of calculations, particularly on proton impact excitation of fine-structure transitions, which broadened to include metastable transitions and ionization processes, all of which participate in determining the ionization structure and the emissivity and energy loss from hot plasmas.

ACKNOWLEDGMENT

I am indebted to Dr. S. Bienstock, Dr. T. G. Heil, and Dr. K. Kirby for comments on this chapter.

References

1. M. J. SEATON, *Proc. Phys. Soc.* **68**, 457 (1955).
2. R. M. PENGELLY and M. J. SEATON, *Mon. Not. R. Astron. Soc.* **127**, 165 (1964).
3. A. BURGESS and H. P. SUMMERS, *Mon. Not. R. Astron. Soc.* **174**, 345 (1976).
4. M. J. SEATON, *Mon. Not. R. Astron. Soc.* **127**, 191 (1964).
5. R. G. H. REID, *J. Phys. B* **6**, 2018 (1973); F. H. MIES, *Phys. Rev. A* **7**, 942 (1973).
6. E. E. NIKITIN, *J. Chem. Phys.* **43**, 744 (1965).
7. F. MASNOU-SEEUWS and R. MCCARROLL, *Astron. Astrophys.* **17**, 441 (1972).
8. D. A. LANDMAN, *Astron. Astrophys.* **43**, 285 (1975).
9. J. G. DOYLE, A. E. KINGSTON, and R. H. G. REID, *Astron. Astrophys.* **90**, 97 (1980).
10. P. FAUCHER, *J. Phys. B* **8**, 1886 (1975).
11. R. G. H. REID and J. H. SCHWARZ, in *VI ICPEAC*, I. Amdur, ed. (MIT Press, Cambridge, Massachusetts), p. 236 (1969).
12. P. FAUCHER and D. A. LANDMAN, *Astron. Astrophys.* **54**, 159 (1977).
13. K. ALDER, A BOHR, T. HUUS, B. MOTTELSON, and A. WINTHER, *Rev. Mod. Phys.* **28**, 432 (1956).
14. O. BELY and P. FAUCHER, *Astron. Astrophys.* **6**, 88 (1970).
15. D. A. LANDMAN, *Solar Phys.* **30**, 371 (1973).
16. K. S. DE BOER, H. OLTHOF, and S. R. POTTASCH, *Astron. Astrophys.* **16**, 417 (1972).
17. S. SAHAL-BRECHOT, *Astron. Astrophys.* **32**, 147 (1974).
18. M. MALINOVSKY, *Astron. Astrophys.* **43**, 101 (1975).
19. J. N. BAHCALL and R. A. WOLF, *Astrophys. J.* **152**, 701 (1968).
20. R. A. CHEVALIER and D. L. LAMBERT, *Solar Phys.* **10**, 115 (1969).
21. R. A. CHEVALIER and D. L. LAMBERT, *Solar Phys.* **11**, 243 (1970).

22. D. A. LANDMAN, *Solar Phys.* **31**, 81 (1973).
23. P. FAUCHER, *Astron. Astrophys.* **54**, 589 (1977); P. FAUCHER, F. MASNOU-SEEUWS and M. PRUDHOMME, *Astron. Astrophys.* **81**, 137 (1980).
24. D. A. LANDMAN, *Astrophys. J.* **220**, 366 (1978); D. A. LANDMAN and T. BROWN, *Astrophys. J.* **232**, 636 (1979).
25. D. A. LANDMAN, *Astron. Astrophys.* **43**, 285 (1975).
26. S. O. KASTNER, *Astron. Astrophys.* **54**, 255 (1977).
27. S. O. KASTNER and A. K. BHATIA, *Astron. Astrophys.* **71**, 211 (1979).
28. E. E. NIKITIN, *Adv. Chem. Phys.* **28**, 317 (1975).
29. C. GAUSSORGUES and F. MASNOU-SEEUWS, *J. Phys. B* **10**, 2125 (1977).
30. E. E. NIKITIN and A. I. RESNIKOV, *J. Phys. B* **13**, L57 (1980).
31. E. E. NIKITIN, *J. Chem. Phys.* **43**, 744 (1965); F. MASNOU-SEEUWS, *J. Phys. B* **3**, 1437 (1970).
32. A. DALGARNO, *Proc. R. Soc. Lond. Ser. A* **262**, 131 (1961); A DALGARNO and M. R. H. RUDGE, *Astrophys. J.* **140**, 800 (1964); S. WOFSY, R. H. G. REID, and A. DALGARNO, *Astrophys. J.* **168**, 161 (1971).
33. R. H. G. REID and R. F. RANKIN, *J. Phys. B* **11**, 55 (1978).
34. C. HAREL, V. LOPEZ, R. MCCARROLL, A. RIERA, and P. WAHNON, *J. Phys. B* **11**, 71 (1978).
35. T. G. HEIL, K. KIRBY, and A. DALGARNO, *Bull. Am. Phys. Soc.* **26**, 1309 (1981).
36. J. H. SCHWARZ, R. H. G. REID, and A. DALGARNO, unpublished (1970).
37. S. BIENSTOCK, T. G. HEIL, and A. DALGARNO, *Bull. Am. Phys. Soc.* **26**, 1301 (1981).
38. T. G. HEIL, S. E. BUTLER, and A. DALGARNO, *Phys. Rev. A* **23**, 1100 (1981).
39. D. R. BATES, H. C. JOHNSTON, and I. STEWART, *Proc. Phys. Soc.* **84**, 517 (1964).
40. B. PEART, D. M. GEE, and K. T. DOLDER, *J. Phys. B* **10**, 2683 (1977).
41. R. W. MCCULLOUGH, W. L. NUTT, and H. B. GILBODY, *J. Phys. B* **12**, 4159 (1979).
42. S. E. BUTLER and A. DALGARNO, *Astrophys. J.* **241**, 838 (1980).
43. S. L. BALIUNAS and S. E. BUTLER, *Astrophys. J. Lett.* **235**, L45 (1980).

5

LONG-RANGE INTERACTIONS IN ATOMS AND DIATOMIC MOLECULES

G. Peach

1. Introduction

It is now many years since Castillejo, Percival, and Seaton[1] published a paper on the theory of elastic collisions between electrons and hydrogen atoms. The Schrödinger equation for the whole system was expressed in terms of the usual infinite set of coupled equations by expanding the total wave function in terms of a complete set of unperturbed states of the hydrogen atom. They then used these equations to show that the leading term in the interaction between the electron and the hydrogen atom at large separations r was $-\alpha_d/(2r^4)$, where here α_d is the static dipole polarizability of hydrogen. Throughout this article atomic units are used, so that all distances are expressed in units of the Bohr radius a_0, and one atomic unit of energy is equal to e^2/a_0. Much more recently, Seaton and Steenman-Clark[2] again used the coupled equations that represent the electron–hydrogen system to show that the next long-range term in the interaction is $-\alpha'/(2r^6)$, where α' is linearly dependent upon the energy of the elastically scattered electron and can be determined analytically. Thus Professor Seaton's interest in the derivation of the form of the long-range potentials spans a period of over 20 years, and I have great pleasure in contributing a chapter on this topic to this special volume in honor of Professor Seaton's 60th birthday. I am indebted to him for originally stimulating my own interest in the electron–atom problem; subsequently I also became interested in two-center problems, and therefore this chapter represents a synthesis of the interests of us both.

G. PEACH • Department of Physics and Astronomy, University College London, Gower Street, London WC1E 6BT, England.

Although in principle all electrons in an atomic or molecular system are indistinguishable, in practice it is well established that many of the chemical and physical properties of the system are determined by the interactions of the outer or *valence* electrons only, whereas the inner or *core* electrons remain closely associated with the nuclei. An equation for a wave function describing the outer electrons can always be obtained from the Schrödinger equation for the whole system. This shows that the outer electrons move in a potential that is nonlocal and depends on the properties of the wave functions of the core and valence electrons in a very complicated way. A numerical study of the nonlocal nature of the effective potentials for electron–hydrogen scattering has been carried out by Seaton and Steenman-Clark.[3] However, effective potentials are only nonlocal at short range, and so when the outer electrons are at large distances from the atomic core the electron–core interaction becomes local, and it is with this region that we shall be concerned.

Over the years quite a few papers have been published in which the problems of the long-range potentials have been discussed. In 1956, Dalgarno and McCarroll[4] derived a general expression for the leading asymptotic correction to the static interaction between any two atoms or atomic ions of arbitrary mass and applied it to the case of a slowly moving charge of arbitrary mass at a large distance from a hydrogen atom. In 1959, Mittleman and Watson[5] showed that for electron–atom scattering, the leading term in the long-range potential is $-\alpha_d/(2r^4)$, where α_d is the static dipole polarizability of the atom. They also showed that if the atom is hydrogen, the next term—the so-called nonadiabatic correction to the static dipole term—is $3\beta_1/r^6$, where β_1 is a positive constant that depends solely on the properties of the atom. This work was subsequently generalized by Öpik[6] in 1967, and by Kleinman, Hahn, and Spruch[7] and Callaway, LaBahn, Pu, and Duxler[8] in 1968. These authors derived an expression for the case of a slowly moving electron interacting with any spherically symmetric atom and showed that the nonadiabatic correction term is also of the form $3\beta_1/r^6$ in the more general case. Later on in 1968, Dalgarno, Drake, and Victor[9] extended the original analysis of Dalgarno and McCarroll to obtain the correction term for any charged or neutral system interacting at long range with a spherically symmetric atom. They reproduced all the earlier results and showed that if two neutral atoms interact, the leading nonadiabatic term is proportional to $1/r^8$.

A very interesting paper was published by Bottcher and Dalgarno[10] in 1974. They examined, in general, the construction and use of model potentials for studies of one-center and two-center systems in which there are many outer electrons that must be treated explicitly. Expressions for the long-range form of these potentials were derived which included certain terms describing three-body interactions that had not previously appeared in the literature. Errors in some of the nonadiabatic correction terms given in this paper were subsequently corrected by Peach,[11] Norcross (private communication), and Valiron, Gayet, McCarroll, Masnou-Seeuws, and Phillipe.[12] In particular, Bottcher and Dalgarno[10] and, in 1976, Wang, Taylor, and Yaris[13] found that the nonadiabatic correction also gave rise to three-body terms involving the coefficient β_1, but Peach and Norcross showed that in fact these terms do not exist. The same mistake was repeated in 1978 in a paper by Laurenzi[14]. However, none of the authors of the period 1959–1976 realized that there could be nonadiabatic corrections to the long-range interaction that are dependent on the energy and are also of order $1/r^6$. Seaton and Steenman-Clark[2] first proved that such a term existed for the case of electron–hydrogen scattering, and their work is extended in this chapter to the general case of interactions in atoms and diatomic molecules. In 1980, Watanabe and Greene[15] also obtained this energy-dependent term but with a coefficient of different magnitude and sign to that obtained by Seaton and Steenman-Clark. In this paper it is shown that it is the result of Watanabe and Greene that is not correct.

Methods that involve model potentials are widely used in physics and chemistry, and in these methods the nonlocal inner region of the potential is replaced by a semiempirical potential that is of local or semilocal form. A semilocal or *l*-dependent potential is one in which the short-range part of the potential may depend on the angular momentum of the outer electrons. A wide variety of model potentials and pseudopotentials have been tried, and detailed discussions of their use can be found in the reviews by Bardsley[16] and Dixon and Robertson.[17] Recently, Peach[18] has examined the relative merits of model and pseudopotential methods and their application to problems of atom–atom scattering. The analytical form of the long-range potentials in one-center and two-center problems is certainly not of purely academic interest. It is well known that, at low energies, the properties of the scattering by atoms of either electrons or other atomic systems are largely determined by the long-range potentials. It is therefore very desirable to choose semiempirical potentials that incorporate the correct long-range interactions, although for practical reasons this is not always done.

The general approach adopted here is essentially the same as the one introduced by Bottcher and Dalgarno.[10] The exact Hamiltonian for the whole system is compared with a model Hamiltonian in which only the outer electrons are represented explicitly, and all the electron–core interactions are replaced by semiempirical model potentials. The difference between the exact and model Hamiltonians is then treated as a small perturbation. A condition is imposed so that, correct to a given order in the perturbation expansion, the change in the energy of any specified state of the system that does not involve an excitation of the core(s) is zero. The long-range forms of the electron–core interactions are obtained from this condition by making multipole expansions of the exact interaction between the valence and core electrons. Since only the long-range nature of the model potentials are of interest, their precise form at short range is of no importance, except that they should have the correct Coulomb form for very small separations. Thus the arguments would not be essentially changed if pseudopotentials rather than model potentials were used in the derivation.

2. General Form of the Model Hamiltonian

We shall consider the electrons in the system to be divided into two groups: the core electrons that essentially remain close to the atomic nucleus or nuclei and the valence electrons whose interaction and excitation determine the appearance of the observed spectra of the atom or molecule. We define $\mathcal{H}$ to be the exact Hamiltonian for the whole one-center or two-center system, and H_c to be the exact Hamiltonian for the core electrons only. If there are two centers, H_c is the sum of the Hamiltonians for the unperturbed separate cores. The kinetic energy operator for the valence electrons is denoted by K, and U is the total potential for all the interactions between pairs of valence electrons and includes any monopole terms that arise when an atomic core is charged. If there are two atomic cores, both of which have an overall charge, this means that U also contains the Coulomb interaction between the two cores themselves. The exact Hamiltonian may then be written as

$$\mathcal{H} = H_c + K + U + \mathcal{V}, \tag{1}$$

where $\mathcal{V}$ is the remaining interaction between the valence electrons and the core electrons together with their nucleus or nuclei. If the system has two centers, $\mathcal{V}$ also includes the interaction between the two sets of core electrons. If the whole system has energy $\mathcal{E}$ and is in a state described by a wave function Ψ, the Schrödinger equation for the system is

$$\mathcal{H}\Psi = \mathcal{E}\Psi, \tag{2}$$

and the eigenfunctions φ_c of the core Hamiltonian H_c satisfy the equation

$$H_c\varphi_c = E_c\varphi_c, \tag{3}$$

where the states of the core have energies denoted by E_c. In the analysis that follows, the ground state of the core will be indicated by setting the subscripts $c = 0$ in Eq. (3). The model Hamiltonian H describes the valence electrons only and is defined by

$$H = K + U + V, \tag{4}$$

where V is a semiempirical model potential that represents the overall perturbation of the valence electrons by the core(s). The replacement of the core by a semiempirical potential means that the spectrum of the model Hamiltonian is restricted to states in which only valence electrons can be excited. The Schrödinger equation for the model system may be written as

$$H\varphi = E\varphi, \tag{5}$$

and the potential V is chosen to be such that for any state of the valence electrons, the corresponding eigenenergy E obtained from Eq. (5) is identical with the exact energy $(\mathcal{E} - E_0)$ obtained from Eqs. (2) and (3). On using Eqs. (1) and (4), we have that

$$\mathcal{K} = H_c + H + \Delta V, \tag{6}$$

where

$$\Delta V \equiv \mathcal{V} - V, \tag{7}$$

and we shall treat ΔV as a small perturbation to the zero-order Hamiltonian $(H_c + H)$.

It is convenient to introduce the projection operators P and Q for the core states defined by

$$P \equiv |\varphi_0\rangle\langle\varphi_0| \tag{8}$$

and

$$Q \equiv 1 - P = \sum_{c \neq 0} |\varphi_c\rangle\langle\varphi_c|; \tag{9}$$

see Feshbach,[19] where the summation in Eq. (9) is over all the states of the core except the ground state. We assume in the following analysis that the ground state of the core is spherically symmetric, and it is easily shown from definitions (8) and (9) that

$$P^2 = P, \quad Q^2 = Q, \quad \text{and} \quad PQ = QP = 0. \tag{10}$$

Equation (2) can then be written in the two forms

$$P(\mathcal{K} - \mathcal{E})(P + Q)|\Psi\rangle = 0 \tag{11}$$

and

$$Q(\mathcal{K} - \mathcal{E})(P + Q)|\Psi\rangle = 0, \tag{12}$$

and on using Eq. (10), Eqs. (11) and (12) become

$$P(\mathcal{K} - \mathcal{E})P|\Psi\rangle + P\mathcal{K}Q|\Psi\rangle = 0 \tag{13}$$

and

$$Q(\mathcal{K} - \mathcal{E})Q|\Psi\rangle + Q\mathcal{K}P|\Psi\rangle = 0. \tag{14}$$

The operators P and Q commute with the model Hamiltonian H, and therefore

$$P\mathcal{K}Q = P\Delta V Q \tag{15}$$

and

$$Q\mathcal{K}P = Q\Delta V P, \tag{16}$$

where $\mathcal{K}$ is given by Eq. (6). Having substituted Eq. (16) into Eq. (14), (14) can be rewritten as an integral equation, i.e.,

$$|\Psi\rangle = -\frac{1}{Q(\mathcal{K} - \mathcal{E})Q} \, Q\Delta V P|\Psi\rangle + P|\Psi\rangle, \tag{17}$$

and then Eqs. (13), (15), and (17) can be combined to give

$$P(\mathcal{K} - \mathcal{E})P|\Psi\rangle - P\Delta V \, Q \frac{1}{Q(\mathcal{K} - \mathcal{E})Q} \, Q\Delta V P|\Psi\rangle = 0. \tag{18}$$

If we define the wave vector $|\psi\rangle$ by

$$|\psi\rangle = \langle\varphi_0|\Psi\rangle, \tag{19}$$

it follows directly from Eqs. (6), (8), and (18) that the wave function ψ satisfies the equation

$$(H + W)\psi = E\psi, \qquad E = \mathcal{E} - E_0, \tag{20}$$

where W is given by

$$W = \langle \varphi_0 | \Delta V | \varphi_0 \rangle - \langle \varphi_0 | \Delta V \, Q \, \frac{1}{Q(\mathcal{H} - \mathcal{E})Q} \, Q \Delta V | \varphi_0 \rangle. \quad (21)$$

So far, no approximations have been made in obtaining Eq. (20) so that it is entirely equivalent to Eqs. (2) and (3). Equation (20) is identical with the wave equation (5) for the model system, if the nonlocal potential W is in fact zero. Therefore, the model potential V in Eq. (7) must satisfy the condition

$$0 = \langle \varphi_0 | \Delta V | \varphi_0 \rangle - \langle \varphi_0 | \Delta V \, Q \, \frac{1}{\mathcal{D} + Q \Delta V \, Q} \, Q \Delta V | \varphi_0 \rangle, \quad (22)$$

where the operator $\mathcal{D}$ is defined by

$$\mathcal{D} \equiv Q(H_c + H - \mathcal{E}) Q. \quad (23)$$

We now introduce the first approximation by treating ΔV as a small perturbation to the Hamiltonian $H_c + H$, so that we write

$$\frac{1}{\mathcal{D} + Q \Delta V \, Q} = \frac{1}{\mathcal{D}} - \frac{1}{\mathcal{D}} \, Q \Delta V \, Q \frac{1}{\mathcal{D}} + O(\Delta V^2); \quad (24)$$

and therefore using Eqs. (22) and (24) we have that

$$0 = \langle \varphi_0 | \Delta V | \varphi_0 \rangle - \langle \varphi_0 | \Delta V \, G \, \Delta V | \varphi_0 \rangle + \langle \varphi_0 | \Delta V \, G \, \Delta V \, G \, \Delta V | \varphi_0 \rangle + O(\Delta V^4), \quad (25)$$

where the Green's operator G is defined by

$$G \equiv Q \frac{1}{\mathcal{D}} \, Q = \sum_{c \neq 0} | \varphi_c \rangle \, \frac{1}{E_c - E_0 + H - E} \, \langle \varphi_c |. \quad (26)$$

Since the model potential V commutes with Q, we find that on using Eq. (7), condition (25) may be reexpressed as

$$V = \langle \varphi_0 | \mathcal{V} | \varphi_0 \rangle - \langle \varphi_0 | \mathcal{V} G \mathcal{V} | \varphi_0 \rangle + \langle \varphi_0 | \mathcal{V} G \Delta V \, G \mathcal{V} | \varphi_0 \rangle \quad (27)$$

correct to third order in ΔV. Implicit in the division of the electrons in the atom or molecule into groups of inner and outer electrons is the idea that the core electrons are very much more tightly bound than the valence electrons, and therefore the energy differences $(E_c - E_0)$ for the core are very much larger than the typical energy differences that occur in the spectrum of the model Hamiltonian in Eq. (5). We define the operator D_c by

$$D_c = Q(H_c - E_0)Q \tag{28}$$

and make the expansion

$$\frac{1}{\mathcal{D}} = \frac{1}{D_c} \sum_{k=0}^{\infty} \left(-D\frac{1}{D_c}\right)^k, \tag{29}$$

where

$$D \equiv Q(H - E)Q. \tag{30}$$

On combining Eqs. (26) and (29),

$$G = G_c \sum_{k=0}^{\infty} \left[-(H-E)G_c\right]^k \tag{31}$$

where G_c is the Green's operator for the core(s) only and is defined by

$$G_c \equiv Q\frac{1}{D_c}Q = \sum_{c \neq 0} |\varphi_c\rangle \frac{1}{E_c - E_0} \langle\varphi_c|. \tag{32}$$

Substituting Eq. (31) into (27), we have that

$$V = V_0^{(1)} + \sum_{k=0}^{\infty} V_k^{(2)} + \sum_{k=0}^{\infty} \sum_{k'=0}^{\infty} V_{kk'}^{(3)}, \tag{33}$$

where

$$V_0^{(1)} \equiv \langle\varphi_0|\mathcal{V}|\varphi_0\rangle, \tag{34}$$

$$V_k^{(2)} \equiv (-1)^{k-1} \langle\varphi_0|\mathcal{V}G_c\left[(H-E)G_c\right]^k \mathcal{V}|\varphi_0\rangle \tag{35}$$

and

$$V_{kk'}^{(3)} \equiv (-1)^{k+k'} \langle\varphi_0|\mathcal{V}G_c\left[(H-E)G_c\right]^k \Delta V G_c\left[(H-E)G_c\right]^{k'}\mathcal{V}|\varphi_0\rangle. \tag{36}$$

Equations (33)–(36) form the basis for our determination of the form of the model potential V at long range.

3. Form of the Model Potential for Atomic Systems

3.1. The Exact Interaction between the Valence Electrons and the Core

We consider an atom or atomic ion that is composed of a core with an overall charge z and containing N electrons, together with n outer or valence electrons. If the position vectors of the core and valence electrons relative to the atomic nucleus are denoted by $\mathbf{r}_i$, $i = 1, 2, \ldots, N$ and $\mathbf{r}_j$, $j = 1, 2, \ldots, n$ respectively, then the interaction $\mathcal{V}$ defined by Eq. (1) can be written as

$$\mathcal{V} = \sum_{i=1}^{N} \sum_{j=1}^{n} \frac{1}{|\mathbf{r}_i - \mathbf{r}_j|} - \sum_{j=1}^{n} \frac{N}{r_j}. \tag{37}$$

We expand $\mathcal{V}$ in terms of the usual normalized spherical harmonics $Y_{\lambda\mu}(\hat{\mathbf{r}})$ [see Edmonds[20]] in the form

$$\mathcal{V} = \sum_{\lambda=1}^{\infty} \sum_{\mu=-\lambda}^{+\lambda} S^*(\lambda, \mu) T(\lambda, \mu), \tag{38}$$

where

$$S(\lambda, \mu) \equiv \left(\frac{4\pi}{2\lambda + 1} \right)^{1/2} \sum_{i=1}^{N} r_i^{\lambda} Y_{\lambda\mu}(\hat{\mathbf{r}}_i) \tag{39}$$

and

$$T(\lambda, \mu) \equiv \left(\frac{4\pi}{2\lambda + 1} \right)^{1/2} \sum_{j=1}^{n} \frac{1}{r_j^{\lambda+1}} Y_{\lambda\mu}(\hat{\mathbf{r}}_j). \tag{40}$$

In making expansion (38), we have assumed that $r_i < r_j$ for all $i = 1, 2, \ldots, N$ and $j = 1, 2, \ldots, n$, since we are only interested in extracting expressions for the long-range form of the model potential. Equation (38) when used in conjunction with (33)–(36) gives rise to errors in V which are appreciable only at short range, i.e., they decay exponentially in the outer region.

3.2. Second-Order Perturbation Theory: The Static Contribution

We now substitute Eq. (38) into (34)–(36) and consider the separate contributions to V. Since the ground state of the core is spherically symmetric, there is no contribution from the first-order term, i.e.,

$$V_0^{(1)} = 0, \tag{41}$$

in (34). In order to write down more explicit expressions for the higher order terms, we suppose that the core state c is described by the angular momentum quantum numbers $L_c M_c$ and that all the other quantum numbers are denoted by Γ_c. Spin quantum numbers are not indicated explicitly, since the spin of the core is always conserved. From Eqs. (32) and (35) we then have that

$$V_0^{(2)} = -\sum_{c \neq 0} \frac{\langle \varphi_0 | \mathcal{V} | \varphi_c \rangle \langle \varphi_c | \mathcal{V} | \varphi_0 \rangle}{E_c - E_0}, \tag{42}$$

and on introducing expansion (38) and the angular momenta of the core states, we obtain

$$V_0^{(2)} = -\sum_{\Gamma_c} \sum_{L_c, M_c} \frac{1}{E_c - E_0} \sum_{\lambda, \mu, \lambda', \mu'} \langle \Gamma_0 00 | S^*(\lambda, \mu) | \Gamma_c L_c M_c \rangle$$

$$\times \langle \Gamma_c L_c M_c | S(\lambda', \mu') | \Gamma_0 00 \rangle T(\lambda, \mu) T^*(\lambda', \mu'), \tag{43}$$

where E_c and E_0 are the energies of the states with quantum numbers $\Gamma_c L_c$ and $\Gamma_0 0$. The standard techniques of Racah algebra can be used to simplify Eq. (43), and therefore we define the reduced matrix elements $\langle \Gamma_c L_c \| S(\lambda) \| \Gamma'_c L'_c \rangle$ by the relation

$$\langle \Gamma_c L_c M_c | S(\lambda, \mu) | \Gamma'_c L'_c M'_c \rangle$$

$$= (-1)^{L_c - M_c} \begin{pmatrix} L_c & \lambda & L'_c \\ -M_c & \mu & M'_c \end{pmatrix} \langle \Gamma_c L_c \| S(\lambda) \| \Gamma'_c L'_c \rangle, \tag{44}$$

where

$$\begin{pmatrix} a & b & c \\ d & e & f \end{pmatrix}$$

denotes a $3 - j$ symbol [see Edmonds[20]]. Equation (43) can then be expressed as

$$V_0^{(2)} = -\frac{1}{2} \sum_{\lambda=1}^{\infty} \alpha(\lambda) \sum_{\mu} |T(\lambda, \mu)|^2, \tag{45}$$

where

$$\alpha(\lambda) = 2 \sum_{\Gamma_c} \frac{1}{2\lambda + 1} \frac{|\langle \Gamma_c \lambda \| \mathbf{S}(\lambda) \| \Gamma_0 0 \rangle|^2}{E_c - E_0} \tag{46}$$

and

$$\sum_{\mu} |T(\lambda, \mu)|^2 = \frac{P_\lambda(\hat{\mathbf{f}}_j \cdot \hat{\mathbf{f}}_{j'})}{(r_j r_{j'})^{\lambda + 1}}, \tag{47}$$

on using definition (40). The coefficient $\alpha(\lambda)$ is the static 2^λ-pole polarizability of the core, so that, in particular, $\alpha(1) \equiv \alpha_d$ and $\alpha(2) \equiv \alpha_q$, where α_d and α_q are the usual static dipole and static quadrupole polarizabilities, respectively.

3.3. The First Nonadiabatic Correction

The first nonadiabatic correction to V comes from the term in Eq. (35) with $k = 1$. Since G_c commutes with $(H - E)$, we obtain

$$V_1^{(2)} = \sum_{c \neq 0} \frac{\langle \varphi_0 | \mathcal{V} | \varphi_c \rangle}{(E_c - E_0)^2} (H - E) \langle \varphi_c | \mathcal{V} | \varphi_0 \rangle, \tag{48}$$

and we see that $V_1^{(2)}$ is an operator. It must be emphasized that $V_1^{(2)}$ is to be used in Eq. (5) as a contribution to the Hamiltonian H, and in fact the operator $V_1^{(2)}$ only has meaning when it is considered in this context. We therefore determine an effective contribution to the potential V, $\overline{V}_1^{(2)}$, by evaluating the matrix element

$$\langle \varphi | V_1^{(2)} | \varphi \rangle \equiv \langle \varphi | \overline{V}_1^{(2)} | \varphi \rangle, \tag{49}$$

where φ is the eigenfunction of Eq. (5) that has the eigenvalue E. If there is more than one state corresponding to this eigenvalue, the matrix elements in (49) should be interpreted as including an average over all the degenerate states. If we now substitute expansion (38) for $\mathcal{V}$ into (48), introduce the notation of (43), and use the definition of a reduced matrix element given in (44), we have

$$\langle \varphi | V_1^{(2)} | \varphi \rangle = \sum_{\lambda = 1}^{\infty} \beta(\lambda) \sum_{\mu} \langle \varphi | T(\lambda, \mu)(H - E) T^*(\lambda, \mu) | \varphi \rangle, \tag{50}$$

where

$$\beta(\lambda) = \sum_{\Gamma_c} \frac{1}{2\lambda + 1} \frac{|\langle \Gamma_c \lambda \| S(\lambda) \| \Gamma_0 0 \rangle|^2}{(E_c - E_0)^2}. \tag{51}$$

The coefficient $\beta(1)$, which represents the first dynamical correction to the static dipole polarizability $\alpha(1)$, is more usually denoted by β_1. The sum of matrix elements in (50) is given by

$$\sum_{\mu} \langle \varphi | T(\lambda, \mu)(H - E) T^*(\lambda, \mu) | \varphi \rangle$$

$$= \frac{4\pi}{2\lambda + 1} \sum_{j'=1}^{n} \sum_{j=1}^{n} \int_{\tau} \sum_{\mu} \varphi^* \frac{1}{r_{j'}^{\lambda+1}} Y_{\lambda\mu}(\hat{\mathbf{r}}_{j'})(H - E)$$

$$\times \frac{1}{r_j^{\lambda+1}} Y_{\lambda\mu}^*(\hat{\mathbf{r}}_j) \varphi \, d\mathbf{r}_1 \, d\mathbf{r}_2 \cdots d\mathbf{r}_n, \tag{52}$$

where $\int_{\tau}$ denotes an integral over $3n$-dimensional space. We define this space τ to be bounded by two multidimensional surfaces, the one at infinity defined by $\mathbf{r}_j = \infty$, $j = 1, 2, \ldots, n$, and an inner one upon which $\varphi \equiv 0$ everywhere. This inner surface encompasses the origin, but is distinct from it, since the divergence in the integrand at the origin must be excluded. For example, in the case where there is only one bound valence electron, it would be reasonable on physical grounds to choose the inner surface to be at $r_1 = r_0$, where r_0 gives the position of the outermost node in the wave function of this outer electron. This is because the core of the atomic system can be effectively contained within the sphere of radius r_0, see Peach[18] and in particular Fig. 5 of that paper. Clearly, the eigenfunctions φ that do not have any nodes at a finite radius for any one of the coordinate vectors $\mathbf{r}_j$, $j = 1, 2, \ldots, n$, cannot be used in this derivation of the long-range effective potential. This is because an appropriate bounding surface cannot be defined that does not include the origin. However, this still means that there exists an infinite number of bound and free states of the atomic system for which definition (49) of $\overline{V}_1^{(2)}$ must be valid. This analysis would be very similar if we chose to work with a pseudopotential for V rather than a model potential. The only difference would be that the eigenfunctions φ would be solutions of an equation of type (5) in which the pseudo-Hamiltonian would depend explicitly on the angular momenta of the outer electrons.

The kinetic energy operator K in Eq. (4) is given by

$$K = -\frac{1}{2} \sum_{j=1}^{n} \nabla_j^2 \tag{53}$$

where ∇_j denotes the del operator for the position vector $\mathbf{r}_j$, and so we have that

$$(H - E)\left[\frac{1}{r_j^{\lambda+1}} Y_{\lambda\mu}^*(\hat{\mathbf{r}}_j)\varphi \right] = -\nabla_j\left[\frac{1}{r_j^{\lambda+1}} Y_{\lambda\mu}^*(\hat{\mathbf{r}}_j) \right] \cdot \nabla_j\varphi, \tag{54}$$

since

$$\nabla_j^2\left[\frac{1}{r_j^{\lambda+1}} Y_{\lambda\mu}^*(\hat{\mathbf{r}}_j) \right] = 0 \tag{55}$$

for all possible values of λ and μ. We now consider separately the contributions to the right-hand side of (52) from terms with $j' = j$ and $j' \neq j$. On using Eq. (54), the first contribution is given by

$$-\frac{4\pi}{2\lambda+1} \sum_{j=1}^{n} \int_\tau \sum_\mu \varphi^* \frac{1}{r_j^{\lambda+1}} Y_{\lambda\mu}(\hat{\mathbf{r}}_j)\nabla_j\left[\frac{1}{r_j^{\lambda+1}} Y_{\lambda\mu}^*(\hat{\mathbf{r}}_j) \right] \cdot \nabla_j\varphi \, d\mathbf{r}_1 \, d\mathbf{r}_2 \cdots d\mathbf{r}_n$$

$$= -\frac{1}{4}\frac{4\pi}{2\lambda+1} \sum_{j=1}^{n} \int_\tau \nabla_j\left[\frac{1}{r_j^{2\lambda+2}} \sum_\mu Y_{\lambda\mu}(\hat{\mathbf{r}}_j) Y_{\lambda\mu}^*(\hat{\mathbf{r}}_j) \right]$$

$$\cdot \nabla_j(\varphi^*\varphi) d\mathbf{r}_1 \, d\mathbf{r}_2 \cdots d\mathbf{r}_n. \tag{56}$$

This result follows directly from the fact that the quantities

$$\sum_\mu \frac{1}{r_j^{\lambda+1}} Y_{\lambda\mu}(\hat{\mathbf{r}}_j)\nabla_j\left[\frac{1}{r_j^{\lambda+1}} Y_{\lambda\mu}^*(\hat{\mathbf{r}}_j) \right], \qquad j = 1, 2, \ldots, n, \tag{57}$$

are real and that the matrix element (49) is also real. Green's theorem can now be applied to the integral in (56) to give

$$\mathcal{I}_1 \equiv -\frac{1}{4} \sum_{j=1}^{n} \int_{\tau_j}\int_{\sigma_j} \varphi^*\varphi \, \nabla_j\left(\frac{1}{r_j^{2\lambda+2}} \right) \cdot d\hat{\mathbf{f}}_j \, d\mathbf{r}_1 \cdots d\mathbf{r}_{j-1} \, d\mathbf{r}_{j+1} \cdots d\mathbf{r}_n$$

$$+ \frac{1}{4} \sum_{j=1}^{n} \int \varphi^*\varphi \, \nabla_j^2\left(\frac{1}{r_j^{2\lambda+2}} \right) d\mathbf{r}_1 \, d\mathbf{r}_2 \cdots d\mathbf{r}_n, \tag{58}$$

where we have also carried out the sum over the spherical harmonics in (56). In Eq. (58), σ_j indicates the bounding surface in the three-dimensional space of the jth electron and τ_j denotes the $3(n-1)$-dimensional volume filled by the other $(n-1)$ electrons. The second contribution to (52) is

$$\mathcal{I}_2 \equiv -\frac{4\pi}{2\lambda+1} \sum_{j'=1}^{n} \sum_{j=1}^{n}{}'$$

$$\times \int_\tau \sum_\mu \varphi^* \frac{1}{r_{j'}^{\lambda+1}} Y_{\lambda\mu}(\hat{\mathbf{r}}_{j'}) \nabla_j \left[\frac{1}{r_j^{\lambda+1}} Y_{\lambda\mu}^*(\hat{\mathbf{r}}_j) \right] \cdot \nabla_j \varphi \, d\mathbf{r}_1 \, d\mathbf{r}_2 \cdots d\mathbf{r}_n, \quad (59)$$

where the prime on the second summation indicates the term $j' = j$ is to be omitted. Then having carried out the sum over μ, the application of Green's theorem to (59) gives

$$\mathcal{I}_2 = -\frac{1}{2} \sum_{j=1}^{n} \sum_{j'=1}^{n}{}' \int_{\tau_j} \int_{\sigma_j} \varphi^* \varphi \nabla_j \left[\frac{P_\lambda(\hat{\mathbf{r}}_j \cdot \hat{\mathbf{r}}_{j'})}{(r_j r_{j'})^{\lambda+1}} \right] \cdot d\hat{\mathbf{r}}_j \, d\mathbf{r}_1 \cdots d\mathbf{r}_{j-1} \, d\mathbf{r}_{j+1} \cdots d\mathbf{r}_n$$

$$+ \frac{1}{2} \sum_{j=1}^{n} \sum_{j'=1}^{n}{}' \int_\tau \varphi^* \varphi \nabla_j^2 \left[\frac{P_\lambda(\hat{\mathbf{r}}_j \cdot \hat{\mathbf{r}}_{j'})}{(r_j r_{j'})^{\lambda+1}} \right] d\mathbf{r}_1 \, d\mathbf{r}_2 \cdots d\mathbf{r}_n. \quad (60)$$

The surface terms in (58) and (60) are zero since the integrand vanishes everywhere on the surface, and by substituting (55) into (60), it can be seen that the second term in (60) is also zero. The work of Bottcher and Dalgarno[10] is therefore incorrect here since they predict that nonadiabatic terms involving three bodies do exist. Thus from Eq. (52) and (58), we have that

$$\sum_\mu \langle \varphi | T(\lambda, \mu)(H - E) T^*(\lambda, \mu) | \varphi \rangle = \tfrac{1}{2}(\lambda+1)(2\lambda+1) \sum_{j=1}^{n} \langle \varphi | \frac{1}{r_j^{2\lambda+4}} | \varphi \rangle,$$

$$(61)$$

and since relation (50) must be true for an infinite number of eigenfunctions φ of the Hamiltonian H, we have from (49), (50), and (61) that the effective potential is

$$\overline{V}_1^{(2)} = \tfrac{1}{2} \sum_{\lambda=1}^{\infty} (\lambda+1)(2\lambda+1)\beta(\lambda) \sum_{j=1}^{n} \frac{1}{r_j^{2\lambda+4}}. \quad (62)$$

Finally, it must be emphasized that this derivation of $\overline{V}_1^{(2)}$, based on definition (49), only has meaning within the context of first-order perturbation theory. Thus if we wish to determine the contribution of these long-range terms to the energy E, we simply evaluate the expectation value of $\bar{V}_1^{(2)}$ given by (49); a higher-order calculation is not justified.

3.4. The Second Nonadiabatic Correction

The second nonadiabatic correction to V comes from the term in Eq. (35) with $k = 2$. This term is given by

$$V_2^{(2)} = - \sum_{c \neq 0} \frac{\langle \varphi_0 | \mathcal{V} | \varphi_c \rangle}{(E_c - E_0)^3} (H - E)^2 \langle \varphi_0 | \mathcal{V} | \varphi_c \rangle \tag{63}$$

and as in the theory of Section 3.3, we wish to determine an effective potential, $\bar{V}_2^{(2)}$ say, such that

$$\langle \varphi | V_2^{(2)} | \varphi \rangle = \langle \varphi | \bar{V}_2^{(2)} | \varphi \rangle, \tag{64}$$

for the eigenfunctions φ of the Hamiltonian H and their corresponding eigenvalues E. Using an analysis similar to that of Section 3.2, we obtain

$$\langle \varphi | V_2^{(2)} | \varphi \rangle = -2 \sum_{\lambda=1}^{\infty} \gamma(\lambda) \sum_{\mu} \langle \varphi | T(\lambda, \mu)(H - E)^2 T^*(\lambda, \mu) | \varphi \rangle, \tag{65}$$

where

$$\gamma(\lambda) = \frac{1}{2} \sum_{\Gamma_c} \frac{1}{2\lambda + 1} \frac{|\langle \Gamma_c \lambda \| \mathbf{S}(\lambda) \| \Gamma_0 0 \rangle|^2}{(E_c - E_0)^3}. \tag{66}$$

The coefficient $\gamma(1)$ represents the second dynamical correction to the static dipole polarizability α_d and has been denoted by γ_1 in the paper by Peach.[18] The definition (66) of $\gamma(\lambda)$ is consistent with the definition of γ_1 given by Seaton and Steenman-Clark[2] for hydrogen. By using the definition (40) of $T(\lambda, \mu)$ and the result (54) it is easily shown that

$$\sum_{\mu} T(\lambda, \mu)(H - E)^2 T^*(\lambda, \mu)\varphi$$

$$= -\frac{4\pi}{2\lambda + 1} \sum_{j=1}^{n} \sum_{j'=1}^{n} \sum_{\mu} \frac{1}{r_{j'}^{\lambda+1}} Y_{\lambda\mu}(\hat{\mathbf{r}}_{j'})(H - E)\nabla_j \left[\frac{1}{r_j^{\lambda+1}} Y_{\lambda\mu}^*(\hat{\mathbf{r}}_j) \right] \cdot \nabla_j \varphi.$$

$$\tag{67}$$

Then on introducing definitions (4) and (53), (67) becomes

$$\sum_{\mu} T(\lambda, \mu)(H - E)^2 T^*(\lambda, \mu)\varphi = (\mathcal{J}_1 + \mathcal{J}_2)\varphi, \tag{68}$$

where $\mathcal{J}_1$ and $\mathcal{J}_2$ are defined by

$$\mathcal{J}_1 \equiv \frac{4\pi}{2\lambda + 1} \sum_{j=1}^{n} \sum_{j'=1}^{n} \sum_{\mu} \frac{1}{r_{j'}^{\lambda+1}} Y_{\lambda\mu}(\hat{\mathbf{r}}_{j'}) \nabla_j \nabla_j \left[\frac{1}{r_j^{\lambda+1}} Y_{\lambda\mu}^*(\hat{\mathbf{r}}_j) \right] : \nabla_j \nabla_j ; \tag{69}$$

$$\mathcal{J}_2 \equiv \frac{4\pi}{2\lambda + 1} \sum_{j=1}^{n} \sum_{j'=1}^{n} \sum_{\mu} \frac{1}{r_{j'}^{\lambda+1}} Y_{\lambda\mu}(\hat{\mathbf{r}}_{j'}) \nabla_j \left[\frac{1}{r_j^{\lambda+1}} Y_{\lambda\mu}^*(\hat{\mathbf{r}}_j) \right] \cdot \nabla_j(U + V). \tag{70}$$

It can be seen that (69) involves the double scalar product of dyadics that depend on the vector operator ∇_j and this product is defined by

$$\mathbf{ab} : \mathbf{cd} \equiv (\mathbf{a} \cdot \mathbf{c})(\mathbf{b} \cdot \mathbf{d}), \tag{71}$$

where $\mathbf{a}$, $\mathbf{b}$, $\mathbf{c}$, and $\mathbf{d}$ are four arbitrary vectors [see Weatherburn[21]].

In order to evaluate expression (69), we need certain general results involving spherical harmonics and the proofs of these results are given in Section 5. If the vector $\mathbf{r}$ in Eq. (272) is interpreted as the vector operator ∇, (272) is an operator equation that can be applied directly in the evaluation of $\mathcal{J}_1$ in (69). In this case, $l = \lambda$, $l' = 2$, and $\mathbf{r}$ is replaced by ∇_j which operates only on φ. The quantity $\mathcal{J}_1$ may be split up into two parts, so that

$$\mathbf{J}_1 \varphi = \mathcal{K}_1 + \mathcal{K}_2, \tag{72}$$

where

$$\mathcal{K}_1 \equiv \frac{4\pi}{2\lambda + 1} \sum_{j=1}^{n} \sum_{\mu} \frac{1}{r_j^{\lambda+1}} Y_{\lambda\mu}(\hat{\mathbf{r}}_j) \nabla_j \nabla_j \left[\frac{1}{r_j^{\lambda+1}} Y_{\lambda\mu}^*(\hat{\mathbf{r}}_j) \right] : \nabla_j \nabla_j \varphi \tag{73}$$

and

$$\mathcal{K}_2 \equiv \sum_{j=1}^{n} \sum_{j'=1}^{n}{}' \frac{1}{r_{j'}^{\lambda+1}} \nabla_j \nabla_j \left[\frac{1}{r_j^{\lambda+1}} P_\lambda(\hat{\mathbf{r}}_j \cdot \hat{\mathbf{r}}_{j'}) \right] : \nabla_j \nabla_j \varphi. \tag{74}$$

Using the result (272) and the relation (271) with $L = 2$, $l = \lambda$, and $l' = \lambda + 2$, (73) may be evaluated to give

$$\mathcal{K}_1 = \sum_{j=1}^{n} \frac{(\lambda + 1)(\lambda + 2)}{r_j^{2\lambda+4}} \nabla_j^2 P_2(\hat{\mathbf{r}}_j \cdot \hat{\nabla}_j)\varphi, \tag{75}$$

where $\mathbf{\nabla}_j$ operates only on φ. Equation (272) can also be applied directly to enable (74) to be rewritten in the form

$$\mathcal{K}_2 = \sum_{j=1}^{n} \sum_{j'=1}^{n}{}' \frac{1}{r_{j'}^{\lambda+1}} \frac{1}{r_j^{\lambda+3}} \frac{4\pi(\lambda+1)(\lambda+2)}{[5(2\lambda+5)]^{1/2}} \left[\begin{pmatrix} \lambda & 2 & \lambda+2 \\ 0 & 0 & 0 \end{pmatrix} \right]^{-1}$$

$$\times \sum_{M} \begin{pmatrix} \lambda & 2 & \lambda+2 \\ 0 & -M & M \end{pmatrix} Y_{\lambda+2\,M}(\hat{\mathbf{r}}_j) \nabla_j^2 Y_{2-M}(\hat{\mathbf{\nabla}}_j)\varphi, \tag{76}$$

where $\mathbf{\nabla}_j$ operates only on φ and the directions of the vectors $\hat{\mathbf{r}}_j$ and $\hat{\mathbf{\nabla}}_j$ are referred to the direction $\hat{\mathbf{r}}_{j'}$ which acts as the z axis. Equation (75) may be rewritten as

$$\mathcal{K}_1 = \sum_{j=1}^{n} \frac{(\lambda+1)(\lambda+2)}{r_j^{2\lambda+4}} \frac{1}{2}\left[3(\hat{\mathbf{r}}_j \cdot \mathbf{\nabla}_j)^2 - \nabla_j^2 \right]\varphi$$

$$= \sum_{j=1}^{n} \frac{(\lambda+1)(\lambda+2)}{r_j^{2\lambda+4}} \left[\nabla_j^2 - \frac{3}{r_j} \frac{\partial}{\partial r_j} + \frac{3}{2} \frac{L_j^2}{r_j^2} \right]\varphi, \tag{77}$$

where $\mathbf{L}_j$ is the angular momentum operator for electron j. Since, by using integration by parts, we have that

$$\int_\tau \varphi^* \frac{1}{r_j^{2\lambda+5}} \frac{\partial\varphi}{\partial r_j} d\mathbf{r}_1 d\mathbf{r}_2 \cdots d\mathbf{r}_n$$

$$= \frac{1}{2} \int_\tau \frac{1}{r_j^{2\lambda+3}} \frac{\partial}{\partial r_j} (\varphi^*\varphi) dr_j\, d\hat{\mathbf{r}}_j\, d\mathbf{r}_1 \cdots d\mathbf{r}_{j-1} d\mathbf{r}_{j+1} \cdots d\mathbf{r}_n$$

$$= \frac{1}{2} (2\lambda+3) \int_\tau \varphi^* \frac{1}{r_j^{2\lambda+6}} \varphi\, d\mathbf{r}_1 d\mathbf{r}_2 \cdots d\mathbf{r}_n, \tag{78}$$

the effective contribution to $\mathcal{K}_1$ from the second term on the right-hand side of (77) is of order $1/r_j^{2\lambda+6}$. The third term in (77) is also clearly of this order. However, by using Eqs. (4), (5), and (53), the Schrödinger equation for the atomic system may be written as

$$\sum_{j=1}^{n} \nabla_j^2 \varphi = 2(U + V - E)\varphi, \tag{79}$$

where the potentials U and V are given by

$$U = -\sum_{j=1}^{n} \frac{z}{r_j} + \sum_{j=1}^{n} \sum_{j'=1}^{j-1} \frac{1}{|\mathbf{r}_j - \mathbf{r}_{j'}|} \tag{80}$$

and

$$V = -\frac{1}{2}\alpha(1)\sum_{j=1}^{n}\sum_{j'=1}^{n}\left\{\frac{P_1(\hat{\mathbf{f}}_j\cdot\hat{\mathbf{f}}_{j'})}{(r_j r_{j'})^2}+O\left[\frac{1}{(r_j r_{j'})^3}\right]\right\},\tag{81}$$

[see Eqs. (45) and (47)]. Thus the first term in (77) clearly gives a contribution of order $1/r_j^{2\lambda+4}$.

For the case in which there is just one valence electron, i.e., $n = 1$, Eqs. (77)–(81) can easily be combined to give

$$\mathcal{K}_1 = -\frac{2(\lambda+1)(\lambda+2)}{r_1^{2\lambda+4}}\left[E+\frac{z}{r_1}+O\left(\frac{1}{r_1^2}\right)\right]\varphi,\tag{82}$$

but in the many-electron case no such simple reduction is possible. However, an approximate expression may be given for $\mathcal{K}_1$, which is plausible from the physical point of view. We suppose that the electrons are labeled by $j = 1, 2, \ldots, n$ in such a way that $r_1 < r_2 < \cdots < r_n$ and we define energies $E_j, j = 1, 2, \ldots, n$, which are such that $-E_j$ is the energy required to remove the jth electron from the system, the electrons $j + 1, j + 2, \ldots, n$ having already been removed. The potential U in (80) is then given by

$$U = -\sum_{j=1}^{n}\frac{z_j}{r_j}+\sum_{j=1}^{n}\sum_{j'=1}^{j-1}\sum_{l=1}^{\infty}\frac{r_{j'}^l}{r_j^{l+1}}P_l(\hat{\mathbf{f}}_j\cdot\hat{\mathbf{f}}_{j'})\tag{83}$$

[see (265)], and $\mathcal{K}_1$ can be approximately expressed as

$$\mathcal{K}_1 \simeq -2\sum_{j=1}^{n}\frac{(\lambda+1)(\lambda+2)}{r_j^{2\lambda+4}}\left[E_j+\frac{z_j}{r_j}+O\left(\frac{1}{r_j^2}\right)\right],\tag{84}$$

where

$$z_j = z - j + 1\tag{85}$$

and

$$\sum_{j=1}^{n}E_j = E,\tag{86}$$

since $-E$ is the total energy required to remove the n valence electrons from the atomic system. We have neglected the contribution to $\mathcal{K}_1$ from the second term in (83), since such terms are unlikely to contribute much when averaged over the inner electrons with $j' = 1, 2, \ldots, j - 1$. Alternatively, if the n valence electrons are equivalent it may be better to choose

$$E_j = E/n, \quad j = 1, 2, \ldots, n, \tag{87}$$

and

$$z_j = z_{\text{eff}}, \quad j = 1, 2, \ldots, n, \tag{88}$$

where z_{eff} is an effective charge that can be obtained by, for example, minimizing the energy using wave functions of the screened Coulomb type. If $n = 1$, z_{eff} is taken to be equal to z. Clearly, in either case, the approximate expression for $\mathcal{K}_1$ in (84) reduces to the exact expression (82) in the one-electron case.

The second contribution, $\mathcal{K}_2$, to $\mathcal{J}_1 \varphi$ in (74) only exists in the many-electron case. It is clear from Eqs. (77) and (78) that terms that involve the components of $\hat{\mathbf{V}}_j$ in the directions perpendicular to $\hat{\mathbf{r}}_j$ only contribute terms of order $1/r_j^2$ compared to the term arising from the component of $\hat{\mathbf{V}}_j$ in the direction of $\hat{\mathbf{r}}_j$. Thus the leading terms in $\mathcal{K}_2$ can be obtained by replacing $\hat{\mathbf{V}}_j$ by $\hat{\mathbf{r}}_j$ in (76), so that

$$\mathcal{K}_2 = \sum_{j=1}^{n} \sum_{j'=1}^{n}{}' \frac{(\lambda + 1)(\lambda + 2)}{r_{j'}^{\lambda+1} r_j^{\lambda+3}} \left[P_\lambda(\hat{\mathbf{r}}_j \cdot \hat{\mathbf{r}}_{j'}) + O\left(\frac{1}{r_j^2}\right) \right] \nabla_j^2 \varphi \tag{89}$$

on using (271) with $L = \lambda$, $l = \lambda + 2$, and $l' = 2$. If we make the same assumptions as those used to obtain (84), we have the final approximate relation

$$\mathcal{K}_2 \simeq -2 \sum_{j=1}^{n} \sum_{j'=1}^{n}{}' \frac{(\lambda + 1)(\lambda + 2)}{r_{j'}^{\lambda+1} r_j^{\lambda+3}} P_\lambda(\hat{\mathbf{r}}_j \cdot \hat{\mathbf{r}}_{j'}) \left[E_j + \frac{z_j}{r_j} + O\left(\frac{1}{r_j^2}\right) \right]. \tag{90}$$

The quantity $\mathcal{J}_2$ defined in (70) is more easily evaluated. Since the leading terms in $(U + V)$ depend on r_j, $j = 1, 2, \ldots, n$, the only terms in (70) that contribute are those arising from the component of $\hat{\mathbf{V}}_j$ in the direction of $\hat{\mathbf{r}}_j$, and we have that

$$\mathcal{J}_2 \simeq - \sum_{j=1}^{n} \sum_{j'=1}^{n}{}' \frac{(\lambda + 1)}{r_{j'}^{\lambda+1} r_j^{\lambda+4}} P_\lambda(\hat{\mathbf{r}}_j \cdot \hat{\mathbf{r}}_{j'}) \left[z_j + O\left(\frac{1}{r_j}\right) \right]. \tag{91}$$

The final expression for the effective potential, $\bar{V}_2^{(2)}$, defined by (64) can now be obtained and is valid only in the context of first-order perturbation theory. Using Eqs. (64), (65), (68), (72), (84), (90), and (91), we find that

$$\bar{V}_2^{(2)} \simeq 4 \sum_{\lambda=1}^{\infty} (\lambda + 1)\gamma(\lambda) \sum_{j=1}^{n} \sum_{j'=1}^{n} \frac{P_\lambda(\hat{\mathbf{r}} \cdot \hat{\mathbf{r}}_{j'})}{r_{j'}^{\lambda+1} r_j^{\lambda+3}}$$

$$\times \left[(\lambda + 2)E_j + \frac{1}{2}(2\lambda + 5)\frac{z_j}{r_j} + O\left(\frac{1}{r_j^2}\right) \right], \qquad (92)$$

and this expression becomes exact in the one-electron case.

3.5. Nonadiabatic Corrections of Higher Order

We shall not examine the potentials $V_k^{(2)}$, $k = 3, 4, \ldots$, in detail. However, some indication of their form can be deduced relatively simply by making use of results (62) and (92). From Eq. (35) we have that

$$V_k^{(2)} = (-1)^{k-1} \sum_{c \neq 0} \frac{\langle \varphi_0 | \mathcal{V} | \varphi_c \rangle}{(E_c - E_0)^{k+1}} (H - E)^k \langle \varphi_c | \mathcal{V} | \varphi_0 \rangle, \qquad (93)$$

and using the analysis described in Section 3.2, we obtain

$$\langle \varphi | V_k^{(2)} | \varphi \rangle = (-2)^{k-1} \sum_{\lambda=1}^{\infty} \delta_k(\lambda) \sum_{\mu} \langle \varphi | T(\lambda, \mu)(H - E)^k T^*(\lambda, \mu) | \varphi \rangle,$$

$$(94)$$

where

$$\delta_k(\lambda) = \left(\frac{1}{2}\right)^{k-1} \sum_{\Gamma_c} \frac{1}{2\lambda + 1} \frac{|\langle \Gamma_c \lambda \| S(\lambda) \| \Gamma_0 0 \rangle|^2}{(E_c - E_0)^{k+1}}. \qquad (95)$$

The coefficients $\delta_0(\lambda)$, $\delta_1(\lambda)$, and $\delta_2(\lambda)$ are identical with the coefficients $\alpha(\lambda)$, $\beta(\lambda)$, and $\gamma(\lambda)$, respectively, as defined by Eqs. (46), (51), and (66). Result (92) may be used for the evaluation of the contribution $V_k^{(2)}$, i.e.,

$$\sum_{\mu} T(\lambda, \mu)(H - E)^k T^*(\lambda, \mu)\varphi$$

$$\simeq -2(\lambda + 1)\frac{4\pi}{2\lambda + 1} \sum_{j=1}^{n} \sum_{j'=1}^{n} \sum_{\mu} \frac{1}{r_{j'}^{\lambda+1}} Y_{\lambda\mu}(\hat{\mathbf{r}}_{j'})(H - E)^{k-2}\frac{1}{r_j^{\lambda+3}}$$

$$\times Y_{\lambda\mu}^*(\hat{\mathbf{r}}_j)\left[(\lambda + 2)E_j + \frac{1}{2}(2\lambda + 5)\frac{z_j}{r_j} + O\left(\frac{1}{r_j^2}\right) \right]\varphi. \qquad (96)$$

For the case of $k = 3$, since

$$(H - E) \frac{1}{r_j^{\lambda+3}} Y_{\lambda\mu}^*(\hat{\mathbf{r}}_j)\varphi \simeq - \nabla_j \left[\frac{1}{r_j^{\lambda+3}} Y_{\lambda\mu}^*(\hat{\mathbf{r}}_j) \right] \cdot \nabla_j \varphi - \frac{2\lambda + 3}{r_j^{\lambda+5}} Y_{\lambda\mu}^*(\hat{\mathbf{r}}_j)\varphi$$

(97)

[see Eq. (54)], Eq. (96) becomes

$$\sum_\mu \langle \varphi | T(\lambda, \mu)(H - E)^3 T^*(\lambda, \mu) | \varphi \rangle$$

$$\simeq - \frac{1}{2}(\lambda + 1)(\lambda + 2) \sum_{j=1}^n E_j \int_\tau \nabla_j \left(\frac{1}{r_j^{2\lambda+4}} \right) \cdot \nabla_j (\varphi^*\varphi) \, d\mathbf{r}_1 \, d\mathbf{r}_2 \cdots d\mathbf{r}_n$$

$$- (\lambda + 1)(\lambda + 2) \sum_{j=1}^n \sideset{}{'}\sum_{j'=1}^n E_j \int_\tau \nabla_j \left[\frac{P_\lambda(\hat{\mathbf{r}} \cdot \hat{\mathbf{r}}_{j'})}{r_{j'}^{\lambda+1} r_j^{\lambda+3}} \right] \cdot \nabla_j (\varphi^*\varphi) \, d\mathbf{r}_1 \, d\mathbf{r}_2 \cdots d\mathbf{r}_n$$

$$- 2(\lambda + 1)(\lambda + 2)(2\lambda + 3) \sum_{j=1}^n \sum_{j'=1}^n E_j \int_\tau \varphi^* \frac{P_\lambda(\hat{\mathbf{r}}_j \cdot \hat{\mathbf{r}}_{j'})}{r_{j'}^{\lambda+1} r_j^{\lambda+5}} \varphi \, d\mathbf{r}_1 \, d\mathbf{r}_2 \cdots d\mathbf{r}_n,$$

(98)

[see Eqs. (56) and (59)]. The first two terms in (98) can be converted to the desired form by an application of Green's theorem, see (58) and (60), so that if we define the effective potential $\overline{V}_3^{(2)}$ by the relation

$$\langle \varphi | V_k^{(2)} | \varphi \rangle = \langle \varphi | \overline{V}_k^{(2)} | \varphi \rangle, \qquad k = 3, 4, \ldots,$$

(99)

we obtain the result

$$\overline{V}_3^{(2)} \simeq 4 \sum_{\lambda=1}^\infty \delta_3(\lambda)\lambda(\lambda + 1)(\lambda + 2)(2\lambda + 3) \sum_{j=1}^n E_j \left[\frac{1}{r_j^{2\lambda+6}} + O\left(\frac{1}{r_j^{2\lambda+7}} \right) \right]$$

(100)

from (97) and (98).

It is clear from the analysis of Section 3.4 that the effective potential derived from the term with $k = 4$ in (93) is of the form

$$\overline{V}_4^{(2)} = \sum_{\lambda=1}^\infty \delta_4(\lambda) \sum_{j=1}^n \sum_{j'=1}^n O\left(\frac{1}{r_{j'}^{\lambda+1} r_j^{\lambda+5}} \right)$$

(101)

but we shall not attempt to evaluate it in detail. It may be conjectured that, in general,

$$\bar{V}_k^{(2)} = \sum_{\lambda=1}^{\infty} \delta_k(\lambda) \sum_{j=1}^{n} \sum_{j'=1}^{n} O\left(\frac{1}{r_{j'}^{\lambda+1} r_j^{\lambda+k+1}} \right) \tag{102}$$

if k is even, and

$$\bar{V}_k^{(2)} = \sum_{\lambda=1}^{\infty} \delta_k(\lambda) \sum_{j=1}^{n} O\left(\frac{1}{r_j^{2\lambda+k+3}} \right) \tag{103}$$

if k is odd.

3.6. Third-Order Perturbation Theory: The Static Contribution

The static contribution to V from the third-order term in Eq. (33) is given by $V_{kk'}^{(3)}$ in (36) with $k = k' = 0$. Equation (36) gives

$$V_{00}^{(3)} = \sum_{c \neq 0} \sum_{c' \neq 0} \frac{\langle \varphi_0 | \mathcal{V} | \varphi_c \rangle \langle \varphi_c | \Delta V | \varphi_{c'} \rangle \langle \varphi_{c'} | \mathcal{V} | \varphi_0 \rangle}{(E_c - E_0)(E_{c'} - E_0)} \tag{104}$$

and since ΔV depends upon V itself, see Eq. (7), it is convenient to write

$$V_{00}^{(3)} = W_1 + W_2, \tag{105}$$

where

$$W_1 \equiv \sum_{c \neq 0} \sum_{c' \neq 0} \frac{\langle \varphi_0 | \mathcal{V} | \varphi_c \rangle \langle \varphi_c | \mathcal{V} | \varphi_{c'} \rangle \langle \varphi_{c'} | \mathcal{V} | \varphi_0 \rangle}{(E_c - E_0)(E_{c'} - E_0)} \tag{106}$$

and

$$W_2 \equiv - \sum_{c \neq 0} \frac{\langle \varphi_0 | \mathcal{V} | \varphi_c \rangle \langle \varphi_c | \mathcal{V} | \varphi_0 \rangle}{(E_c - E_0)^2} V, \tag{107}$$

since V only depends upon the position vectors of the valence electrons. On introducing definitions (38)–(40) and the notation of Section 3.2, we obtain

$$W_1 = \sum_{\Gamma_c} \sum_{\Gamma_{c'}} \sum_{L_c, M_c} \sum_{L_{c'}, M_{c'}} \frac{1}{(E_c - E_0)(E_{c'} - E_0)}$$

$$\times \sum_{\lambda, \mu, \lambda', \mu', \lambda'', \mu''} \langle \Gamma_0 00 | S(\lambda, \mu) | \Gamma_c L_c M_c \rangle \langle \Gamma_c L_c M_c | S(\lambda, \mu) | \Gamma_{c'} L_{c'} M_{c'} \rangle$$

$$\times \langle \Gamma_{c'} L_{c'} M_{c'} | S(\lambda, \mu) | \Gamma_0 00 \rangle T^*(\lambda, \mu) T^*(\lambda', \mu') T^*(\lambda'', \mu''), \tag{108}$$

and on using (44), (108) can be simplified to give

$$W_1 = \sum_{\lambda=1}^{\infty} \sum_{\lambda'=1}^{\infty} \sum_{\lambda''=1}^{\infty} \varepsilon(\lambda,\lambda',\lambda'') \sum_{\mu,\mu',\mu''} \left[\begin{pmatrix} \lambda & \lambda' & \lambda'' \\ 0 & 0 & 0 \end{pmatrix} \right]^{-1} \begin{pmatrix} \lambda & \lambda' & \lambda'' \\ \mu & \mu' & \mu'' \end{pmatrix}$$

$$\times T^*(\lambda, \mu)T^*(\lambda', \mu')T^*(\lambda'', \mu''), \tag{109}$$

where

$$\varepsilon(\lambda,\lambda',\lambda'') \equiv \sum_{\Gamma_c} \sum_{\Gamma_{c'}} \left[(2\lambda + 1)(2\lambda'' + 1) \right]^{-1/2} \begin{pmatrix} \lambda & \lambda' & \lambda'' \\ 0 & 0 & 0 \end{pmatrix}$$

$$\times \frac{\langle \Gamma_0 0 \| \mathbf{S}(\lambda) \| \Gamma_c \lambda \rangle \langle \Gamma_c \lambda \| \mathbf{S}(\lambda') \| \Gamma_{c'} \lambda'' \rangle \langle \Gamma_{c'} \lambda'' \| \mathbf{S}(\lambda'') \| \Gamma_0 0 \rangle}{(E_{c'} - E_0)(E_c - E_0)}.$$

$$\tag{110}$$

If we neglect the contributions to W_1 that involve sets of three distinct electrons, the sum over μ, μ', and μ'' in (109) can be evaluated to give

$$\sum_{\mu,\mu',\mu''} \left[\begin{pmatrix} \lambda & \lambda' & \lambda'' \\ 0 & 0 & 0 \end{pmatrix} \right]^{-1} \begin{pmatrix} \lambda & \lambda' & \lambda'' \\ \mu & \mu' & \mu'' \end{pmatrix} T^*(\lambda, \mu)T^*(\lambda', \mu')T^*(\lambda'', \mu'')$$

$$= 3 \sum_{j=1}^{n} \sum_{j'=1}^{n}{}' \frac{1}{r_{j'}^{\lambda'+1}} \frac{1}{r_j^{\lambda+\lambda''+2}} P_\lambda(\hat{\mathbf{f}}_j \cdot \hat{\mathbf{f}}_{j'}) + \sum_{j=1}^{n} \frac{1}{r_j^{\lambda+\lambda'+\lambda''+3}} \tag{111}$$

on using (271). If there is only one outer electron, the leading term in (109) is of order $1/r^7$, and this term has recently been discussed by Drachman[22] for the case of electrons scattered by hydrogen. Note that both here and in Eq. (114) below, r_1 has been replaced by r for the case of $n = 1$. Also by comparing (107) with (48), it can be shown directly that

$$W_2 = \frac{1}{2}\alpha(1) \sum_{\lambda=1}^{\infty} \beta(\lambda) \left[\sum_{j=1}^{n} \sum_{j'=1}^{n} \frac{P_\lambda(\hat{\mathbf{f}}_j \cdot \hat{\mathbf{f}}_{j'})}{(r_j r_{j'})^{\lambda+1}} \right]$$

$$\times \left\{ \sum_{j=1}^{n} \sum_{j'=1}^{n} \frac{P_\lambda(\hat{\mathbf{f}}_j \cdot \hat{\mathbf{f}}_{j'})}{(r_j r_{j'})^2} + O\left[\frac{1}{(r_j r_{j'})^3} \right] \right\}, \tag{112}$$

in which we have made use of (47) and the expression (81) for V.

The nonadiabatic corrections in third-order perturbation theory derived from (36) with nonzero values of k and/or k' clearly give contributions of higher order than the corresponding terms in (109) and (112) and will not be considered further.

3.7. Summary of Results

We have taken the case of a general atomic system and have considered the model in which the nucleus and inner electrons are replaced by a spherically symmetric core of overall charge z. Only the n outer or valence electrons of the system are treated explicitly, and these electrons may or may not be bound to the core. The difference between the exact interaction of the core with the valence electrons, $\mathcal{V}$, and the corresponding effective interaction, V, is treated as a perturbation on the model system. Terms up to third order in this perturbation have been examined. On collecting together the results given in (41), (45), (47), (62), (92), (100), (101), (105), (109), (111), and (112), we have from (33) that the interaction with the core at long range (excluding the pure Coulomb terms) is given by

$$V \simeq \sum_{j=1}^{n} \sum_{j'=1}^{n} \left\{ -\frac{1}{2}\alpha(1) \frac{P_1(\hat{\mathbf{r}}_j \cdot \hat{\mathbf{r}}_{j'})}{r_j^2 r_{j'}^2} - \frac{1}{2}\alpha(2) \frac{P_2(\hat{\mathbf{r}}_j \cdot \hat{\mathbf{r}}_{j'})}{r_j^3 r_{j'}^3} + \frac{3\beta(1)}{r_j^6} \delta_{jj'} \right.$$

$$+ 8\gamma(1) \frac{P_1(\hat{\mathbf{r}}_j \cdot \hat{\mathbf{r}}_{j'})}{r_j^2 r_{j'}^4} \left(3E_j + \frac{7}{2} \frac{z_j}{r_j} \right) + (3 - 2\delta_{jj'})$$

$$\left. \times \left[2\varepsilon(2,1,1) \frac{P_1(\hat{\mathbf{r}}_j \cdot \hat{\mathbf{r}}_{j'})}{r_j^2 r_{j'}^5} + \varepsilon(1,2,1) \frac{P_2(\hat{\mathbf{r}}_j \cdot \hat{\mathbf{r}}_{j'})}{r_j^3 r_{j'}^4} \right] + O\left(\frac{1}{r_j^8}\right) \right\}$$

(113)

since $\varepsilon(2,1,1) = \varepsilon(1,1,2)$. In (113) it has been assumed that $r_1 < r_2 < \cdots < r_n$, and we have also neglected all contributions that involve more than three distinct bodies. For the important special case of a single electron interacting with a core, we have shown that

$$V = -\frac{1}{2}\alpha(1) \frac{1}{r^4} + \left[-\frac{1}{2}\alpha(2) + 3\beta(1) + 24\gamma(1)E \right] \frac{1}{r^6}$$

$$+ \left[28z\gamma(1) + 2\varepsilon(2,1,1) + \varepsilon(1,2,1) \right] \frac{1}{r^7} + O\left(\frac{1}{r^8}\right) \qquad (114)$$

since $E_1 = E$ [see (86) and (87)].

4. Form of the Model Potential for Diatomic Systems

4.1. The Exact Interaction between the Valence Electrons and the Cores

Throughout this section, we assume that the electronic and nuclear motions in the system may be treated separately, and therefore $\mathcal{H}$ in Eq. (1)

is taken to represent the exact electronic Hamiltonian for the molecule. We introduce a natural extension of the notation of Section 3, and we consider a diatomic system that is composed of two spherically symmetric cores A and B with overall charges z_a and z_b and containing N_a and N_b electrons, respectively, together with n outer electrons. The Hamiltonian H_c in Eq. (1) may then be written

$$H_c = H_{ac} + H_{bc},$$
(115)

where H_{ac} and H_{bc} are the unperturbed Hamiltonians for the cores A and B, respectively. The eigenfunctions φ_{ac} and φ_{bc} for states c of the cores A and B satisfy the equations

$$H_{ac}\varphi_{ac} = E_{ac}\varphi_{ac} \quad \text{and} \quad H_{bc}\varphi_{bc} = E_{bc}\varphi_{bc},$$
(116)

where E_{ac} and E_{bc} denote the energies of these states. Thus the wave function φ_c in (3) can be written explicitly as

$$\varphi_c = \varphi_{ac}\varphi_{bc},$$
(117)

and the corresponding energy as

$$E_c = E_{ac} + E_{bc}.$$
(118)

The position vectors of the core and valence electrons relative to the atomic nuclei A and B are given by $\mathbf{r}_{ai}$, $i = 1, 2, \ldots, N_a$, $\mathbf{r}_{aj}$, $j = 1, 2, \ldots, n$ and $\mathbf{r}_{bi}$, $i = 1, 2, \ldots, N_b$, $\mathbf{r}_{bj}$, $j = 1, 2, \ldots, n$; as in Section 3 subscripts i and j denote core and valence electrons, respectively. The exact interaction $\mathcal{V}$ defined in Eq. (1) is then given by

$$\mathcal{V} = \sum_{i=1}^{N_a} \sum_{j=1}^{n} \frac{1}{|\mathbf{r}_{ai} - \mathbf{r}_{aj}|} - \sum_{j=1}^{n} \frac{N_a}{r_{aj}} + \sum_{i=1}^{N_b} \sum_{j=1}^{n} \frac{1}{|\mathbf{r}_{bi} - \mathbf{r}_{bj}|} - \sum_{j=1}^{n} \frac{N_b}{r_{bj}}$$

$$+ \sum_{i=1}^{N_a} \sum_{i'=1}^{N_b} \frac{1}{|\mathbf{R} - \mathbf{r}_{ai} + \mathbf{r}_{bi'}|} - \sum_{i=1}^{N_b} \frac{z_a + N_a}{|\mathbf{R} + \mathbf{r}_{bi}|}$$

$$- \sum_{i=1}^{N_a} \frac{(z_b + N_b)}{|\mathbf{R} - \mathbf{r}_{ai}|} + \frac{z_a N_b}{R} + \frac{z_b N_a}{R} - \frac{N_a N_b}{R}$$
(119)

where $\mathbf{R} = \overrightarrow{AB}$ is the position vector of nucleus B relative to nucleus A.

In order to carry out a complete expansion of $\mathcal{V}$ in (119) in terms of spherical harmonics, we need result (276) for expansion of the terms of type $|\mathbf{R} - \mathbf{r}_1 + \mathbf{r}_2|^{-1}$. Throughout the analysis that follows, we shall choose the coordinate system to be such that the z axis is fixed in the direction $\hat{\mathbf{R}}$. Having made this choice, (276) reduces to

$$\frac{1}{|\mathbf{R} - \mathbf{r}_1 + \mathbf{r}_2|} = \sum_{l,l',m} (-1)^{l'+m} (4\pi) \frac{(l+l')!}{l!\,l'!}$$

$$\times \left[(2l+1)(2l'+1) \right]^{-1/2} \left[\begin{pmatrix} l & l' & l+l' \\ 0 & 0 & 0 \end{pmatrix} \right]^{-1}$$

$$\times \begin{pmatrix} l & l' & l+l' \\ m & -m & 0 \end{pmatrix} \frac{r_1^l r_2^{l'}}{R^{l+l'+1}} Y_{lm}(\hat{\mathbf{r}}_1) Y_{l'm}^*(\hat{\mathbf{r}}_2). \quad (120)$$

Using a similar notation to that introduced in Section 3.1, we define the quantities

$$S_a(\lambda, \mu) \equiv \left(\frac{4\pi}{2\lambda+1} \right)^{1/2} \sum_{i=1}^{N_a} r_{ai}^\lambda Y_{\lambda\mu}(\hat{\mathbf{r}}_{ai});$$

$$\quad (121)$$

$$S_b(\lambda, \mu) \equiv \left(\frac{4\pi}{2\lambda+1} \right)^{1/2} \sum_{i=1}^{N_b} r_{bi}^\lambda Y_{\lambda\mu}(\hat{\mathbf{r}}_{bi})$$

and

$$T_a(\lambda, \mu) \equiv \left(\frac{4\pi}{2\lambda+1} \right)^{1/2} \sum_{j=1}^{n} \frac{1}{r_{aj}^{\lambda+1}} Y_{\lambda\mu}(\hat{\mathbf{r}}_{aj});$$

$$\quad (122)$$

$$T_b(\lambda, \mu) \equiv \left(\frac{4\pi}{2\lambda+1} \right)^{1/2} \sum_{j=1}^{n} \frac{1}{r_{bj}^{\lambda+1}} Y_{\lambda\mu}(\hat{\mathbf{r}}_{bj}).$$

Then on using (120) and (265), we have that

$$\mathcal{V} = \mathcal{V}_a + \mathcal{V}_b + \mathcal{W}_a + \mathcal{W}_b + \mathcal{W}_{ab}, \quad (123)$$

where $\mathcal{V}_a$, $\mathcal{V}_b$, $\mathcal{W}_a$, $\mathcal{W}_b$, and $\mathcal{W}_{ab}$ are given by

$$\mathcal{V}_a \equiv \sum_{\lambda=1}^{\infty} \sum_{\mu=-\lambda}^{+\lambda} S_a^*(\lambda, \mu) T_a(\lambda, \mu); \qquad \mathcal{V}_b \equiv \sum_{\lambda=1}^{\infty} \sum_{\mu=-\lambda}^{+\lambda} S_b^*(\lambda, \mu) T_b(\lambda, \mu);$$

$$\quad (124)$$

$$\mathcal{W}_a \equiv -\sum_{\lambda=1}^{\infty} \frac{z_b}{R^{\lambda+1}} S_a(\lambda, 0); \qquad \mathcal{W}_b \equiv -\sum_{\lambda=1}^{\infty} \frac{z_a(-1)^\lambda}{R^{\lambda+1}} S_b(\lambda, 0) \quad (125)$$

and

$$\mathcal{W}_{ab} \equiv \sum_{\lambda=1}^{\infty} \sum_{\lambda'=1}^{\infty} \sum_{\mu=-\lambda_0}^{\lambda_0} (-1)^{\lambda} S_a(\lambda, \mu) S_b^*(\lambda', \mu) W(\lambda, \lambda', \mu),$$

$$\lambda_0 \equiv \min(\lambda, \lambda'), \quad (126)$$

where

$$W(\lambda, \lambda', \mu) \equiv (-1)^{\mu} \frac{(\lambda + \lambda')!}{\lambda! \lambda'!} \left[\begin{pmatrix} \lambda & \lambda' & \lambda + \lambda' \\ 0 & 0 & 0 \end{pmatrix} \right]^{-1}$$

$$\times \begin{pmatrix} \lambda & \lambda' & \lambda + \lambda' \\ \mu & -\mu & 0 \end{pmatrix} \frac{1}{R^{\lambda+\lambda'+1}}. \quad (127)$$

In making expansions (124), we have assumed that $r_{ai} < r_{aj}$ and $r_{bi'} < r_{bj}$ for all $i = 1, 2, \ldots, N_a$, $i' = 1, 2, \ldots, N_b$ and $j = 1, 2, \ldots, n$; the expansions in (125) and (126) are valid provided that $r_{ai} < R$ and $r_{bi'} < R$ for all values of i and i'.

4.2. Second-Order Perturbation Theory: The Static Contribution

The first nonzero contribution to V arises from the second-order terms $V_k^{(2)}$, $k = 0, 1, \ldots$, in Eq. (33). The static term is given by putting $k = 0$ in Eq. (35), and then on using the relations (123)–(127) that specify V, we find that

$$V_0^{(2)} = V_0^{(2)}(a, a) + V_0^{(2)}(b, b) + V_0^{(2)}(a, b), \quad (128)$$

where

$$V_0^{(2)}(a, a) = -\sum_{c \neq 0} \frac{\langle \varphi_{a0} | (\mathcal{V}_a + \mathcal{W}_a) | \varphi_{ac} \rangle \langle \varphi_{ac} | (\mathcal{V}_a + \mathcal{W}_a) | \varphi_{a0} \rangle}{E_{ac} - E_{a0}}, \quad (129)$$

$$V_0^{(2)}(b, b) = -\sum_{c \neq 0} \frac{\langle \varphi_{b0} | (\mathcal{V}_b + \mathcal{W}_b) | \varphi_{bc} \rangle \langle \varphi_{bc} | (\mathcal{V}_b + \mathcal{W}_b) | \varphi_{b0} \rangle}{E_{bc} - E_{b0}}, \quad (130)$$

and

$$V_0^{(2)}(a, b) = -\sum_{c \neq 0} \sum_{c' \neq 0} \frac{\langle \varphi_{a0} \varphi_{b0} | \mathcal{W}_{ab} | \varphi_{ac} \varphi_{bc'} \rangle \langle \varphi_{ac} \varphi_{bc'} | \mathcal{W}_{ab} | \varphi_{a0} \varphi_{b0} \rangle}{E_{ac} + E_{bc'} - E_{a0} - E_{b0}}. \quad (131)$$

All the other terms arising from substituting (123) into (35) vanish as a result of the selection rules imposed on the matrix elements by the angular integrations involved.

Following the notation of Section 3.2, we suppose the core states of A and B to be described by the quantum numbers $\Gamma_{ac} L_{ac} M_{ac}$ and $\Gamma_{bc} L_{bc} M_{bc}$, respectively. Then, on introducing the reduced matrix elements of $S_a(\lambda, \mu)$ and $S_b(\lambda, \mu)$ by using (44), we obtain straightforwardly

$$V_0^{(2)}(a,a)$$

$$= -\frac{1}{2} \sum_{\lambda=1}^{\infty} \alpha_a(\lambda) \left[\sum_{j=1}^{n} \sum_{j'=1}^{n} \frac{P_\lambda(\hat{\mathbf{f}}_{aj} \cdot \hat{\mathbf{f}}_{aj'})}{(r_{aj} r_{aj'})^{\lambda+1}} - 2z_b \sum_{j=1}^{n} \frac{P_\lambda(\hat{\mathbf{f}}_{aj} \cdot \hat{\mathbf{R}})}{(r_{aj} R)^{\lambda+1}} + \frac{z_b^2}{R^{2\lambda+2}} \right]$$

$$(132)$$

and

$$V_0^{(2)}(b,b) = -\frac{1}{2} \sum_{\lambda=1}^{\infty} \alpha_b(\lambda) \left[\sum_{j=1}^{n} \sum_{j'=1}^{n} \frac{P_\lambda(\hat{\mathbf{f}}_{bj} \cdot \hat{\mathbf{f}}_{bj'})}{(r_{bj} r_{bj'})^{\lambda+1}} \right.$$

$$\left. - 2(-1)^\lambda z_a \sum_{j=1}^{n} \frac{P_\lambda(\hat{\mathbf{f}}_{bj} \cdot \hat{\mathbf{R}})}{(r_{bj} R)^{\lambda+1}} + \frac{z_a^2}{R^{2\lambda+2}} \right]; \quad (133)$$

compare Eqs. (45) and (47), where $\alpha_a(\lambda)$ and $\alpha_b(\lambda)$ are the static 2^λ-pole polarizabilities of the cores A and B and are given by

$$\alpha_a(\lambda) = 2 \sum_{\Gamma_{ac}} \frac{1}{2\lambda+1} \frac{|\langle \Gamma_{ac} \lambda \| S_a(\lambda) \| \Gamma_a 0 \rangle|^2}{E_{ac} - E_{a0}} \qquad (134)$$

and

$$\alpha_b(\lambda) = 2 \sum_{\Gamma_{bc}} \frac{1}{2\lambda+1} \frac{|\langle \Gamma_{bc} \lambda \| S_b(\lambda) \| \Gamma_b 0 \rangle|^2}{E_{bc} - E_{b0}}, \qquad (135)$$

[see Eq. (46)]. On using (126), (127), and (131), the third contribution, $V_0^{(2)}(a,b)$ in (128), can be simplified in a similar manner to yield

$$V_0^{(2)}(a,b) = -\frac{1}{4} \sum_{\lambda=1}^{\infty} \sum_{\lambda'=1}^{\infty} \frac{1}{2\lambda+2\lambda'+1} \left[\frac{(\lambda+\lambda')!}{\lambda! \lambda'!} \right]^2$$

$$\times \left[\begin{pmatrix} \lambda & \lambda' & \lambda+\lambda' \\ 0 & 0 & 0 \end{pmatrix} \right]^{-2} \frac{\alpha_{ab}(\lambda, \lambda')}{R^{2\lambda+2\lambda'+2}}$$

$$\equiv - \sum_{\lambda=1}^{\infty} \sum_{\lambda'=1}^{\infty} \frac{\varepsilon_2(\lambda, \lambda')}{R^{2\lambda+2\lambda'+2}}, \qquad (136)$$

where the coefficient $\alpha_{ab}(\lambda, \lambda')$ is given by

$$\alpha_{ab}(\lambda, \lambda') = 4 \sum_{\Gamma_{ac}} \sum_{\Gamma_{bc'}} \frac{1}{(2\lambda + 1)(2\lambda' + 1)}$$

$$\times \frac{|\langle \Gamma_{ac}\lambda \| S_a(\lambda) \| \Gamma_{a}0 \rangle|^2 |\langle \Gamma_{bc'}\lambda' \| S_b(\lambda') \| \Gamma_{b0}0 \rangle^2}{E_{ac} + E_{bc'} - E_{a0} - E_{b0}}. \quad (137)$$

The coefficient $\varepsilon_2(\lambda, \lambda')$ is the one defined by Dalgarno and Davison,[23] and $\varepsilon_2(1, 1)$ is the exact Van der Waals coefficient for the interaction of core A with core B.

4.3. The First Nonadiabatic Correction

The first nonadiabatic correction to V is given by $V_1^{(2)}$ in Eq. (35). Since the model Hamiltonian H given by Eq. (4) involves the kinetic energy operator K given by (53), which depends only on the coordinates of the valence electrons, the operator $(H - E)$ commutes with pure functions of R. Therefore, using (35) with $k = 1$ and (123)–(126) for $\mathcal{V}$, we obtain

$$V_1^{(2)} = V_1^{(2)}(a, a) + V_1^{(2)}(b, b), \quad (138)$$

where

$$V_1^{(2)}(a, a) = \sum_{c \neq 0} \frac{\langle \varphi_{a0} | (\mathcal{V}_a + \mathcal{W}_a) | \varphi_{ac} \rangle}{(E_{ac} - E_{a0})^2} (H - E) \langle \varphi_{ac} | \mathcal{V}_a | \varphi_{a0} \rangle \quad (139)$$

and

$$V_1^{(2)}(b, b) = \sum_{c \neq 0} \frac{\langle \varphi_{b0} | (\mathcal{V}_b + \mathcal{W}_b) | \varphi_{bc} \rangle}{(E_{bc} - E_{b0})^2} (H - E) \langle \varphi_{bc} | \mathcal{V}_b | \varphi_{b0} \rangle. \quad (140)$$

As in Section 3.3, $V_1^{(2)}$ is an operator, and we introduce an effective contribution $\overline{V}_1^{(2)}$, defined in (49), by evaluating the matrix elements

$$\langle \varphi | V_1^{(2)}(a, a) | \varphi \rangle = \langle \varphi | \overline{V}_1^{(2)}(a, a) | \varphi \rangle;$$

$$\langle \varphi | V_1^{(2)}(b, b) | \varphi \rangle = \langle \varphi | \overline{V}_1^{(2)}(b, b) | \varphi \rangle. \quad (141)$$

It is easily shown that

$$\langle \varphi | V_1^{(2)}(a, a) | \varphi \rangle = \sum_{\lambda=1}^{\infty} \beta_a(\lambda) \left[\sum_{\mu} \langle \varphi | T_a(\lambda, \mu)(H - E) T_a^*(\lambda, \mu) | \varphi \rangle \right.$$

$$\left. - \frac{z_b}{R^{\lambda+1}} \langle \varphi | (H - E) T_a^*(\lambda, 0) | \varphi \rangle \right] \quad (142)$$

and

$$\langle\varphi|V_1^{(2)}(b,b)|\varphi\rangle = \sum_{\lambda=1}^{\infty} \beta_b(\lambda)\left[\sum_{\mu}\langle\varphi|T_b(\lambda,\mu)(H-E)T_b^*(\lambda,\mu)|\varphi\rangle\right.$$
$$\left. - \frac{(-1)^\lambda z_a}{R^{\lambda+1}}\langle\varphi|(H-E)T_b^*(\lambda,0)|\varphi\rangle\right], \quad (143)$$

where

$$\beta_a(\lambda) = \sum_{\Gamma_{ac}} \frac{1}{2\lambda+1} \frac{|\langle\Gamma_{ac}\lambda\|S_a(\lambda)\|\Gamma_{a0}0\rangle|^2}{(E_{ac}-E_{a0})^2} \quad (144)$$

and

$$\beta_b(\lambda) = \sum_{\Gamma_{bc}} \frac{1}{2\lambda+1} \frac{|\langle\Gamma_{bc}\lambda\|S_b(\lambda)\|\Gamma_{b0}0\rangle|^2}{(E_{bc}-E_{b0})^2} \quad (145)$$

are dynamical correction coefficients for the cores A and B [see Eq. (51)]. The sums over μ in (142) and (143) are identical to the corresponding sum in (50), and so the analysis of Section 3.3 applies directly and shows that there are no contributions to these terms that involve two distinct electrons. The second terms in (142) and (143) that involve the factor $1/R^{\lambda+1}$ can also be shown to vanish, by using equations similar to (59) and (60) in which $r_{j'}$ has been replaced by R. Thus again, contrary to the predictions of Bottcher and Dalgarno,[10] no three-body terms contribute to the first nonadiabatic correction, and we obtain

$$\bar{V}_1^{(2)}(a,a) = \frac{1}{2}\sum_{\lambda=1}^{\infty}(\lambda+1)(2\lambda+1)\beta_a(\lambda)\sum_{j=1}^{n}\frac{1}{r_{aj}^{2\lambda+4}} \quad (146)$$

and

$$\bar{V}_1^{(2)}(b,b) = \frac{1}{2}\sum_{\lambda=1}^{\infty}(\lambda+1)(2\lambda+1)\beta_b(\lambda)\sum_{j=1}^{n}\frac{1}{r_{bj}^{2\lambda+4}} \quad (147)$$

[see Eq. (62)].

4.4. The Second Nonadiabatic Correction

The second nonadiabatic correction is given by Eq. (35) with $k=2$, i.e.,

$$V_2^{(2)} = V_2^{(2)}(a,a) + V_2^{(2)}(b,b), \quad (148)$$

where

$$V_2^{(2)}(a,a) = -\sum_{c \neq 0} \frac{\langle \varphi_{a0} | (\mathcal{V}_a + \mathcal{U}_a) | \varphi_{ac} \rangle}{(E_{ac} - E_{a0})^3} (H - E)^2 \langle \varphi_{ac} | \mathcal{V}_a | \varphi_{a0} \rangle \quad (149)$$

and

$$V_2^{(2)}(b,b) = -\sum_{c \neq 0} \frac{\langle \varphi_{b0} | (\mathcal{V}_b + \mathcal{U}_b) | \varphi_{bc} \rangle}{(E_{bc} - E_{b0})^3} (H - E)^2 \langle \varphi_{bc} | \mathcal{V}_b | \varphi_{b0} \rangle \quad (150)$$

[see Eqs. (139) and (140)]. The corresponding effective contribution $\overline{V}_2^{(2)}$, defined in (64), can be obtained by evaluating the matrix elements

$$\langle \varphi | V_2^{(2)}(a,a) | \varphi \rangle \equiv \langle \varphi | \overline{V}_2^{(2)}(a,a) | \varphi \rangle;$$

$$\langle \varphi | V_2^{(2)}(b,b) | \varphi \rangle \equiv \langle \varphi | \overline{V}_2^{(2)}(b,b) | \varphi \rangle, \quad (151)$$

where

$$\langle \varphi | V_2^{(2)}(a,a) | \varphi \rangle \equiv -2 \sum_{\lambda=1}^{\infty} \gamma_a(\lambda) \left[\sum_{\mu} \langle \varphi | T_a(\lambda, \mu)(H - E)^2 T_a^*(\lambda, \mu) | \varphi \rangle \right.$$

$$\left. - \frac{z_b}{R^{\lambda+1}} \langle \varphi | (H - E)^2 T_a^*(\lambda, 0) | \varphi \rangle \right] \quad (152)$$

and

$$\langle \varphi | V_2^{(2)}(b,b) | \varphi \rangle \equiv -2 \sum_{\lambda=1}^{\infty} \gamma_b(\lambda) \left[\sum_{\mu} \langle \varphi | T_b(\lambda, \mu)(H - E)^2 T_b^*(\lambda, \mu) | \varphi \rangle \right.$$

$$\left. - \frac{(-1)^\lambda z_a}{R^{\lambda+1}} \langle \varphi | (H - E)^2 T_b^*(\lambda, 0) | \varphi \rangle \right], \quad (153)$$

[see Eqs. (142) and (143)]. The coefficients $\gamma_a(\lambda)$ and $\gamma_b(\lambda)$ represent the second-order dynamical corrections to the 2^λ-pole polarizabilities of cores A and B, and are given by

$$\gamma_a(\lambda) = \frac{1}{2} \sum_{\Gamma_{ac}} \frac{1}{2\lambda + 1} \frac{|\langle \Gamma_{ac} \lambda \| \mathbf{S}_a(\lambda) \| \Gamma_{a0} 0 \rangle|^2}{(E_{ac} - E_{a0})^3} \quad (154)$$

and

$$\gamma_b(\lambda) = \frac{1}{2} \sum_{\Gamma_{bc}} \frac{1}{2\lambda + 1} \frac{|\langle \Gamma_{bc}\lambda \| S_b(\lambda) \| \Gamma_{b0} 0 \rangle|^2}{(E_{bc} - E_{b0})^3} \tag{155}$$

[see Eq. (66)]. The analysis required in order to simplify (152) and (153) is essentially the same as that presented in Section 3.4, but some differences arise because of the explicit appearance in the expressions of the potential U, which has a different form in this case. In line with the definitions (68)–(70), we write

$$\sum_\mu T_a(\lambda, \mu)(H - E)^2 T_a^*(\lambda, \mu)\varphi = (\mathcal{I}_{a1} + \mathcal{I}_{a2})\varphi \tag{156}$$

and

$$\sum_\mu T_b(\lambda, \mu)(H - E)^2 T_b^*(\lambda, \mu)\varphi = (\mathcal{I}_{b1} + \mathcal{I}_{b2})\varphi, \tag{157}$$

where

$$\mathcal{I}_{a1} \equiv \frac{4\pi}{2\lambda + 1} \sum_{j=1}^n \sum_{j'=1}^n \sum_\mu \frac{1}{r_{aj'}^{\lambda+1}} Y_{\lambda\mu}(\hat{\mathbf{r}}_{aj'}) \boldsymbol{\nabla}_j \boldsymbol{\nabla}_j \left[\frac{1}{r_{aj}^{\lambda+1}} Y_{\lambda\mu}^*(\hat{\mathbf{r}}_{aj}) \right] : \boldsymbol{\nabla}_j \boldsymbol{\nabla}_j ; \tag{158}$$

$$\mathcal{I}_{b1} \equiv \frac{4\pi}{2\lambda + 1} \sum_{j=1}^n \sum_{j'=1}^n \sum_\mu \frac{1}{r_{bj'}^{\lambda+1}} Y_{\lambda\mu}(\hat{\mathbf{r}}_{bj'}) \boldsymbol{\nabla}_j \boldsymbol{\nabla}_j \left[\frac{1}{r_{bj}^{\lambda+1}} Y_{\lambda\mu}^*(\hat{\mathbf{r}}_{bj}) \right] : \boldsymbol{\nabla}_j \boldsymbol{\nabla}_j ; \tag{159}$$

$$\mathcal{I}_{a2} \equiv \frac{4\pi}{2\lambda + 1} \sum_{j=1}^n \sum_{j'=1}^n \sum_\mu \frac{1}{r_{aj'}^{\lambda+1}} Y_{\lambda\mu}(\hat{\mathbf{r}}_{aj'}) \boldsymbol{\nabla}_j \left[\frac{1}{r_{aj}^{\lambda+1}} Y_{\lambda\mu}^*(\hat{\mathbf{r}}_{aj}) \right] \cdot \boldsymbol{\nabla}_j(U + V);$$

$$\tag{160}$$

$$\mathcal{I}_{b2} \equiv \frac{4\pi}{2\lambda + 1} \sum_{j=1}^n \sum_{j'=1}^n \sum_\mu \frac{1}{r_{bj'}^{\lambda+1}} Y_{\lambda\mu}(\hat{\mathbf{r}}_{bj'}) \boldsymbol{\nabla}_j \left[\frac{1}{r_{bj}^{\lambda+1}} Y_{\lambda\mu}^*(\hat{\mathbf{r}}_{bj}) \right] \cdot \boldsymbol{\nabla}_j(U + V).$$

$$\tag{161}$$

Since the operator $\boldsymbol{\nabla}_j$ is invariant with respect to the choice of the origin of coordinates, the origin can be chosen to be at either A or B in (158)–(161) according to which is most convenient.

Using Eqs. (4), (5), and (53), the Schrödinger equation for the diatomic system may be written as

$$\sum_{j=1}^{n} \nabla_{aj}^2 \varphi = \sum_{j=1}^{n} \nabla_{bj}^2 \varphi = 2(U + V - E)\varphi, \tag{162}$$

where V is given by

$$V \simeq -\frac{1}{2}\alpha_a(1)\left[\sum_{j=1}^{n}\sum_{j'=1}^{n}\frac{P_1(\hat{\mathbf{r}}_{aj}\cdot\hat{\mathbf{r}}_{aj'})}{(r_{aj}r_{aj'})^2} - 2z_b\sum_{j=1}^{n}\frac{P_1(\hat{\mathbf{r}}_{aj}\cdot\hat{\mathbf{R}})}{(r_{aj}R)^2} + \frac{z_b^2}{R^4}\right]$$

$$-\frac{1}{2}\alpha_b(1)\left[\sum_{j=1}^{n}\sum_{j'=1}^{n}\frac{P_1(\hat{\mathbf{r}}_{bj}\cdot\hat{\mathbf{r}}_{bj'})}{(r_{bj}r_{bj'})^2} + 2z_a\sum_{j=1}^{n}\frac{P_1(\hat{\mathbf{r}}_{bj}\cdot\hat{\mathbf{R}})}{(r_{bj}R)^2} + \frac{z_a^2}{R^4}\right], \tag{163}$$

see Eqs. (132) and (133). If the n valence electrons are divided into two groups, so that there are n_a electrons for which $r_{aj} < r_{bj}$ and n_b electrons for which $r_{bj} < r_{aj}$, the potential U can be written in the form

$$U = -\sum_{j=1}^{n}\left(\frac{z_a}{r_{aj}} + \frac{z_b}{r_{bj}}\right) + \frac{z_a z_b}{R} + \sum_{j=1}^{n_a}\sum_{j'=1}^{j-1}\frac{1}{|\mathbf{r}_{aj} - \mathbf{r}_{aj'}|} + \sum_{j=1}^{n_b}\sum_{j'=1}^{j-1}\frac{1}{|\mathbf{r}_{bj} - \mathbf{r}_{bj'}|}$$

$$+ \sum_{j=1}^{n_a}\sum_{j'=1}^{n_b}\frac{1}{|\mathbf{R} - \mathbf{r}_{aj} + \mathbf{r}_{bj'}|}. \tag{164}$$

We shall assume that in the limit of infinite separation, n_a electrons are attached to core A and n_b electrons to core B. Furthermore, if the electrons are labeled by $j = 1, 2, \ldots, n_a$ in such a way that $r_{a1} < r_{a2} < \cdots < r_{an_a}$, and by $j = 1, 2, \ldots, n_b$ so that $r_{b1} < r_{b2} < \cdots < r_{bn_b}$, (164) becomes

$$U = -\sum_{j=1}^{n_a}\left[\frac{(z_a - j + 1)}{r_{aj}} + \frac{(z_b - n_b)}{r_{bj}}\right] - \sum_{j=1}^{n_b}\left[\frac{(z_b - j + 1)}{r_{bj}} + \frac{(z_a - n_a)}{r_{aj}}\right]$$

$$+ \frac{(z_a z_b - n_a n_b)}{R} + \sum_{j=1}^{n_a}\sum_{j'=1}^{j-1}\sum_{l=1}^{\infty}\frac{r_{aj'}^l}{r_{aj}^{l+1}}P_l(\hat{\mathbf{r}}_{aj}\cdot\hat{\mathbf{r}}_{aj'})$$

$$+ \sum_{j=1}^{n_b}\sum_{j'=1}^{j-1}\sum_{l=1}^{\infty}\frac{r_{bj'}^l}{r_{bj}^{l+1}}P_l(\hat{\mathbf{r}}_{bj}\cdot\hat{\mathbf{r}}_{bj'}) + O\left(\frac{1}{R^3}\right) \tag{165}$$

on using (120) and (265), and where by definition, $r_{aj} = |\mathbf{R} + \mathbf{r}_{bj}|$ and

$r_{bj} = |\mathbf{R} - \mathbf{r}_{aj}|$. When R becomes large, we may write U as

$$U = -\sum_{j=1}^{n_a} \frac{(z_a - j + 1)}{r_{aj}} - \sum_{j=1}^{n_b} \frac{(z_b - j + 1)}{r_{bj}}$$

$$- (z_b - n_b) \sum_{j=1}^{n_a} \frac{\mathbf{r}_{aj} \cdot \hat{\mathbf{R}}}{R^2} + (z_a - n_a) \sum_{j=1}^{n_b} \frac{\mathbf{r}_{bj} \cdot \hat{\mathbf{R}}}{R^2}$$

$$+ \sum_{j=1}^{n_a} \sum_{j'=1}^{j-1} \sum_{l=1}^{\infty} \frac{r_{aj'}^l}{r_{aj}^{l+1}} P_l(\hat{\mathbf{r}}_{aj} \cdot \hat{\mathbf{r}}_{aj'}) + \sum_{j=1}^{n_b} \sum_{j'=1}^{j-1} \sum_{l=1}^{\infty} \frac{r_{bj'}^l}{r_{bj}^{l+1}} P_l(\hat{\mathbf{r}}_{bj} \cdot \hat{\mathbf{r}}_{bj'})$$

$$+ \frac{(z_a - n_a)(z_b - n_b)}{R} + O\left(\frac{1}{R^3}\right). \tag{166}$$

As in Section 3.4, we define energies E_{aj}, $j = 1, 2, \ldots, n_a$ which are such that $-E_{aj}$ is the energy required to remove the jth electron from the atom A, the electrons $j + 1, j + 2, \ldots, n_a$ having already been removed. Similarly, we define the corresponding energies E_{bj}, $j = 1, 2, \ldots, n_b$ for the atom B. The energy of the diatomic system in the separated atom limit is then given by

$$E(\infty) = E_a + E_b \tag{167}$$

where

$$E_a = \sum_{j=1}^{n_a} E_{aj} \quad \text{and} \quad E_b = \sum_{j=1}^{n_b} E_{bj}, \tag{168}$$

and also

$$E \equiv E(R) = E(\infty) + \frac{(z_a - n_a)(z_b - n_b)}{R} + O\left(\frac{1}{R^4}\right), \tag{169}$$

[see Eq. (163)]. The charges z_{aj} and z_{bj} are defined by

$$z_{aj} = z_a - j + 1 \quad \text{and} \quad z_{bj} = z_b - j + 1 \tag{170}$$

[see Eq. (85)]. Alternatively, if in the limit of infinite separation, the n_a electrons on core A are equivalent, then possibly a better choice for E_{aj} and z_{aj} is given by

$$E_{aj} = E_a/n_a \quad \text{and} \quad z_{aj} = z_{\text{eff}}^a, \quad j = 1, 2, \ldots, n_a, \tag{171}$$

and similarly, if the n_b electrons on core B are equivalent, E_{bj} and z_{bj} may be chosen to be

$$E_{bj} = E_b/n_b \quad \text{and} \quad z_{bj} = z_{\text{eff}}^b, \quad j = 1, 2, \ldots, n_b, \quad (172)$$

where z_{eff}^a and z_{eff}^b are appropriate effective charges [see Eq. (88)]. In order to be consistent, U in (166) should be expressed as

$$U = -\sum_{j=1}^{n_a} \frac{z_{aj}}{r_{aj}} - \sum_{j=1}^{n_b} \frac{z_{bj}}{r_{bj}} - (z_b - n_b) \sum_{j=1}^{n_a} \frac{\mathbf{r}_{aj} \cdot \hat{\mathbf{R}}}{R^2} + (z_a - n_a) \sum_{j=1}^{n_b} \frac{\mathbf{r}_{bj} \cdot \hat{\mathbf{R}}}{R^2}$$

$$+ O\left(\frac{1}{r_{aj}^2}\right) + O\left(\frac{1}{r_{bj}^2}\right) + O\left(\frac{1}{R^3}\right), \quad (173)$$

where $z_{aj}, j = 1, 2, \ldots, n_a$ and $z_{bj}, j = 1, 2, \ldots, n_b$ are given by either (170) or by (171) and (172).

Then if we use (162), (163), (167), (168), and (170) and make the approximations

$$\mathcal{L}_{aj} \varphi \equiv \left[E_{aj} + \frac{z_{aj}}{r_{aj}} + (z_b - n_b) \frac{\mathbf{r}_{aj} \cdot \hat{\mathbf{R}}}{R^2} \right] \varphi \simeq -\frac{1}{2} \nabla_{aj}^2 \varphi, \quad j = 1, 2, \ldots, n_a ;$$

$$(174)$$

and

$$\mathcal{L}_{bj} \varphi \equiv \left[E_{bj} + \frac{z_{bj}}{r_{bj}} - (z_a - n_a) \frac{\mathbf{r}_{bj} \cdot \hat{\mathbf{R}}}{R^2} \right] \varphi \simeq -\frac{1}{2} \nabla_{bj}^2 \varphi, \quad j = 1, 2, \ldots, n_b ;$$

$$(175)$$

where $\mathcal{L}_{aj}$ and $\mathcal{L}_{bj}$ are defined by (174) and (175); eqs. (158), (159), (174), and (175) can be combined to give

$$\mathcal{G}_{a1} \varphi \simeq -2(\lambda + 1)(\lambda + 2) \sum_{j'=1}^{n} \frac{1}{r_{aj'}^{\lambda+1}}$$

$$\times \left[\sum_{j=1}^{n_a} \frac{1}{r_{aj}^{\lambda+3}} P_\lambda(\hat{\mathbf{f}}_{aj} \cdot \hat{\mathbf{f}}_{aj'}) \mathcal{L}_{aj} + \sum_{j=1}^{n_b} \frac{1}{r_{aj}^{\lambda+3}} P_\lambda(\hat{\mathbf{f}}_{aj} \cdot \hat{\mathbf{f}}_{aj'}) \mathcal{L}_{bj} \right] \varphi; \quad (176)$$

$$\mathcal{G}_{b1} \varphi \simeq -2(\lambda + 1)(\lambda + 2) \sum_{j'=1}^{n} \frac{1}{r_{bj'}^{\lambda+1}}$$

$$\times \left[\sum_{j=1}^{n_b} \frac{1}{r_{bj}^{\lambda+3}} P_\lambda(\hat{\mathbf{f}}_{bj} \cdot \hat{\mathbf{f}}_{bj'}) \mathcal{L}_{bj} + \sum_{j=1}^{n_a} \frac{1}{r_{bj}^{\lambda+3}} P_\lambda(\hat{\mathbf{f}}_{bj} \cdot \hat{\mathbf{f}}_{bj'}) \mathcal{L}_{aj} \right] \varphi. \quad (177)$$

The reduction of (158) and (159) has been carried out in exactly the same way as the reduction of $\mathcal{G}_1 \varphi$ in Section 3.4.

Expressions (160) and (161) involve terms that have no counterpart in the treatment of the atomic system given in Section 3.4. These terms are of the type

$$\mathfrak{M}_{ajj'} \equiv - \frac{4\pi}{(\lambda + 1)(2\lambda + 1)} r_{aj}^{\lambda+2} \sum_{\mu} Y_{\lambda\mu}(\hat{\mathbf{f}}_{aj'}) \hat{\mathbf{R}} \cdot \boldsymbol{\nabla}_{aj} \left[\frac{1}{r_{aj}^{\lambda+1}} Y_{\lambda\mu}^*(\hat{\mathbf{f}}_{aj}) \right], \quad (178)$$

$$\mathfrak{M}_{bjj'} \equiv \frac{4\pi}{(\lambda + 1)(2\lambda + 1)} r_{bj}^{\lambda+2} \sum_{\mu} Y_{\lambda\mu}(\hat{\mathbf{f}}_{bj'}) \hat{\mathbf{R}} \cdot \boldsymbol{\nabla}_{bj} \left[\frac{1}{r_{bj}^{\lambda+1}} Y_{\lambda\mu}^*(\hat{\mathbf{f}}_{bj}) \right], \quad (179)$$

and can be simplified by using (272) with $\mathbf{r} = \hat{\mathbf{R}}$; we obtain

$$\mathfrak{M}_{ajj'} \equiv \frac{4\pi}{[(2\lambda + 1)(2\lambda + 3)]^{1/2}} \left[\begin{pmatrix} \lambda & 1 & \lambda + 1 \\ 0 & 0 & 0 \end{pmatrix} \right]^{-1}$$

$$\times \sum_{\mu} \begin{pmatrix} \lambda & 1 & \lambda + 1 \\ \mu & 0 & -\mu \end{pmatrix} Y_{\lambda\mu}(\hat{\mathbf{f}}_{aj'}) Y_{\lambda+1-\mu}(\hat{\mathbf{f}}_{aj}), \quad (180)$$

$$\mathfrak{M}_{bjj'} \equiv - \frac{4\pi}{[(2\lambda + 1)(2\lambda + 3)]^{1/2}} \left[\begin{pmatrix} \lambda & 1 & \lambda + 1 \\ 0 & 0 & 0 \end{pmatrix} \right]^{-1}$$

$$\times \sum_{\mu} \begin{pmatrix} \lambda & 1 & \lambda + 1 \\ \mu & 0 & -\mu \end{pmatrix} Y_{\lambda\mu}(\hat{\mathbf{f}}_{bj'}) Y_{\lambda+1-\mu}(\hat{\mathbf{f}}_{bj}). \quad (181)$$

If $j = j'$, (271) can be used to show that

$$\mathfrak{M}_{ajj} = P_1(\hat{\mathbf{f}}_{aj} \cdot \hat{\mathbf{R}}) \quad \text{and} \quad \mathfrak{M}_{bjj} = - P_1(\hat{\mathbf{f}}_{bj} \cdot \hat{\mathbf{R}}). \quad (182)$$

We can now write down the final expressions for $\mathcal{G}_{a2}$ and $\mathcal{G}_{b2}$ in (160) and (161). We collect the formulas (160), (161), (163), (173), (180), and (181) together and we find that

$$\mathcal{G}_{a2} = -(\lambda + 1) \sum_{j'=1}^{n} \frac{1}{r_{aj'}^{\lambda+1}} \left(\sum_{j=1}^{n_a} \frac{1}{r_{aj}^{\lambda+3}} \mathfrak{M}_{abjj'} + \sum_{j=1}^{n_b} \frac{1}{r_{aj}^{\lambda+3}} \mathscr{P}_{abjj'} \right) \quad (183)$$

and

$$\mathcal{G}_{b2} = -(\lambda + 1) \sum_{j'=1}^{n} \frac{1}{r_{bj'}^{\lambda+1}} \left(\sum_{j=1}^{n_b} \frac{1}{r_{bj}^{\lambda+3}} \mathfrak{M}_{bajj'} + \sum_{j=1}^{n_a} \frac{1}{r_{bj}^{\lambda+3}} \mathscr{P}_{bajj'} \right), \quad (184)$$

where $\mathfrak{N}_{abjj'}$ and $\mathscr{P}_{abjj'}$ are defined by

$$\mathfrak{N}_{abjj'} \equiv \frac{z_{aj}}{r_{aj}} P_\lambda(\hat{\mathbf{f}}_{aj} \cdot \hat{\mathbf{f}}_{aj'}) - (z_b - n_b) \frac{r_{aj}}{R^2} \mathfrak{M}_{ajj'} \tag{185}$$

and

$$\mathscr{P}_{abjj'} \equiv z_{bj} \frac{r_{aj}^2}{r_{bj}^3} P_\lambda(\hat{\mathbf{f}}_{aj} \cdot \hat{\mathbf{f}}_{aj'}) + \left[(z_a - n_a) \frac{1}{R^2} - z_{bj} \frac{R}{r_{bj}^3} \right] r_{aj} \mathfrak{M}_{ajj'} . \tag{186}$$

In order to get the final expressions for $\bar{V}_2^{(2)}(a, a)$ and $\bar{V}_2^{(2)}(b, b)$, we need to evaluate the second terms on the right-hand side of (152) and (153). We write

$$\frac{1}{R^{\lambda+1}} (H - E)^2 T_a^*(\lambda, 0) = (\mathcal{K}_{a1} + \mathcal{K}_{a2})\varphi \tag{187}$$

and

$$\frac{1}{R^{\lambda+1}} (H - E)^2 T_b^*(\lambda, 0) = (\mathcal{K}_{b1} + \mathcal{K}_{b2})\varphi, \tag{188}$$

where

$$\mathcal{K}_{a1} \equiv \frac{1}{R^{\lambda+1}} \sum_{j=1}^n \nabla_j \nabla_j \left[\frac{1}{r_{aj}^{\lambda+1}} P_\lambda(\hat{\mathbf{f}}_{aj} \cdot \hat{\mathbf{R}}) \right] : \nabla_j \nabla_j , \tag{189}$$

$$\mathcal{K}_{b1} \equiv \frac{1}{R^{\lambda+1}} \sum_{j=1}^n \nabla_j \nabla_j \left[\frac{1}{r_{bj}^{\lambda+1}} P_\lambda(\hat{\mathbf{f}}_{bj} \cdot \hat{\mathbf{R}}) \right] : \nabla_j \nabla_j , \tag{190}$$

$$\mathcal{K}_{a2} \equiv \frac{1}{R^{\lambda+1}} \sum_{j=1}^n \nabla_j \left[\frac{1}{r_{aj}^{\lambda+1}} P_\lambda(\hat{\mathbf{f}}_{aj} \cdot \hat{\mathbf{R}}) \right] \cdot \nabla_j(U + V), \tag{191}$$

and

$$\mathcal{K}_{b2} \equiv \frac{1}{R^{\lambda+1}} \sum_{j=1}^n \nabla_j \left[\frac{1}{r_{bj}^{\lambda+1}} P_\lambda(\hat{\mathbf{f}}_{bj} \cdot \hat{\mathbf{R}}) \right] \cdot \nabla_j(U + V) \tag{192}$$

by analogy with (158)–(161). We have that

$$\mathcal{K}_{a1} \varphi \simeq -2(\lambda + 1)(\lambda + 2) \frac{1}{R^{\lambda+1}}$$

$$\times \left[\sum_{j=1}^{n_a} \frac{1}{r_{aj}^{\lambda+3}} P_\lambda(\hat{\mathbf{f}}_{aj} \cdot \hat{\mathbf{R}})\mathcal{L}_{aj} + \sum_{j=1}^{n_b} \frac{1}{r_{aj}^{\lambda+3}} P_\lambda(\hat{\mathbf{f}}_{aj} \cdot \hat{\mathbf{R}})\mathcal{L}_{bj} \right] \varphi, \tag{193}$$

and

$$\mathcal{K}_{b1}\varphi \simeq -2(\lambda + 1)(\lambda + 2)\frac{1}{R^{\lambda+1}}$$

$$\times\left[\sum_{j=1}^{n_b}\frac{1}{r_{bj}^{\lambda+3}}P_\lambda(\hat{\mathbf{f}}_{bj}\cdot\hat{\mathbf{R}})\mathcal{L}_{bj} + \sum_{j=1}^{n_a}\frac{1}{r_{bj}^{\lambda+3}}P_\lambda(\hat{\mathbf{f}}_{bj}\cdot\hat{\mathbf{R}})\mathcal{L}_{aj}\right]\varphi, \quad (194)$$

and Eqs. (193) and (194) can be obtained directly from (176) and (177) by replacing $\mathbf{r}_a$ and $\mathbf{r}_{bj'}$ by $\mathbf{R}$ and dropping the summation involving j'. Similarly, by analogy with (183) and (184), $\mathcal{K}_{a2}$ and $\mathcal{K}_{b2}$ are given by

$$\mathcal{K}_{a2} = -(\lambda + 1)\frac{1}{R^{\lambda+1}}\left(\sum_{j=1}^{n_a}\frac{1}{r_{aj}^{\lambda+3}}\mathfrak{N}_{abj}^0 + \sum_{j=1}^{n_b}\frac{1}{r_{aj}^{\lambda+3}}\mathcal{P}_{abj}^0\right) \quad (195)$$

and

$$\mathcal{K}_{b2} = -(\lambda + 1)\frac{1}{R^{\lambda+1}}\left(\sum_{j=1}^{n_b}\frac{1}{r_{bj}^{\lambda+3}}\mathfrak{N}_{baj}^0 + \sum_{j=1}^{n_a}\frac{1}{r_{bj}^{\lambda+3}}\mathcal{P}_{baj}^0\right), \quad (196)$$

where $\mathfrak{N}_{abj}^0$ and $\mathcal{P}_{abj}^0$ are defined by

$$\mathfrak{N}_{abj}^0 \equiv \frac{z_{aj}}{r_{aj}}P_\lambda(\hat{\mathbf{f}}_{aj}\cdot\hat{\mathbf{R}}) - (z_b - n_b)\frac{r_{aj}}{R^2}\mathfrak{N}_{aj}^0 \quad (197)$$

and

$$\mathcal{P}_{abj}^0 \equiv z_{bj}\frac{r_{aj}^2}{r_{bj}^3}P_\lambda(\hat{\mathbf{f}}_{aj}\cdot\hat{\mathbf{R}}) + \left[(z_a - n_a)\frac{1}{R^2} - z_{bj}\frac{R}{r_{bj}^3}\right]r_{aj}\mathfrak{N}_{aj}^0. \quad (198)$$

The quantities $\mathfrak{N}_{aj}^0$ and $\mathfrak{N}_{bj}^0$ are given by replacing $\hat{\mathbf{f}}_{aj'}$ and $\hat{\mathbf{f}}_{bj'}$ in (180) and (181) by $\hat{\mathbf{R}}$, and so we have that

$$\mathfrak{N}_{aj}^0 = P_{\lambda+1}(\hat{\mathbf{f}}_{aj}\cdot\hat{\mathbf{R}}) \quad \text{and} \quad \mathfrak{N}_{bj}^0 = -P_{\lambda+1}(\hat{\mathbf{f}}_{bj}\cdot\hat{\mathbf{R}}). \quad (199)$$

Thus on combining the results of (176), (177), (183), (184), and (193)–(196), we obtain

$$\bar{V}_2^{(2)}(a,a) = 4\sum_{\lambda=1}^{\infty}(\lambda + 1)\gamma_a(\lambda)\left[\sum_{j'=1}^{n}\frac{1}{r_{aj'}^{\lambda+1}}\left(\sum_{j=1}^{n_a}\frac{1}{r_{aj}^{\lambda+3}}\mathcal{Q}_{abjj'} + \sum_{j=1}^{n_b}\frac{1}{r_{aj}^{\lambda+3}}\mathcal{R}_{abjj'}\right)\right.$$

$$\left. - \frac{z_b}{R^{\lambda+1}}\left(\sum_{j=1}^{n_a}\frac{1}{r_{aj}^{\lambda+3}}\mathcal{Q}_{abj}^0 + \sum_{j=1}^{n_b}\frac{1}{r_{aj}^{\lambda+3}}\mathcal{R}_{abj}^0\right)\right]$$

$$(200)$$

and

$$\overline{V}_2^{(2)}(b,b) = 4 \sum_{\lambda=1}^{\infty} (\lambda + 1)\gamma_b(\lambda)$$

$$\times \left[\sum_{j'=1}^{n} \frac{1}{r_{bj'}^{\lambda+1}} \left(\sum_{j=1}^{n_b} \frac{1}{r_{bj}^{\lambda+3}} \mathcal{Q}_{bajj'} + \sum_{j=1}^{n_a} \frac{1}{r_{bj}^{\lambda+3}} \mathcal{R}_{bajj'} \right) \right.$$

$$\left. - \frac{z_a(-1)^{\lambda}}{R^{\lambda+1}} \left(\sum_{j=1}^{n_b} \frac{1}{r_{bj}^{\lambda+3}} \mathcal{Q}_{baj}^0 + \sum_{j=1}^{n_a} \frac{1}{r_{bj}^{\lambda+3}} \mathcal{R}_{baj}^0 \right) \right],$$

(201)

where

$$\mathcal{Q}_{abjj'} = (\lambda + 2)P_{\lambda}(\hat{\mathbf{r}}_{aj} \cdot \hat{\mathbf{r}}_{aj'})\mathcal{L}_{aj} + \tfrac{1}{2}\mathcal{N}_{abjj'},$$

(202)

$$\mathcal{R}_{abjj'} = (\lambda + 2)P_{\lambda}(\hat{\mathbf{r}}_{aj} \cdot \hat{\mathbf{r}}_{aj'})\mathcal{L}_{bj} + \tfrac{1}{2}\mathcal{P}_{abjj'},$$

(203)

$$\mathcal{Q}_{abj}^0 = (\lambda + 2)P_{\lambda}(\hat{\mathbf{r}}_{aj} \cdot \hat{\mathbf{R}})\mathcal{L}_{aj} + \tfrac{1}{2}\mathcal{N}_{abj}^0,$$

(204)

and

$$\mathcal{R}_{abj}^0 = (\lambda + 2)P_{\lambda}(\hat{\mathbf{r}}_{aj} \cdot \hat{\mathbf{R}})\mathcal{L}_{bj} + \tfrac{1}{2}\mathcal{P}_{abj}^0.$$

(205)

If we compare the results (200)–(205) with (92), it can be seen that it is the first term in $\mathcal{Q}^{abjj'}$ and $\mathcal{Q}_{bajj'}$ that gives the purely atomic contribution to $\overline{V}_2^{(2)}(a,a)$ and $\overline{V}_2^{(2)}(b,b)$.

In the case where there is only one valence electron, these expressions simplify considerably and also become exact. If in the limit of infinite separation, the electron is on core A, then $n = n_a = 1$ and $n_b = 0$, and so (200) and (201) reduce to

$$\overline{V}_2^{(2)}(a,a) = 4 \sum_{\lambda=1}^{\infty} (\lambda + 1)\gamma_a(\lambda)\left(\frac{\mathcal{Q}_{ab11}^0}{r_a^{2\lambda+4}} - \frac{z_b \mathcal{Q}_{ab1}^0}{R^{\lambda+1}r_a^{\lambda+3}} \right)$$

(206)

and

$$\overline{V}_2^{(2)}(b,b) = 4 \sum_{\lambda=1}^{\infty} (\lambda + 1)\gamma_b(\lambda)\left(\frac{\mathcal{R}_{ba11}^0}{r_b^{2\lambda+4}} - \frac{z_a(-1)^{\lambda}\mathcal{R}_{ba1}^0}{R^{\lambda+1}r_b^{\lambda+3}} \right);$$

(207)

compare Eq. (82), where the subscript 1 has been dropped so that $r_{a1} = r_a$ and $r_{b1} = r_b$.

4.5. Nonadiabatic Corrections of Higher Order

In this section, we briefly examine the form of the potentials $V_k^{(2)}$, $k = 3, 4, \ldots$. From Eq. (35), we have that

$$V_k^{(2)} = V_k^{(2)}(a, a) + V_k^{(2)}(b, b), \qquad (208)$$

where

$$V_k^{(2)}(a, a) = (-1)^{k-1} \sum_{c \neq 0} \frac{\langle \varphi_{a0} | (\mathcal{V}_a + \mathcal{U}_a) | \varphi_{ac} \rangle}{(E_{ac} - E_{a0})^{k+1}} (H - E)^k \langle \varphi_{ac} | \mathcal{V}_a | \varphi_{a0} \rangle$$

$$\qquad (209)$$

and

$$V_k^{(2)}(b, b) = (-1)^{k-1} \sum_{c \neq 0} \frac{\langle \varphi_{b0} | (\mathcal{V}_b + \mathcal{U}_b) | \varphi_{bc} \rangle}{(E_{bc} - E_{b0})^{k+1}} (H - E)^k \langle \varphi_{bc} | \mathcal{V}_b | \varphi_{b0} \rangle,$$

$$\qquad (210)$$

[see Eqs. (149) and (150)]. The corresponding effective potential $\overline{V}_k^{(2)}$ defined in (99) can be obtained by evaluating the matrix elements

$$\langle \varphi | V_k^{(2)}(a, a) | \varphi \rangle \equiv \langle \varphi | \overline{V}_k^{(2)}(a, a) | \varphi \rangle,$$

$$\langle \varphi | V_k^{(2)}(b, b) | \varphi \rangle \equiv \langle \varphi | \overline{V}_k^{(2)}(b, b) | \varphi \rangle, \qquad (211)$$

where

$$\langle \varphi | V_k^{(2)}(a, a) | \varphi \rangle = (-2)^{k-1} \sum_{\lambda=1}^{\infty} \delta_{ak}(\lambda)$$

$$\times \left[\sum_{\mu} \langle \varphi | T_a(\lambda, \mu)(H - E)^k T_a^*(\lambda, \mu) | \varphi \rangle \right.$$

$$\left. - \frac{z_b}{R^{\lambda+1}} \langle \varphi | (H - E)^k T_a^*(\lambda, 0) | \varphi \rangle \right] \qquad (212)$$

and

$$\langle \varphi | V_k^{(2)}(b,b) | \varphi \rangle = (-2)^{k-1} \sum_{\lambda=1}^{\infty} \delta_{bk}(\lambda)$$

$$\times \left[\sum_{\mu} \langle \varphi | T_b(\lambda, \mu)(H - E)^k T_b^*(\lambda, \mu) | \varphi \rangle \right.$$

$$\left. - \frac{(-1)^{\lambda} z_a}{R^{\lambda+1}} \langle \varphi | (H - E)^k T_b^*(\lambda, 0) | \varphi \rangle \right] \quad (213)$$

[see Eqs. (152) and (153)]. The coefficients $\delta_{ak}(\lambda)$ and $\delta_{bk}(\lambda)$ are given by

$$\delta_{ak}(\lambda) = 2^{1-k} \sum_{\Gamma_{ac}} \frac{1}{2\lambda+1} \frac{|\langle \Gamma_{ac} \lambda \| S_a(\lambda) \| \Gamma_{a0} 0 \rangle|^2}{(E_{ac} - E_{a0})^{k+1}} \quad (214)$$

and

$$\delta_{bk}(\lambda) = 2^{1-k} \sum_{\Gamma_{bc}} \frac{1}{2\lambda+1} \frac{|\langle \Gamma_{bc} \lambda \| S_b(\lambda) \| \Gamma_{b0} 0 \rangle|^2}{(E_{bc} - E_{b0})^{k+1}}, \quad (215)$$

and they are identical with $\alpha_a(\lambda)$, $\beta_a(\lambda)$, $\gamma_a(\lambda)$ and $\alpha_b(\lambda)$, $\beta_b(\lambda)$, $\gamma_b(\lambda)$, respectively for $k = 0$, 1, and 2. The matrix elements (212) and (213) can be analyzed in a similar way to that described in Section 3.5; this involves only retaining the energy-dependent terms from (202)–(205) in (200) and (201).

It is then easily shown that

$$\overline{V}_3^{(2)}(a,a) \simeq 4 \sum_{\lambda=1}^{\infty} \delta_{a3}(\lambda)\lambda(\lambda+1)(\lambda+2)(\lambda+3)$$

$$\times \left[\sum_{j=1}^{n_a} \frac{E_{aj}}{r_{aj}^{2\lambda+6}} + \sum_{j=1}^{n_b} \frac{E_{bj}}{r_{aj}^{2\lambda+6}} + O\left(\frac{1}{r_{aj}^{2\lambda+7}} \right) \right] \quad (216)$$

and

$$\overline{V}_3^{(2)}(b,b) \simeq 4 \sum_{\lambda=1}^{\infty} \delta_{b3}(\lambda)\lambda(\lambda+1)(\lambda+2)(\lambda+3)$$

$$\times \left[\sum_{j=1}^{n_b} \frac{E_{bj}}{r_{bj}^{2\lambda+6}} + \sum_{j=1}^{n_a} \frac{E_{aj}}{r_{bj}^{2\lambda+6}} + O\left(\frac{1}{r_{bj}^{2\lambda+7}} \right) \right] \quad (217)$$

[see Eq. (100)]. Clearly if $k = 4$, the effective potentials have the forms

$$\bar{V}_4^{(2)}(a,a) = \sum_{\lambda=1}^{\infty} \delta_{a4}(\lambda) \left(\sum_{j'=1}^{n} \frac{1}{r_{aj'}^{\lambda+1}} - \frac{z_b}{R^{\lambda+1}} \right) \sum_{j=1}^{n} O\left(\frac{1}{r_{aj}^{\lambda+5}} \right) \quad (218)$$

and

$$\bar{V}_4^{(2)}(b,b) = \sum_{\lambda=1}^{\infty} \delta_{b4}(\lambda) \left[\sum_{j'=1}^{n} \frac{1}{r_{bj'}^{\lambda+1}} - \frac{(-1)^{\lambda} z_a}{R^{\lambda+1}} \right] \sum_{j=1}^{n} O\left(\frac{1}{r_{bj}^{\lambda+5}} \right). \quad (219)$$

In general, it may be conjectured that

$$\bar{V}_k^{(2)}(a,a) = \sum_{\lambda=1}^{\infty} \delta_{ak}(\lambda) \left(\sum_{j'=1}^{n} \frac{1}{r_{aj'}^{\lambda+1}} - \frac{z_b}{R^{\lambda+1}} \right) \sum_{j=1}^{n} O\left(\frac{1}{r_{aj}^{\lambda+k+1}} \right) \quad (220)$$

and

$$\bar{V}_k^{(2)}(b,b) = \sum_{\lambda=1}^{\infty} \delta_{bk}(\lambda) \left[\sum_{j'=1}^{n} \frac{1}{r_{bj'}^{\lambda+1}} - \frac{(-1)^{\lambda} z_a}{R^{\lambda+1}} \right] \sum_{j=1}^{n} O\left(\frac{1}{r_{bj}^{\lambda+k+1}} \right) \quad (221)$$

if k is even [see Eq. (102)], and also that

$$\bar{V}_k^{(2)}(a,a) = \sum_{\lambda=1}^{\infty} \delta_{ak}(\lambda) \sum_{j=1}^{n} O\left(\frac{1}{r_{aj}^{2\lambda+k+3}} \right) \quad (222)$$

and

$$\bar{V}_k^{(2)}(b,b) = \sum_{\lambda=1}^{\infty} \delta_{bk}(\lambda) \sum_{j=1}^{n} O\left(\frac{1}{r_{bj}^{2\lambda+k+3}} \right) \quad (223)$$

if k is odd [see Eq. (103)].

4.6. Third-Order Perturbation Theory: The Static Contribution

The static contribution to V from the third-order term in Eq. (33) is given by

$$V_0^{(3)} = V_{00}^{(3)}(a,a) + V_{00}^{(3)}(b,b) + V_{00}^{(3)}(a,b), \quad (224)$$

where

$$V_{00}^{(3)}(a,a) = W_{a1} + W_{a2} \tag{225}$$

and

$$V_{00}^{(3)}(b,b) = W_{b1} + W_{b2} \tag{226}$$

[see Eq. (105)]. The quantities W_{a1}, W_{b1}, W_{a2}, and W_{b2} are defined by

$$W_{a1} \equiv \sum_{c \neq 0} \sum_{c' \neq 0} \frac{\langle \varphi_{a0} | (\mathcal{V}_a + \mathcal{U}_a) | \varphi_{ac} \rangle \langle \varphi_{ac} | (\mathcal{V}_a + \mathcal{U}_a) | \varphi_{ac'} \rangle \langle \varphi_{ac'} | (\mathcal{V}_a + \mathcal{U}_a) | \varphi_{a0} \rangle}{(E_{ac} - E_{a0})(E_{ac'} - E_{a0})},$$

$$\tag{227}$$

$$W_{b1} \equiv \sum_{c \neq 0} \sum_{c' \neq 0} \frac{\langle \varphi_{b0} | (\mathcal{V}_b + \mathcal{U}_b) | \varphi_{bc} \rangle \langle \varphi_{bc} | (\mathcal{V}_b + \mathcal{U}_b) | \varphi_{bc'} \rangle \langle \varphi_{bc'} | (\mathcal{V}_b + \mathcal{U}_b) | \varphi_{b0} \rangle}{(E_{bc} - E_{b0})(E_{bc'} - E_{b0})},$$

$$\tag{228}$$

$$W_{a2} \equiv - \sum_{c \neq 0} \frac{\langle \varphi_{a0} | (\mathcal{V}_a + \mathcal{U}_a) | \varphi_{ac} \rangle \langle \varphi_{ac} | (\mathcal{V}_a + \mathcal{U}_a) | \varphi_{a0} \rangle}{(E_{ac} - E_{a0})^2} V \tag{229}$$

and

$$W_{b2} \equiv - \sum_{c \neq 0} \frac{\langle \varphi_{b0} | (\mathcal{V}_b + \mathcal{U}_b) | \varphi_{bc} \rangle \langle \varphi_{bc} | (\mathcal{V}_b + \mathcal{U}_b) | \varphi_{b0} \rangle}{(E_{bc} - E_{b0})^2} V \tag{230}$$

[see Eqs. (106) and (107)]. The expression for $V_{00}^{(3)}(a,b)$ is very lengthy and is made up of several different types of terms, all of which may be written in the form

$$\mathcal{Y}(1,2,3)$$

$$\equiv \sum_{cd \neq 00} \sum_{c'd' \neq 00} \frac{\langle \varphi_{a0} \varphi_{b0} | \mathcal{U}_1 | \varphi_{ac} \varphi_{bd} \rangle \langle \varphi_{ac} \varphi_{bd} | \mathcal{U}_2 | \varphi_{ac'} \varphi_{bd'} \rangle \langle \varphi_{ac'} \varphi_{bd'} | \mathcal{U}_3 | \varphi_{a0} \varphi_{b0} \rangle}{(E_{ac} + E_{bd} - E_{a0} - E_{b0})(E_{ac'} + E_{bd'} - E_{a0} - E_{b0})},$$

$$\tag{231}$$

where $\mathcal{U}_1$, $\mathcal{U}_2$, and $\mathcal{U}_3$ can be equal to the potentials $(\mathcal{V}_a + \mathcal{U}_a)$, $(\mathcal{V}_b +$

TABLE 1

Nonzero Terms $\mathcal{U}(1, 2, 3)^a$

| | Indices[b] | |
1	2	3
a	b	ab
a	ab	b
b	a	ab
b	ab	a
ab	a	b
ab	b	a
a	ab	ab
b	ab	ab
ab	a	ab
ab	b	ab
ab	ab	a
ab	ab	b
ab	ab	ab

[a] See Eq. (231).
[b] For example, if $(1, 2, 3) = (a, b, ab)$, the potentials $\mathcal{U}_1$, $\mathcal{U}_2$, and $\mathcal{U}_3$ in Eq. (231) are equal to $(\mathcal{V}_a + \mathcal{W}_a)$, $(\mathcal{V}_b + \mathcal{W}_b)$, and $\mathcal{W}_{ab}$, respectively.

$\mathcal{W}_b$), and $\mathcal{W}_{ab}$ in any permutation. In Table 1 we list all the terms that give a nonzero contribution to $V_{00}^{(3)}(a, b)$.

Expressions for W_{a1} and W_{b1} can be written down immediately by using an analysis similar to that of Section 3.6. If, as before, we neglect all terms that involve four distinct bodies, i.e., three electrons and a core or two electrons and two cores, it can be shown that

$$W_{a1} = \sum_{\lambda=1}^{\infty} \sum_{\lambda'=1}^{\infty} \sum_{\lambda''=1}^{\infty} \varepsilon_a(\lambda, \lambda', \lambda'')$$

$$\times \left[3 \sum_{j=1}^{n} \sum_{j'=1}^{n}{}' \frac{1}{r_{aj'}^{\lambda'+1}} \frac{1}{r_{aj}^{\lambda+\lambda''+2}} P_{\lambda'}(\hat{\mathbf{f}}_{aj} \cdot \hat{\mathbf{f}}_{aj'}) \right.$$

$$+ \frac{3z_b^2}{R^{\lambda+\lambda''+2}} \sum_{j=1}^{n} \frac{1}{r_{aj}^{\lambda'+1}} P_{\lambda'}(\hat{\mathbf{f}}_{aj} \cdot \hat{\mathbf{R}}) + \sum_{j=1}^{n} \frac{1}{r_{aj}^{\lambda+\lambda'+\lambda''+3}}$$

$$\left. - \frac{3z_b}{R^{\lambda'+1}} \sum_{j=1}^{n} \frac{1}{r_{aj}^{\lambda+\lambda''+2}} P_{\lambda'}(\hat{\mathbf{f}}_{aj} \cdot \hat{\mathbf{R}}) - \frac{z_b^3}{R^{\lambda+\lambda'+\lambda''+3}} \right] \quad (232)$$

and

$$W_{b1} = \sum_{\lambda=1}^{\infty} \sum_{\lambda'=1}^{\infty} \sum_{\lambda''=1}^{\infty} \varepsilon_b(\lambda, \lambda', \lambda'')$$

$$\times \left[3 \sum_{j=1}^{n} \sum_{j'=1}^{n} \frac{1}{r_{bj'}^{\lambda'+1}} \frac{1}{r_{bj}^{\lambda+\lambda''+2}} P_{\lambda'}(\hat{\mathbf{f}}_{bj} \cdot \hat{\mathbf{f}}_{bj'}) \right.$$

$$+ \frac{3(-1)^{\lambda'} z_a^2}{R^{\lambda+\lambda''+2}} \sum_{j=1}^{n} \frac{1}{r_{bj}^{\lambda'+1}} P_{\lambda'}(\hat{\mathbf{f}}_{bj} \cdot \hat{\mathbf{R}}) + \sum_{j=1}^{n} \frac{1}{r_{bj}^{\lambda+\lambda'+\lambda''+3}}$$

$$\left. - \frac{3(-1)^{\lambda'} z_a}{R^{\lambda'+1}} \sum_{j=1}^{n} \frac{1}{r_{bj}^{\lambda+\lambda''+2}} P_{\lambda'}(\hat{\mathbf{f}}_{bj} \cdot \hat{\mathbf{R}}) - \frac{z_a^3}{R^{\lambda+\lambda'+\lambda''+3}} \right], \quad (233)$$

where

$$\varepsilon_a(\lambda, \lambda', \lambda'')$$

$$\equiv \sum_{\Gamma_{ac}} \sum_{\Gamma_{ac'}} \left[(2\lambda+1)(2\lambda''+1) \right]^{-1/2} \begin{pmatrix} \lambda & \lambda' & \lambda'' \\ 0 & 0 & 0 \end{pmatrix}$$

$$\times \frac{\langle \Gamma_{a0} 0 \| S_a(\lambda) \| \Gamma_{ac} \lambda \rangle \langle \Gamma_{ac} \lambda \| S_a(\lambda') \| \Gamma_{ac'} \lambda'' \rangle \langle \Gamma_{ac'} \lambda'' \| S_a(\lambda'') \| \Gamma_{a0} 0 \rangle}{(E_{ac} - E_{a0})(E_{ac'} - E_{a0})}$$

$$(234)$$

and

$$\varepsilon_b(\lambda, \lambda', \lambda'')$$

$$\equiv \sum_{\Gamma_{bc}} \sum_{\Gamma_{bc'}} \left[(2\lambda+1)(2\lambda''+1) \right]^{-1/2} \begin{pmatrix} \lambda & \lambda' & \lambda'' \\ 0 & 0 & 0 \end{pmatrix}$$

$$\times \frac{\langle \Gamma_{b0} 0 \| S_b(\lambda) \| \Gamma_{bc} \lambda \rangle \langle \Gamma_{bc} \lambda \| S_b(\lambda') \| \Gamma_{bc'} \lambda'' \rangle \langle \Gamma_{bc'} \lambda'' \| S_b(\lambda'') \| \Gamma_{b0} 0 \rangle}{(E_{bc} - E_{b0})(E_{bc'} - E_{b0})}$$

$$(235)$$

[see Eqs. (109)–(111)]. Also by comparing (229) and (230) with (129) and (130), it can be seen directly that

$$W_{a2} = - V \sum_{\lambda=1}^{\infty} \beta_a(\lambda) \left[\sum_{j=1}^{n} \sum_{j'=1}^{n} \frac{P_\lambda(\hat{\mathbf{f}}_{aj} \cdot \hat{\mathbf{f}}_{aj'})}{(r_{aj} r_{aj'})^{\lambda+1}} - 2z_b \sum_{j=1}^{n} \frac{P_\lambda(\hat{\mathbf{f}}_{aj} \cdot \hat{\mathbf{R}})}{(r_{aj} R)^{\lambda+1}} + \frac{z_b^2}{R^{2\lambda+2}} \right]$$

$$(236)$$

and

$$W_{b2} = -V \sum_{\lambda=1}^{\infty} \beta_b(\lambda) \left[\sum_{j=1}^{n} \sum_{j'=1}^{n} \frac{P_\lambda(\hat{\mathbf{r}}_{bj} \cdot \hat{\mathbf{r}}_{bj'})}{(r_{bj} r_{bj'})^{\lambda+1}} \right.$$

$$\left. - 2(-1)^\lambda z_a \sum_{j=1}^{n} \frac{P_\lambda(\hat{\mathbf{r}}_{bj} \cdot \hat{\mathbf{R}})}{(r_{bj} R)^{\lambda+1}} + \frac{z_a^2}{R^{2\lambda+2}} \right] \quad (237)$$

where V is given by (163), and we have used the results (132) and (133).

We shall not give algebraic expressions for all the individual terms $\mathcal{Y}(1,2,3)$ in Table 1, and indeed it is easily shown that the second and third blocks of terms are of higher order than those in the first block. Therefore, retaining only these first six terms, and carrying out the angular algebra by introducing the reduced matrix elements defined in (44), we obtain the result

$$V_{00}^{(3)}(a,b)$$

$$\simeq 2 \sum_{\lambda=1}^{\infty} \sum_{\lambda'=1}^{\infty} \frac{(\lambda + \lambda')!}{\lambda! \lambda'!} \beta_{ab}(\lambda, \lambda') \frac{1}{R^{\lambda+\lambda'+1}}$$

$$\times \left[\sum_\mu (-1)^{\lambda'} \frac{W(\lambda, \lambda', \mu)}{W(\lambda, \lambda', 0)} T_a^*(\lambda, \mu) T_b^*(\lambda', \mu) - \frac{(-1)^{\lambda'} z_b}{R^{\lambda+1}} T_b^*(\lambda', 0) \right.$$

$$\left. - \frac{z_a}{R^{\lambda'+1}} T_a(\lambda, 0) + \frac{z_a z_b}{R^{\lambda+\lambda'+2}} \right], \quad (238)$$

where

$$\beta_{ab}(\lambda, \lambda') \equiv \sum_{\Gamma_{ac}} \sum_{\Gamma_{bc'}} \frac{|\langle \Gamma_{ac} \lambda \| \mathbf{S}_a(\lambda) \| \Gamma_{a0} 0 \rangle|^2 |\langle \Gamma_{bc'} \lambda' \| \mathbf{S}_b(\lambda') \| \Gamma_{b0} 0 \rangle|^2}{(2\lambda + 1)(2\lambda' + 1)(\Delta E)^2} \quad (239)$$

[see Eq. (137)], and

$$(\Delta E)^{-2} \equiv (E_{ac} + E_{bc'} - E_{a0} - E_{b0})^{-1} \left[(E_{ac} - E_{a0})^{-1} + (E_{bc'} - E_{b0})^{-1} \right]$$

$$+ (E_{ac} - E_{a0})^{-1} (E_{bc'} - E_{b0})^{-1}. \quad (240)$$

The nonadiabatic corrections in third-order perturbation theory also give terms of higher order than the corresponding terms in (232), (233), (236), (237), and (238), and we do not attempt to consider them further.

4.7. Summary of Results and the Separated Atom Limit

We have considered a model of a general diatomic system, in which the two nuclei and the electrons in the inner shells are replaced by two spherically symmetric cores A and B with overall charges z_a and z_b. The outer or valence electrons are treated explicitly and are divided into two groups so that n_a electrons are associated with core A and n_b electrons with core B. In the limit of infinite separation of the cores, the n_a and n_b electrons may or may not be bound to cores A and B, respectively. Terms up to third order in the perturbation ($\mathcal{V} - V$) on the model system have been studied. We may collect up all contributions to V that are of order less than or equal to $1/r^7$, and as the complete expression is rather cumbersome, we shall not write it down explicitly, but instead details of each contribution are listed in Table 2. In the many-electron case, the result is approximate insofar as the energy-dependent terms in $\overline{V}_2^{(2)}(a,a)$ and $\overline{V}_2^{(2)}(b,b)$ cannot be uniquely determined, although some plausible assumptions can be made by considering the physical picture. We have also neglected contributions to $V_{00}^{(3)}$ that involve more than three distinct bodies.

In the one-electron case, however, the results become exact, and the expressions for the various contributions simplify considerably. If we take the case of a neutral core interacting with a neutral atom that has one valence electron, $z_a = 1$, $z_b = 0$, $n = n_a = 1$, and $n_b = 0$, and from the

TABLE 2
Model Potential V for a Diatomic System[a]

Contribution	Equation	λ	λ'	λ''
$V_0^{(2)}(a,a)$	(132)	1, 2	—	—
$V_0^{(2)}(b,b)$	(133)	1, 2	—	—
$V_0^{(2)}(a,b)$	(136)	1	1	—
$V_1^{(2)}(a,a)$	(146)	1	—	—
$V_1^{(2)}(b,b)$	(147)	1	—	—
$V_2^{(2)}(a,a)$	(200)	1	—	—
$V_2^{(2)}(b,b)$	(201)	1	—	—
$V_{00}^{(3)}(a,a) \simeq W_{a1}$	(232)	2	1	1
		1	2	1
		1	1	2
$V_{00}^{(3)}(b,b) \simeq W_{b1}$	(233)	2	1	1
		1	2	1
		1	1	2
$V_{00}^{(3)}(a,b)$	(238)	1	1	—

[a] All contributions to V are listed that are of order less than or equal to $1/r^7$.

equations listed in Table 2, we find that

$$V = V_a + V_b + V_{ab} + V_{\text{int}} \qquad (241)$$

say, where

$$V_a = -\frac{1}{2}\alpha_a(1)\frac{1}{r_a^4} + \left[-\frac{1}{2}\alpha_a(2) + 3\beta_a(1) + 24\gamma_a(1)E_a\right]\frac{1}{r_a^6}$$

$$+ \left[28\gamma_a(1) + 2\varepsilon_a(2,1,1) + \varepsilon_a(1,2,1)\right]\frac{1}{r_a^7} + O\left(\frac{1}{r_a^8}\right) \qquad (242)$$

and

$$V_b = -\frac{1}{2}\alpha_b(1)\frac{1}{r_b^4} + \left[-\frac{1}{2}\alpha_b(2) + 3\beta_b(1) + 24\gamma_b(1)E_a\right]\frac{1}{r_b^6}$$

$$+ \left[2\varepsilon_b(2,1,1) + \varepsilon_b(1,2,1)\right]\frac{1}{r_b^7} + O\left(\frac{1}{r_b^8}\right), \qquad (243)$$

[see Eq. (114)]. The term V_{ab} in (241) represents the pure core–core interaction and V_{int} is a three-body term; they are given by

$$V_{ab} = -\frac{1}{2}\alpha_b(1)\frac{1}{R^4} - \frac{1}{2}\left[\alpha_b(2) + 3\alpha_{ab}(1,1)\right]\frac{1}{R^6}$$

$$- \left[2\varepsilon_b(2,1,1) + \varepsilon_b(1,2,1)\right]\frac{1}{R^7} + O\left(\frac{1}{R^8}\right) \qquad (244)$$

and

$$V_{\text{int}} = -\alpha_b(1)\frac{P_1(\hat{\mathbf{r}}_b \cdot \hat{\mathbf{R}})}{r_b^2 R^2} + \alpha_b(2)\frac{P_2(\hat{\mathbf{r}}_b \cdot \hat{\mathbf{R}})}{r_b^3 R^3}$$

$$+ 4\gamma_b(1)\left[\frac{6}{r_a r_b^6} + \frac{1}{r_a^3 r_b^4} + \left(\frac{6}{r_a r_b^4 R^2} + \frac{1}{r_a^3 r_b^2 R^2} + \frac{R}{r_a^3 r_b^5} + \frac{6E_a}{r_b^4 R^2}\right)\right.$$

$$\left. \times P_1(\hat{\mathbf{r}}_b \cdot \hat{\mathbf{R}}) + \frac{P_2(\hat{\mathbf{r}}_b \cdot \hat{\mathbf{R}})}{r_a^3 r_b^3 R}\right]$$

$$+ 6\varepsilon_b(2,1,1)\left(\frac{1}{r_b^3} - \frac{1}{R^3}\right)\frac{P_1(\hat{\mathbf{r}}_b \cdot \hat{\mathbf{R}})}{r_b^2 R^2} - 3\varepsilon_b(1,2,1)\left(\frac{1}{r_b} - \frac{1}{R}\right)\frac{P_2(\hat{\mathbf{r}}_b \cdot \hat{\mathbf{R}})}{r_b^3 R^3}$$

$$- 4\beta_{ab}(1,1)\left[\frac{P_1(\hat{\mathbf{r}}_a \cdot \hat{\mathbf{r}}_b)}{r_b^2} + \frac{P_1(\hat{\mathbf{r}}_a \cdot \hat{\mathbf{R}})}{R^2}\right]\frac{1}{r_a^2 R^3} + O\left(\frac{1}{r^8}\right). \qquad (245)$$

When the cores are well separated and $r_a \ll r_b$, we may write

$$r_b = \left(r_a^2 - 2\mathbf{r}_a \cdot \mathbf{R} + R^2 \right)^{1/2}$$

$$\simeq R \left\{ 1 - \frac{r_a}{R} P_1(\hat{\mathbf{r}}_a \cdot \hat{\mathbf{R}}) + \frac{r_a^2}{3R^2} \left[1 - P_2(\hat{\mathbf{r}}_a \cdot \hat{\mathbf{R}}) \right] \right\}, \qquad (246)$$

and on using (243)–(246), it can be shown that the perturbation of the neutral atom A by the core B is given by

$$V_b + V_{ab} + V_{\mathrm{int}}$$

$$\simeq - \left\{ \alpha_b(1) \left[1 + P_2(\hat{\mathbf{r}}_a \cdot \hat{\mathbf{R}}) \right] r_a^2 - 3\beta_b(1) + 4\gamma_b(1) \left[1 + P_2(\hat{\mathbf{r}}_a \cdot \hat{\mathbf{R}}) \right] \frac{1}{r_a} \right.$$

$$\left. + \frac{3}{2} \alpha_{ab}(1,1) - 8\beta_{ab}(1,1) P_2(\hat{\mathbf{r}}_a \cdot \hat{\mathbf{R}}) \frac{1}{r_a} \right\} \frac{1}{R^6} + O\left(\frac{1}{R^7} \right). \quad (247)$$

If the unperturbed atom occupies a valence state labeled by v, whose wave function is denoted by φ_{av} and whose energy is E_{av}, the Schrödinger equation for the atom is

$$\left(-\frac{1}{2} \nabla_a^2 - \frac{1}{r_a} + V_a \right) \varphi_{av} = E_{av} \varphi_{av}, \qquad (248)$$

where V_a is given by (242) when r_a is greater than the radius of core A. The average change in the energy of the ground state of the atom produced by the presence of core B is then approximately

$$\langle \varphi_{a0} | V_b + V_{ab} + V_{\mathrm{int}} | \varphi_{a0} \rangle \simeq - \frac{C_6}{R^6}, \qquad (249)$$

where the ground state is denoted by $v = 0$. Note that the energy of this state, E_{a0}, can be identified with the energy E_a defined by (168). The van der Waals constant, C_6, is given by

$$C_6 = \alpha_b(1) \langle \varphi_{a0} | r_a^2 | \varphi_{a0} \rangle - 3\beta_b(1) + 4\gamma_b(1) \langle \varphi_{a0} | \frac{1}{r_a} | \varphi_{a0} \rangle + \frac{3}{2} \alpha_{ab}(1,1).$$

$$(250)$$

In writing down the result (249), we have assumed that an average has been carried out over the degenerate magnetic sublevels of the ground state, and thus the terms involving $P_2(\hat{\mathbf{r}}_a \cdot \hat{\mathbf{R}})$ in (247) do not contribute to C_6. The first term in (250) just involves the mean square radius of the state $v = 0$, and the third term can be evaluated by using the virial theorem. This may be written as

$$\langle \varphi_{a0} | \nabla_a^2 | \varphi_{a0} \rangle \simeq \langle \varphi_{a0} | - \frac{1}{r_a} | \varphi_{a0} \rangle, \tag{251}$$

and so (248) and (251) give

$$2E_{a0} = \langle \varphi_{a0} | - \frac{1}{r_a} + 2V_a | \varphi_{a0} \rangle \simeq \langle \varphi_{a0} | - \frac{1}{r_a} | \varphi_{a0} \rangle, \tag{252}$$

where the approximation made in (252) is consistent with the approximations made in Section 4.4 for the expansions of the potentials. Thus (250) and (252) give the final result for C_6, i.e.,

$$C_6 = \alpha_b(1) \langle \varphi_{a0} | r_a^2 | \varphi_{a0} \rangle - 3\beta_b(1) - 8\gamma_b(1)E_{a0} + \tfrac{3}{2}\alpha_{ab}(1,1). \tag{253}$$

As a check, we may compare (253) with the result for C_6 that is given by[23]

$$C_6 = \frac{3}{2} \sum_{v \neq 0} \sum_{c \neq 0} \frac{f_a(0,v)f_b(0,c)}{(E_{bc} - E_{b0})(E_{av} - E_{a0})(E_{bc} - E_{b0} + E_{av} - E_{a0})}, \tag{254}$$

where $f_a(0,v)$ is the oscillator strength for the transition $0 \to v$ of the valence electron in the neutral atom, and $f_b(0,c)$ is the oscillator strength for a transition $0 \to c$ in the core B. Expression (254) is almost exact; we have only neglected a small contribution from transitions within core A, and it is precisely this contribution that is given by the term $\tfrac{3}{2}\alpha_{ab}(1,1)$ in (253). The oscillator strengths in (254) may be written as

$$f_a(0,v) = \tfrac{2}{3}(E_{av} - E_{a0})|\langle \Gamma_{av} 1 \| S_a(1) \| \Gamma_{a0} 0 \rangle|^2 \tag{255}$$

and

$$f_b(0,c) = \tfrac{2}{3}(E_{bc} - E_{b0})|\langle \Gamma_{bc} 1 \| S_b(1) \| \Gamma_{b0} 0 \rangle|^2, \tag{256}$$

where $S_a(1)$ is the dipole operator for the valence electron and the quantum numbers of state v are specified by Γ_{av} and $\lambda = 1$. Equations (255) and (256) provide the link between the notation of (254) and that of (137). Expressions (135), (145) and (155) for $\alpha_b(1)$, $\beta_b(1)$, and $\gamma_b(1)$ can be rewritten as

$$\alpha_b(1) = \sum_{c \neq 0} \frac{f_b(0,c)}{(E_{bc} - E_{b0})^2} , \tag{257}$$

$$\beta_b(1) = 2 \sum_{c \neq 0} \frac{f_b(0,c)}{(E_{bc} - E_{b0})^3} \tag{258}$$

and

$$\gamma_b(1) = 4 \sum_{c \neq 0} \frac{f_b(0,c)}{(E_{bc} - E_{b0})^4} \tag{259}$$

by using (256). We shall now assume that $(E_{av} - E_{a0}) \ll (E_{bc} - E_{b0})$ in (254), so that

$$C_6 \simeq \frac{3}{2} \alpha_b(1) \sum_{v \neq 0} \frac{f_a(0,v)}{(E_{av} - E_{a0})} - 3\beta_b(1) \sum_{v \neq 0} f_a(0,v)$$

$$+ 6\gamma_b(1) \sum_{v \neq 0} (E_{av} - E_{a0}) f_a(0,v). \tag{260}$$

This is a very reasonable assumption, and indeed if it is not valid it also means that the internal structure of the core B cannot be neglected. Finally, we can make use of the sum rules for oscillator strengths first derived by Vinti[24] and later stated in the form given here by Dalgarno and Lynn.[25] For our case the rules state that

$$\sum_{v \neq 0} \frac{f_a(0,v)}{(E_{av} - E_{a0})} = \frac{2}{3} \langle \varphi_{a0} | r_a^2 | \varphi_{a0} \rangle, \tag{261}$$

$$\sum_{v \neq 0} f_a(0,v) = 1, \tag{262}$$

and

$$\sum_{v \neq 0} (E_{av} - E_{a0}) f_a(0,v) = -\frac{4}{3} E_{a0}, \tag{263}$$

and by substituting (261)–(263) into (260), we obtain complete agreement with (253).

5. Addition Theorems for Solid Harmonics

In this section, we prove some general results involving the solid harmonics $r^l Y_{lm}(\hat{r})$ that are required in Sections 3 and 4. We begin by considering the Taylor series expansion of $|\mathbf{R} - \mathbf{r}|^{-1}$, i.e.,

$$\frac{1}{|\mathbf{R} - \mathbf{r}|} = \sum_{l=0}^{\infty} \frac{(-1)^l}{l!} (\mathbf{r} \cdot \nabla_R)^l \frac{1}{R} = \exp(-\mathbf{r} \cdot \nabla_R) \frac{1}{R} \qquad (264)$$

formally, where ∇_R operates only on functions of R. If $R > r$, we may also expand $|\mathbf{R} - \mathbf{r}|^{-1}$ in terms of Legendre polynomials so that

$$\frac{1}{|\mathbf{R} - \mathbf{r}|} = \sum_{l=0}^{\infty} \frac{r^l}{R^{l+1}} P_l(\hat{r} \cdot \hat{R}), \qquad (265)$$

and on equating the corresponding terms in (264) and (265), we have that

$$(\mathbf{r} \cdot \nabla_R)^l \frac{1}{R} = \frac{l!}{(-1)^l} \frac{r^l}{R^{l+1}} P_l(\hat{r} \cdot \hat{R}). \qquad (266)$$

Also since from (266)

$$(\mathbf{r} \cdot \nabla_R)^{l+l'} \frac{1}{R} = \frac{(l+l')!}{(-1)^{l+l'}} \frac{r^{l+l'}}{R^{l+l'+1}} P_{l+l'}(\hat{r} \cdot \hat{R}), \qquad (267)$$

(266) and (267) may be combined to give

$$(\mathbf{r} \cdot \nabla_R)^{l'} \frac{1}{R^{l+1}} P_l(\hat{r} \cdot \hat{R}) = \frac{(l+l')!}{(-1)^{l'} l!} \frac{r^{l'}}{R^{l+l'+1}} P_{l+l'}(\hat{r} \cdot \hat{R}). \qquad (268)$$

On expanding the Legendre polynomials in (268) using the addition theorem for spherical harmonics, i.e.,

$$P_l(\hat{r} \cdot \hat{R}) = \frac{4\pi}{2l+1} \sum_m Y_{lm}^*(\hat{r}) Y_{lm}(\hat{R}), \qquad (269)$$

we obtain

$$\sum_m Y_{lm}(\hat{\mathbf{r}})(\mathbf{r} \cdot \nabla_R)^{l'} \frac{1}{R^{l+1}} Y_{lm}^*(\hat{\mathbf{R}})$$

$$= \frac{(l+l')!}{(-1)^{l'} l!} \frac{(2l+1)}{(2l+2l'+1)} \frac{r^{l'}}{R^{l+l'+1}} \sum_M Y_{l+l'M}^*(\hat{\mathbf{r}}) Y_{l+l'M}(\hat{\mathbf{R}}).$$

(270)

The spherical harmonic $Y_{l+l'M}^*(\hat{\mathbf{r}})$ can be expanded using the relation

$$Y_{LM}^*(\hat{\mathbf{r}}) = \left[\frac{4\pi(2L+1)}{(2l+1)(2l'+1)} \right]^{1/2} \left[\begin{pmatrix} l & l' & L \\ 0 & 0 & 0 \end{pmatrix} \right]^{-1}$$

$$\times \sum_{m,m'} \begin{pmatrix} l & l' & L \\ m & m' & M \end{pmatrix} Y_{lm}(\hat{\mathbf{r}}) Y_{l'm'}(\hat{\mathbf{r}})$$

(271)

with $L = l + l'$; this result can easily be deduced from the relation (179) given by Edmonds.[20] Then having substituted result (271) into (270), our key result may be obtained by equating the coefficients of $Y_{lm}(\hat{\mathbf{r}})$ on each side of the equation, i.e.,

$$(\mathbf{r} \cdot \nabla_R)^{l'} \frac{1}{R^{l+1}} Y_{lm}^*(\hat{\mathbf{R}})$$

$$= \frac{(l+l')!}{(-1)^{l'} l!} \left[\frac{4\pi(2l+1)}{(2l+2l'+1)(2l'+1)} \right]^{1/2} \left[\begin{pmatrix} l & l' & l+l' \\ 0 & 0 & 0 \end{pmatrix} \right]^{-1}$$

$$\times \sum_{M,m'} \begin{pmatrix} l & l' & l+l' \\ m & m' & M \end{pmatrix} \frac{1}{R^{l+l'+1}} Y_{l+l'M}(\hat{\mathbf{R}}) r^{l'} Y_{l'm'}(\hat{\mathbf{r}}).$$

(272)

Formulas (264) and (272) can now be used to obtain an expansion for quantities of the type $|\mathbf{R} - \mathbf{r}_1 + \mathbf{r}_2|^{-1}$, which is required in Section 4.1. By setting

$$\mathbf{r} = \mathbf{r}_1 - \mathbf{r}_2$$

(273)

in (264), we have that

$$\frac{1}{|\mathbf{R} - \mathbf{r}_1 + \mathbf{r}_2|} = \exp\left[-(\mathbf{r}_1 - \mathbf{r}_2) \cdot \nabla_R\right] \frac{1}{R}$$

$$= \sum_{l'=0}^{\infty} \frac{1}{l'!} (\mathbf{r}_2 \cdot \nabla_R)^{l'} \sum_{l=0}^{\infty} \frac{(-1)^l}{l!} (\mathbf{r}_1 \cdot \nabla_R)^l \frac{1}{R}, \quad (274)$$

and so by using (266), (269), and (272), we obtain

$$\frac{1}{|\mathbf{R} - \mathbf{r}_1 + \mathbf{r}_2|} = \sum_{l,l',m,m',M} (-1)^{l'} (4\pi)^{3/2} \frac{(l+l')!}{l!\, l'!}$$

$$\times \left[(2l+1)(2l'+1)(2l+2l'+1)\right]^{-1/2}$$

$$\times \left[\begin{pmatrix} l & l' & l+l' \\ 0 & 0 & 0 \end{pmatrix}\right]^{-1} \begin{pmatrix} l & l' & l+l' \\ m & m' & M \end{pmatrix}$$

$$\times \frac{r_1^l r_2^{l'}}{R^{l+l'+1}} Y_{lm}(\hat{\mathbf{r}}_1) Y_{l'm'}(\hat{\mathbf{r}}_2) Y_{l+l'M}(\hat{\mathbf{R}}). \quad (275)$$

Alternatively, (275) may be written as

$$\frac{1}{|\mathbf{R} - \mathbf{r}_1 + \mathbf{r}_2|} = \sum_{l,L,m,m',M} (-1)^{L-l} (4\pi)^{3/2} \frac{L!}{l!\,(L-l)!}$$

$$\times \left[(2l+1)(2L-2l+1)(2L+1)\right]^{-1/2}$$

$$\times \left[\begin{pmatrix} l & L-l & L \\ 0 & 0 & 0 \end{pmatrix}\right]^{-1} \begin{pmatrix} l & L-l & L \\ m & m' & M \end{pmatrix}$$

$$\times \frac{r_1^l r_2^{L-l}}{R^{L+1}} Y_{lm}(\hat{\mathbf{r}}_1) Y_{L-,m'}(\hat{\mathbf{r}}_2) Y_{LM}(\hat{\mathbf{R}}). \quad (276)$$

Finally, two interesting results can be extracted from (276). By using (265), (269), and (273) to expand $|\mathbf{R} - \mathbf{r}_1 + \mathbf{r}_2|^{-1}$ in terms of the spherical harmonics $Y_{LM}(\hat{\mathbf{r}})$ and $Y_{LM}(\hat{\mathbf{R}})$, the coefficients of $Y_{LM}(\hat{\mathbf{R}})$ on each side of (276) may then be equated to give

$$r^L Y_{LM}^*(\hat{\mathbf{r}}) = \sum_{l,m,m'} (-1)^{L-l} \frac{L!}{l!(L-l)!} \left[\frac{4\pi(2L+1)}{(2l+1)(2L-2l+1)} \right]^{1/2}$$

$$\times \left[\begin{pmatrix} l & L-l & L \\ 0 & 0 & 0 \end{pmatrix} \right]^{-1} \begin{pmatrix} l & L-l & L \\ m & m' & M \end{pmatrix} r_1^l r_2^{L-l} Y_{lm}(\hat{\mathbf{r}}_1) Y_{L-,m'}(\hat{\mathbf{r}}_2).$$

$$(277)$$

However, if we write

$$\mathbf{R}' = \mathbf{R} + \mathbf{r}_2, \tag{278}$$

then $|\mathbf{R} - \mathbf{r}_1 + \mathbf{r}_2|^{-1}$ can be expanded in terms of $Y_{lm}(\hat{\mathbf{r}}_1)$ and $Y_{lm}(\hat{\mathbf{R}}')$, and provided that $R' > r_1$, the coefficients of $Y_{lm}(\hat{\mathbf{r}}_1)$ on each side of (276) can also be equated so that

$$\frac{1}{R'^{l+1}} Y_{lm}^*(\hat{\mathbf{R}}') = \sum_{L,m',M} (-1)^{L-l} \frac{L!}{l!(L-l)!} \left[\frac{4\pi(2l+1)}{(2L+1)(2L-2l+1)} \right]^{1/2}$$

$$\times \left[\begin{pmatrix} l & L-l & L \\ 0 & 0 & 0 \end{pmatrix} \right]^{-1} \begin{pmatrix} l & L-l & L \\ m & m' & M \end{pmatrix}$$

$$\times \frac{r_2^{L-l}}{R^{L+1}} Y_{L-,m'}(\hat{\mathbf{r}}_2) Y_{LM}(\hat{\mathbf{R}}). \tag{279}$$

If we substitute explicit expressions for the $3-j$ symbols into (275) [see Edmonds[20]], we obtain the result for the expansion of $|\mathbf{R} - \mathbf{r}_1 + \mathbf{r}_2|^{-1}$ as given by Margenau and Kestner,[26] for example. Rose[27] was the first person to derive results (275) and (277) in essentially the same form, although the more compact proofs presented in this chapter reveal the presence of the factor

$$\left[\begin{pmatrix} l & l' & l+l' \\ 0 & 0 & 0 \end{pmatrix} \right]^{-1}$$

in (275), and this leads to a more pleasing symmetry in the results. It is interesting to note that the formulas (273)–(277) can be easily generalized to find the corresponding expansions of $|\mathbf{R} - \mathbf{r}_1 - \mathbf{r}_2 \cdots - \mathbf{r}_n|^{-1}$.

ACKNOWLEDGMENTS

I should like to thank Dr. D. W. Norcross for some helpful comments
on the original manuscript.

References

1. L. CASTILLEJO, I. C. PERCIVAL, and M. J. SEATON, On the theory of elastic collisions
    between electrons and hydrogen atoms, *Proc. R. Soc. Lond. Ser. A* **254**, 259–272 (1960).
2. M. J. SEATON and L. STEENMAN-CLARK, Effective potentials for electron-atom scattering
    below inelastic thresholds I. Asymptotic forms, *J. Phys. B* **10**, 2639–2647 (1977).
3. M. J. SEATON and L. STEENMAN-CLARK, Effective potentials for electron-atom scattering
    below inelastic thresholds II. The finite-range problem, *J. Phys. B* **11**, 293–307 (1978).
4. A. DALGARNO and R. W. MCCARROLL, Adiabatic coupling between electronic and
    nuclear motion in molecules, *Proc. R. Soc. Lond. Ser. A* **237**, 383–394 (1956).
5. M. H. MITTLEMAN and K. M. WATSON, Scattering of charged particles by neutral atoms,
    Phys. Rev. **113**, 198–211 (1959).
6. U. OPIK, A simple method for calculating van der Waals interactions, *Proc. Phys. Soc.*
    92, 573–576 (1967).
7. C. J. KLEINMAN, Y. HAHN, and L. SPRUCH, Dominant nonadiabatic contribution to the
    long-range electron-atom interaction, *Phys. Rev.* **165**, 53–62 (1968).
8. J. CALLAWAY, R. W. LABAHN, R. T. PU, and W. M. DUXLER, Extended polarisation
    potential: applications to atomic scattering, *Phys. Rev.* **168**, 12–21 (1968).
9. A. DALGARNO, G. W. F. DRAKE, and G. A. VICTOR, Nonadiabatic long-range forces,
    Phys. Rev. **176**, 194–197 (1968).
10. C. BOTTCHER and A. DALGARNO, A constructive model potential method for atomic
    interactions, *Proc. R. Soc. Lond. Ser. A* **340**, 187–198 (1974).
11. G. PEACH, Low-energy scattering of excited helium atoms by rare gases I: The model
    potentials, *J. Phys. B* **11**, 2107–2131 (1978).
12. P. VALIRON, R. GAYET, R. MCCARROLL, F. MASNOU-SEEUWS, and M. PHILLIPE, Model-
    potential methods for the calculation of atom–rare-gas interactions: application to the
    H–He system, *J. Phys. B* **12**, 53–67 (1979).
13. S. W. WANG, H. S. TAYLOR, and R. YARIS, An *ab initio* effective potential for two
    particles in the field of the core: a reduction of the $(N + 2)$-electron problem to the
    two-electron problem, *Chem. Phys.* **14**, 53–71 (1976).
14. B. J. LAURENZI, Atomic and molecular model potentials, *J. Chem. Phys.* **69**, 4838–4850
    (1978).
15. S. WATANABE and C. H. GREENE, Atomic polarisability in negative-ion photodetachment,
    Phys. Rev. A **22**, 158–169 (1980).
16. J. N. BARDSLEY, in *Case Studies in Atomic Physics*, Vol. 4, E. W. McDaniel and M. R. C.
    McDowell, Eds., North-Holland, Amsterdam, 1974, pp. 299–368.
17. R. N. DIXON and I. L. ROBERTSON, in *Specialist Periodical Reports, Theoretical Chemistry*,
    Vol. 3, R. N. Dixon and C. Thomson, Eds., The Chemical Society, London, 1978, pp.
    100–134.
18. G. PEACH, Model potentials or pseudopotentials; whats the difference? *Comments At.
    Mol. Phys.* **11**, 101–118 (1982).
19. H. FESHBACH, A unified theory of nuclear reactions. II, *Ann. Phys. N.Y.* **19**, 287–313
    (1962).
20. A. R. EDMONDS, *Angular Momentum in Quantum Mechanics*, 2nd ed., Princeton Univer-
    sity Press, Princeton, N.J., 1960.
21. C. E. WEATHERBURN, *Advanced Vector Analysis*, Bell, London, 1951.
22. R. J. DRACHMAN, Asymptotic effective potentials for electron–hydrogen scattering, *J.
    Phys. B.* **11**, L699–702 (1979).

23. A. DALGARNO and W. D. DAVISON, The calculation of van der Waals interactions, *Adv. At. Mol. Phys.* **2**, 1–32 (1966).
24. J. P. VINTI, Sum rules for atomic transition probabilities, *Phys. Rev.* **41**, 432–442 (1932).
25. A. DALGARNO and N. LYNN, Properties of the helium atom, *Proc. Phys. Soc. Lond.* **70A**, 802–808 (1957).
26. H. MARGENAU and N. R. KESTNER, *Theory of Intermolecular Forces*, 2nd ed., Pergamon, Oxford, 1971.
27. M. E. ROSE, The electrostatic interaction of two arbitrary charge distributions, *J. Math. & Phys.* **37**, 215–222 (1958).

6

APPLICATIONS OF QUANTUM DEFECT THEORY

DAVID L. MOORES AND HANNELORE E. SARAPH

Introduction

We follow the evolution of quantum defect theory and its applications from the 1960's to the present. The merits and practical usefulness of quantum defect methods are discussed. On most topics, we only give a general discussion and refer the reader to the literature, but some illustrative examples are given in great detail.

The emphasis of this chapter is on work inspired by the developments at University College London; no attempt is made to present a comprehensive review.

Part 1 outlines the historical development; Part 2 quotes those mathematical formulas that are required for this chapter; Part 3 discusses the independent particle approximation in connection with single channel quantum defect theory; Part 4 presents some detailed examples of the use of multichannel quantum defect methods; Part 5 discusses extrapolation techniques based on quantum defect theory and their application to the calculation of cross sections.

1. Historical Survey

Quantum defect theory (QDT) is concerned with the relationship between the bound and continuum states of atomic systems composed of an electron moving in the field of a positive atomic or molecular ion. The theory is applied to problems in atomic structure and in electron–ion scattering, including radiative transition probabilities and dielectronic recombination.

DAVID L. MOORES and HANNELORE E. SARAPH ● University College London, Gower Street, London WC1 E6BT, England.

The concept of the quantum defect originates in the work of Rydberg in the last century on the analysis of alkali spectra. Observations of hydrogenic one-electron spectra had revealed series of lines, the wavelengths λ_n of which were given by

$$\lambda_n^{-1} = \left[(-E_0 - z^2)/n^2 \right] R, \tag{1}$$

where z is the residual charge on the ion and R is the Rydberg constant. The Bohr theory showed that the integer n, now known as the principal quantum number, was of fundamental importance in the theoretical description of an ion, and that E_0 was the energy (in rydbergs) of the lower level. For the spectra of the alkalis, Rydberg found empirically that a similar expression could be used:

$$\lambda_n^{-1} = \left[(-E_0 - z^2)/n^{*2} \right] R. \tag{2}$$

Here, $n^* = n - \mu_n(l)$ is the effective quantum number (also denoted by v) and is not in general equal to an integer. The quantity μ is called the quantum defect, it depends strongly on l but varies slowly with n. This latter property made μ an attractive parameter for spectral analysis. It was soon discovered that most spectra could be described quite well by the Rydberg formula (2), at least for the higher members of the series. To a good approximation the quantum defect could be fitted to a power series in the energy with only two or three terms:

$$\mu = \mu_0 + \sum_{i=1}^{M} \mu_i E^i, \tag{3}$$

which is known as the Ritz formula. In view of the practical importance of these empirical formulas a sound theoretical justification was required.

In hydrogenic systems the electron moves in a pure Coulomb potential, whereas in nonhydrogenic systems an electron experiences a modified potential which only assumes the Coulomb form at large distances. The quantum defect represents the effects of the non-Coulomb part of the potential. The Bohr theory for hydrogenic ions verified formula (1); a justification for (2) and (3) has been given by Hartree.[1] Using Hartree's notation the radial part of the wave function for a valence electron satifies

$$\left[\frac{d^2}{dr^2} + \frac{2Y_0(r)}{r} - \frac{l(l+1)}{r^2} - E_{nl,nl} \right] P_{nl}(r) + \frac{2}{r} \int_0^\infty K(r,s) P_{nl}(s)\, ds = 0.$$

$$\tag{4}$$

Suppose r_0 is such that $K(r,s) = 0$ and $Y_0(r) = C$ for $r > r_0$, then Eq. (4) becomes

$$\left[\frac{d^2}{dr^2} + \frac{2C}{r} - \frac{C^2}{n^{*2}} - \frac{l(l+1)}{r^2} \right] P_{nl}(r) = 0, \qquad r > r_0 \qquad (5)$$

where we have put

$$E = C^2/n^{*2} = C^2 \varepsilon^*. \qquad (6)$$

Equation (5) may be rewritten, putting $\sigma = 2Cr$

$$\left[\frac{d^2}{d\sigma^2} + \frac{1}{\sigma} - \frac{\varepsilon^*}{4} - \frac{l(l+1)}{\sigma^2} \right] P_{nl} = 0. \qquad (7)$$

Hartree studied the behavior of the solutions of this equation for fixed σ and integer l. He found that in the power series expansion for the regular and irregular solutions, i.e.,

$$(2l+1)! \, G_l(\sigma, \varepsilon^*) = \sigma^{l+1}(1 + a_1 \sigma + a_2 \sigma^2 \cdots) \qquad (8)$$

$$H_l(\sigma, \varepsilon^*) = \sigma^l(b_0 + b_1 \sigma + b_2 \sigma^2 \cdots) + \alpha G_l(\sigma, \varepsilon^*) \log \sigma, \qquad (9)$$

the recurrence relations for the expansion coefficients involve integer powers of ε^* only, so that G_l and H_l are regular in ε^* for given σ. The condition that the solution of (4) has no exponentially increasing part for $r > r_0$ determines the discrete energy eigenvalues as the roots of the simultaneous equations

$$\sin \pi \left[n^*(E) + \mu(E) \right] = 0 \quad \text{and} \quad E = I - \frac{z^2}{n^{*2}(E)}, \qquad (10)$$

giving the Rydberg formula

$$E_n = I - \frac{z^2}{\left[n - \mu(E_n) \right]^2}. \qquad (11)$$

The asymptotic form of the wave function may then be expressed in terms of the Whittaker function as

$$\frac{(-n^*)^{l+1}}{\Gamma(n^* + l + 1)} W_{n^*, l+1/2}(\sigma/n^*) = G_l(\sigma, \varepsilon^*) \cos \pi n^* + H_l(\sigma, \varepsilon^*) \sin \pi n^*.$$

$$(12)$$

The condition that the solutions of (4) and (5) should match at $r = r_0$ gives an expression for $\mu = n - n^*$, with $\sigma_0 = 2Cr_0$, i.e.,

$$\frac{P'(r_0)}{P(r_0)} = \frac{2C\left[G'_l(\sigma_0,\varepsilon^*)\cos \pi n^* + H'_l(\sigma_0,\varepsilon^*)\sin \pi n^*\right]}{\left[G_l(\sigma_0,\varepsilon^*)\cos \pi n^* + H_l(\sigma_0,\varepsilon^*)\sin \pi n^*\right]}, \tag{13}$$

On rearranging (13),

$$\tan \pi\mu = \frac{2CP(r_0)G'_l(2Cr_0,\varepsilon^*) - P'(r_0)G_l(2Cr_0,\varepsilon^*)}{2CP(r_0)H'_l(2Cr_0,\varepsilon^*) - P'(r_0)H_l(2Cr_0,\varepsilon^*)}. \tag{14}$$

If P and P' are expressible as power series in ε^* we obtain the Ritz formula

$$\mu = \mu_0 + \mu_1\varepsilon^* + \mu_2\varepsilon^{*2} + \cdots. \tag{15}$$

Hartree's work not only gave a theoretical confirmation of the Ritz formula but more importantly provided the incentive for the development of quantum defect theory. Ham[2] developed a form of the theory for application to solid-state problems, in which known quantum defects of valence electron states of a free atom were used to obtain wave functions and energy levels for an atom in a finite enclosure. In the course of this work, he laid down a rigorous mathematical framework that has been the basis of all subsequent studies. However, it was through the early work of Seaton, who extended the theory to the continuum states, that the power and versatility of the theory became apparent.[3] This paper will be referred to as QDM. Its principal conclusions are as follows:

1. The asymptotic form of the wave function for any member of a Rydberg series can be expressed as a linear combination of Coulomb functions analytic in the energy and is therefore known exactly if μ is known from spectral analysis for a few members of the series.

2. For continuum states the asymptotic phase shift δ represents the effects of the non-Coulomb part of the potential and hence plays the same role as the quantum defect for bound states. Seaton was able to derive the exact relationship between the phase shift δ and the quantum defect μ:

$$\delta(\varepsilon = 0) = \pi\mu(n^* = \infty). \tag{16}$$

Thus Eq. (3) can be used to estimate phases and asymptotic forms of continuum wave functions, using extrapolated quantum defects.

3. Theoretical calculations can be interpolated and extrapolated utilizing the known dependence of the solutions upon μ.

These results provided a basis for general formulas for radiative transition probabilities, although they applied only to singly excited states, as described by Eqs. (4) and (5).

In spite of the success of the one-channel theory, especially in providing a fast, simple, and generally reliable method for calculation of photoabsorption rates it soon became clear that, in order to interpret a wider range of phenomena, a multichannel form of the theory, applicable to systems with more than one electron in open shells, for example, was required. Nowadays, complicated interactions between electrons in different configurations are taken into account by solving numerically the relevant coupled integro-differential equations on large computers, but this is expensive. Many-channel quantum defect theory provides the means for an analytical study of phases and quantum defects affected by these interactions. From the study of the asymptotic solutions of the close-coupling equations, and in particular the analytic properties of the scattering matrix, Seaton and collaborators in a series of papers[4-16] (referred to as QDT I–XIII) derived powerful interpolation and extrapolation techniques that can describe perturbed Rydberg series and complicated resonance structures. Both authors of this article, having shared in the enthusiasm and success that encouraged the development of QDT over the past 22 years, are glad to be given this opportunity to express their appreciation.

The first paper on multichannel quantum defect theory (MQDT) appeared in the proceedings of the 1963 ICPEAC in London.[17] The titles of the QDT papers give some idea of the scope of the theory. QDT I, by Seaton,[4] gives the general formulation of the many-channel theory. In QDT II,[5] Seaton gave some illustrative examples of applications to one- and two-channel problems. QDT III[6] by Bely deals with the scattering of electrons by He^+, with an attempt to include the effects of the dipole coupling potential in the analytic description. In QDT IV[7] by Moores, the many-channel theory is applied to the calculation of radiative transition probabilities and photoionization cross sections of Ca, including the effects of autoionizing states. A semiempirical method was used, employing experimental data for the perturbed series. QDT V[8] by Moores is concerned with the autoionizing and bound states of Be. In QDT VI[9] by Doughty, Seaton, and Sheorey, the theory is extended to extrapolations along isoelectronic sequences. In QDT VII[10] Seaton developed the theory for the analysis of complex resonance structures in inelastic electron–ion scattering. Expressions are obtained for the widths and positions of resonances and for the cross sections averaged over resonances in regions just below excitation thresholds. The expressions are similar to those obtained by Gailitis.[18] QDT VIII[11] by Martins and Seaton is one of many papers that present applications of the methods developed in QDT VII. In QDT IX[12] by Norcross and Seaton, the concept of a complex quantum defect is introduced, and the method applied to an analysis of the Be spectrum. In QDT X[13] by Dubau and Wells, the many-channel theory is applied to the problem of the calculation of photoionization cross sections, including detailed resonance analysis. QDT XI[14] by Seaton is a summary and clarification of the development so far. QDT XII[15] by Dubau further develops the theory for cases of dipole coupling potentials, overcoming the problems encountered in QDT III. QDT XIII[16] by Dubau and Seaton deals further with autoionizing processes in excitation and photoionization processes of great importance in the study of dielectronic recombination. This latest paper in the series was submitted in 1982. From QDM to QDT XIII[3-16], this work has spanned 24 years with fruitful contributions to a very demanding but highly rewarding subject that stimulated research into a wide range of applications. Fano and collaborators[19-22] have developed simple methods for describing the properties of entire spectra (including autoionization line profiles), with very few parameters that are found to be slowly varying functions of the energy: the quantum defects, oscillator strength density, and the eigenvectors of the scattering-phase matrix. Using frame transformations, the nature of the coupling between the electron and the ionic core can be identified, and this greatly helps in the analysis of experimental data. The methods are restricted to cases where the coupling is not very strong and where there are no overlapping resonances. They have been successfully applied to the spectra of rare gases and alkaline earths; see, for example, Geiger,[23] Armstrong et al.,[24] and Clark and Taylor.[25]

The Fano approach has also been applied with particular success to the analysis of molecular spectra and molecular processes. Following some pioneering work by Fano on l uncoupling in H_2,[19] Herzberg and Jungen[26] carried out an extensive analysis of the H_2 spectrum, using the same formalism. In Section 4 we shall discuss the application of Seaton's formalism to the same problem. Fano's technique has subsequently been extended to include the effects of vibration by Herzberg and Jungen,[26] Atabek et al.,[27] Atabek and Jungen,[28] and Jungen and Atabek,[29] and to include dissociative channels, by Lee[30], and by Giusti,[31] and to include rotational–vibrational preionization, by Jungen and Dill,[32] and electronic predissociation, by Giusti and Jungen.[33]

For heavier, highly stripped ions ($z > 20$), relativistic effects can no longer be neglected. Multichannel quantum defect theory was reformulated for solutions of the Dirac equations by Johnson, Cheng, Lee, and others.[34-37] Here, Sommerfeld's relativistic formula for the level energy is used to define a relativistic quantum defect. As in the nonrelativistic theory the analytic properties of the solutions of the Dirac equations make possible extrapolations from bound to continuum states. The computer power available now has made numerical solutions of the Dirac equations for heavy, highly stripped ions feasible. The extension of quantum defect theory to the relativistic regime will be of great help in the description of these ions and the assessment of the accuracy of calculated data.

2. Mathematical Background to Quantum Defect Theory

2.1. Properties of Coulomb Wave Functions

Quantum defect methods are based on the properties of the confluent hypergeometric funtions which are the solutions of the radial equation for an electron moving in a Coulomb potential. In order to provide a rigorous justification for the extrapolation procedures that are the essence of the method, it is necessary to investigate in some detail the analytic properties of these functions with respect to the energy. In the following analysis we adopt the notation of QDT XI[14] and summarize the formulas required for an exposition of the theory.

The radial Schrödinger equation for an electron in a Coulomb potential is

$$\left[\frac{d^2}{dr^2} - \frac{l(l+1)}{r^2} + \frac{2z}{r} + E\right]F = 0, \tag{17}$$

where E is the energy in rydbergs. Writing $\rho = zr$, $\varepsilon = E/z^2 = -1/\kappa^2$, $l = \lambda - \frac{1}{2}$ (note that in QDM[3] and QDT I[4] the notation $l = \lambda$ is used), Eq. (17) becomes

$$\left(\frac{d^2}{d\rho^2} - \frac{\lambda^2 - \frac{1}{4}}{\rho^2} + \frac{2}{\rho} - \frac{1}{\kappa^2}\right)F = 0. \tag{18}$$

In (18) λ and κ are defined quite generally and may assume complex values. In QDT I Seaton defines two solutions of (18): $y_1(\kappa,\lambda;\rho) = f(\kappa,l;\rho)$ and $y_4(\kappa,\lambda;\rho) = g(\kappa,l;\rho)$, which are linearly independent for $\lambda = l + \frac{1}{2}$, l being a positive integer. They are both analytic in ε, that is, they can be expressed as convergent power series in that variable. The functions f and g may be expressed as linear combinations of functions $\phi^{\pm}$ [defined in Eq.

(2.36) of QDT XI]. All we require here are their asymptotic forms:

$$\phi^\pm \sim \left(\frac{\kappa}{i}\right)^{1/2}\left(\frac{2\rho}{\kappa}\right)^{\pm\kappa}\exp\left\{\pm\left[-\frac{\rho}{\kappa}+\frac{i\pi}{2}(\kappa-l)\right]\right\}. \tag{19}$$

The functions f and g are given by

$$f(\kappa,l;\rho) = i^{1/2}\kappa^{l+1/2}e^{(1/2)i\pi(\kappa-l)}\left[\frac{-\phi^+}{\Gamma(\kappa+l+1)}+\frac{\phi^-}{\Gamma(-\kappa+l+1)}\right] \tag{20}$$

$$g(\kappa,l;\rho) = i^{1/2}\kappa^{l+1/2}e^{(1/2)i\pi(\kappa-l)}\left[\frac{(iA+G)\phi^+}{\Gamma(\kappa+l+1)}-\frac{(A\cot\pi\kappa+G)\phi^-}{\Gamma(-\kappa+l+1)}\right]; \tag{21}$$

these are Eqs. (2.38) and (2.39) of QDT XI. The functions $A(\kappa,\lambda)$ and $G(\kappa,\lambda)$ are discussed in detail in QDM[3] and QDT I[4]. They are defined by

$$A(\kappa,\lambda) = \frac{\Gamma(\kappa+\lambda+\frac{1}{2})}{\kappa^{2\lambda}\Gamma(\kappa-\lambda+\frac{1}{2})} \tag{22}$$

and

$$G(\kappa,\lambda) = \frac{1}{2\pi}\frac{d}{d\lambda}A(\kappa,\lambda). \tag{23}$$

Seaton also defines a third solution which has an asymptotic expansion in powers of ε

$$h(\varepsilon,l;\rho) = -g - \mathcal{G}f \tag{24}$$

where $\mathcal{G}$ is the real part of $G(\kappa,l+\frac{1}{2})$. For real ε one writes

$$\kappa = \begin{cases} \nu, & \varepsilon < 0 \\ i\gamma, & \varepsilon > 0 \end{cases} \tag{25}$$

with ν and γ real and positive. For negative energies, $\kappa = \nu$, Seaton defines solutions $\theta(\varepsilon,l;\rho)$ and $\xi(\varepsilon,l;\rho)$ [QDT XI, Eqs. (2.47) and (2.48)] with asymptotic forms

$$\left.\begin{array}{c}\theta(\varepsilon,l;\rho)\\\xi(\varepsilon,l;\rho)\end{array}\right\} \sim \left(\frac{2\rho}{\nu}\right)^{\pm\nu}e^{\mp\rho/\nu} \tag{26}$$

The analytic functions f and h may be expressed as linear combinations of θ and ξ, for $\varepsilon < 0$:

$$f = B^{-1/2}(-1)^l\left[\frac{\sin\pi\nu}{\nu^{1/2}K}\xi - \nu^{3/2}\cos(\pi\nu)K\theta\right], \qquad (27)$$

$$h = B^{1/2}(-1)^l\left[\frac{\cos\pi\nu}{\nu^{1/2}K}\xi + \nu^{3/2}\sin(\pi\nu)K\theta\right], \qquad (28)$$

where

$$K(\nu,l) = \left[\nu^2\Gamma(\nu + l + 1)\Gamma(\nu - l)\right]^{-1/2} \qquad (29)$$

is the normalization factor for bound-state functions first introduced in QDM, and

$$B(\varepsilon,l) = \begin{cases} A(\nu, l + \tfrac{1}{2}), & \varepsilon < 0 & (30) \\[2mm] \dfrac{A(i\gamma, l + \tfrac{1}{2})}{1 - \exp(-2\pi\gamma)}, & \varepsilon > 0. & (31) \end{cases}$$

$$B = \prod_{m=0}^{l}(1 + m^2\varepsilon), \qquad 0 \leqslant \varepsilon \ll 4\pi^2 \qquad (32)$$

For $\varepsilon > 0$ the asymptotic forms are

$$\left.\begin{array}{c} f \\ h \end{array}\right\} \sim B^{\mp 1/2}\left(\frac{2\gamma}{\pi}\right)^{1/2}\begin{array}{c}\sin\zeta \\ \cos\zeta\end{array}, \qquad (33)$$

where

$$\zeta = \frac{\rho}{\gamma} - \tfrac{1}{2}l\pi + \gamma\ln\left(\frac{2\rho}{\gamma}\right) + \sigma(l,\gamma) \qquad (34)$$

and

$$\sigma(l,\gamma) = \arg\Gamma(l + 1 - i/\gamma). \qquad (35)$$

It is convenient to define functions that asymptotically behave like sine and cosine of the Coulomb phase, i.e.,

$$\left.\begin{array}{c} s \\ c \end{array}\right\} = \frac{1}{2^{1/2}}B^{\pm 1/2}\left.\begin{array}{c} f \\ h \end{array}\right\} \sim \left(\frac{\gamma}{\pi}\right)^{1/2}\begin{array}{c}\sin\zeta \\ \cos\zeta\end{array}. \qquad (36)$$

Writing

$$\phi_{\pm} = c \pm is, \tag{37}$$

we have

$$\phi_{\pm} \sim \left(\frac{\gamma}{\pi}\right)^{1/2} \exp(\pm i\zeta), \qquad \varepsilon > 0 \tag{38}$$

and

$$\phi_{\pm} = (-1)^l e^{\pm i\pi\nu} \left[\frac{\xi}{(2\nu)^{1/2} K} \mp i\left(\frac{\nu^3}{2}\right)^{1/2} K\theta\right], \qquad \varepsilon < 0. \tag{39}$$

The functions $\phi_{\pm}$ differ from $\phi^{\pm}$ in that for $\varepsilon > 0$ the asymptotic forms of $\phi_{\pm}$ contain the Coulomb phase $\sigma(l, \gamma)$, whereas those of $\phi^{\pm}$ do not.

2.2. Solutions of the Coupled Equations

The application of the close-coupling approximation to the problem of an electron moving in the field of a positive ion leads to a set of M coupled integro-differential equations of the form

$$\left[\frac{d^2}{d\rho^2} - \frac{l_i(l_i + 1)}{\rho^2} + \frac{2}{\rho} + \varepsilon_i\right] F_{ij}(\rho) + \sum_{i'=1}^{M} U_{ii'}(\rho) F_{i'j}(\rho) = 0,$$

$$i, j = 1, \dots, M \tag{40}$$

where i is the channel index. The total energy is

$$E = E_i + z^2\varepsilon_i, \tag{41}$$

E_i being the energy of the ion state belonging to channel i. The potential matrix $\mathbf{U}$ is real and symmetric. These equations have M linearly independent sets of solutions satisfying

$$F_{ij}(\rho) \to 0 \qquad \text{as} \quad \rho \to 0. \tag{42}$$

The subscript j specifies a particular set of solutions. The solutions F_{ij} form the elements of an $M \times M$ matrix $\mathbf{F}$. QDT considers a model problem in which the potential $\mathbf{U}(\rho)$ satisfies certain conditions which are not necessarily satisfied, however, by the potential matrix of the electron–ion scattering problem. The model is a generalization of the one introduced by Ham[2] and is given by Seaton in QDT XI, Section 3.2: For ρ_0 and ρ_1 such that $0 < \rho_0 < \rho_1 < \infty$

(i) For $0 \leqslant \rho \leqslant \rho_0$, $\rho U_{ii}(\rho)$ is an analytic function of ρ; and for $i \neq j$, $\rho^{-q_{ij}} U_{ij}(\rho)$ is an analytic function of ρ, where $q_{ij} \geqslant |l_i - l_j|$.

(ii) For $\rho_0 \leqslant \rho \leqslant \rho_1$, $U(\rho)$ is piecewise continuous.

(iii) $U(\rho)$ is of finite range: $U(\rho) = 0$ for $\rho > \rho_1$. $\qquad$ (43)

Condition (iii) is strictly never satisfied but in many problems the long-range part of U is sufficiently small compared to the Coulomb potential and to the short-range part that the following derivations are applicable in practice over a limited range of energies. Assuming that conditions (43) are satisfied, it can be shown that there exists a solution $F(E, \rho)$ of the scattering equations satisfying the boundary condition*

$$\lim_{\rho \to 0} \left[\rho^{-l_i - 1} F_{ij}(E, \rho) \right] = \delta_{ij} \qquad (44)$$

which is an entire analytic function of the energy E for all $|E| < \infty$. Also $F(E, \rho)$, together with its derivative, is a continuous function of the real variable ρ, so that

$$F(E, \rho) = \sum_{m=0}^{\infty} F_{(m)}(\rho) E^m. \qquad (45)$$

Furthermore, for $\rho > \rho_1$ the solution can be expressed as the linear combination of the Coulomb functions f and g:

$$F(E, \rho) = f(\varepsilon, l; \rho) I(E) + g(\varepsilon, l; \rho) J(E). \qquad (46)$$

Here, f and g are diagonal matrices with elements $f(\varepsilon_i, l_i; \rho)$ and $g(\varepsilon_i, l_i; \rho)$, respectively. $I(E)$ and $J(E)$ may be determined by matching F and $dF/d\rho$ to (46). Since f, g, and F are analytic functions with respect to E, $I(E)$ and $J(E)$ must be analytic, and so

$$I(E) = \sum_{m=0}^{\infty} I_{(m)} E^m \quad \text{and} \quad J(E) = \sum_{m=0}^{\infty} J_{(m)} E^m. \qquad (47)$$

The following expressions for the solutions, based on the properties of I and J are of importance in the applications of QDT:

(i) $\qquad F(S_I, \rho) = \phi^- - \phi^+ S_I, \qquad \rho > \rho_1 \qquad (48)$

---

*Letters in italic type without subscripts in matrix equations denote diagonal matrices.

introduces a generalized scattering matrix in terms of matrices $\mathbf{I}$ and $\mathbf{J}$, for $-\frac{1}{2}\pi < \arg(\kappa) < \frac{3}{2}\pi$, which is given by

$$\mathbf{S}_l = \frac{\kappa^{l+1/2}\exp\left[\frac{1}{2}i\pi(\kappa - l)\right]}{\Gamma(\kappa + l + 1)}\left[\mathbf{I} - (i\mathbf{A} + \mathbf{G})\mathbf{J}\right]$$

$$* \left[\mathbf{I} - (A\cot\pi\kappa + \mathbf{G})\mathbf{J}\right]^{-1} * \frac{\Gamma(-\kappa + l + 1)}{\kappa^{l+1/2}\exp\left[\frac{1}{2}i\pi(\kappa - l)\right]}. \tag{49}$$

Note that for $\varepsilon > 0$, $\mathbf{S}_l$ is the full scattering matrix, including the phase due to the Coulomb field.

(ii) $$\mathbf{Y} = \mathbf{J}(\mathcal{G}\mathbf{J} - \mathbf{I})^{-1} \tag{50}$$

In most applications, one has real ε and works with real functions. The real, symmetric matrix $\mathbf{Y}$ is a slowly varying function of E but has poles at the energies where $(\mathcal{G}\mathbf{J} - \mathbf{I})$ is zero. Because of (47) we can write

$$\mathbf{Y}(E) = \mathbf{Q}(E)\mathbf{P}(E)^{-1}, \tag{51}$$

where $\mathbf{Q}(E)$ and $\mathbf{P}(E)$ are power series in E. For a single-channel system with $\varepsilon < 0$, $Y = \tan\pi\eta$. This "η defect" was first introduced by Ham.[2] For many-channel solutions a diagonal matrix $\bar{\eta}$ was first introduced by Seaton,[4] from $\mathbf{Y} = \mathbf{X}\bar{Y}\mathbf{X}^{-1}$, $\tan\pi\bar{\eta} = \bar{Y}$ for all channels closed.

(iii) $$\mathbf{F}(\mathcal{R}, \rho) = s + c\mathcal{R}, \qquad \rho > \rho_1. \tag{52}$$

For $\varepsilon > 0$, $\mathcal{R} = \mathbf{R}$, the usual reactance matrix. One has

$$\mathcal{R} = B^{1/2}\mathbf{Y}B^{1/2}, \tag{53}$$

where $B(\varepsilon_i, l_i)$ is defined by (30) and (31). For $\nu \simeq l$, power series expressions for $\mathbf{Y}$ are more reliable than those for $\mathcal{R}$. At energies $\varepsilon \ll 4\pi$ and $\nu > l$, $\mathcal{R}(E)$ can be fitted to expressions of the form

$$\mathcal{R}(E) = \mathcal{Q}(E)\mathcal{P}^{-1}(E) \tag{54}$$

where $\mathcal{Q}$ and $\mathcal{P}$ can be expanded as power series in E. For a single channel

$$\mathcal{R} = \begin{cases} \tan \delta, & \varepsilon > 0 \\ \tan \pi\mu, & \varepsilon < 0 \end{cases}. \tag{54a}$$

(iv) $$F(\chi, \rho) = \phi_- - \phi_+ \chi, \qquad \rho > \rho_1. \tag{55}$$

When all channels are open we have

$$\mathbf{S}_l = e^{i\sigma} \chi e^{i\sigma}. \tag{56}$$

From (55), (52) and (37) one obtains

$$\chi = (1 + i\mathcal{R})(1 - i\mathcal{R})^{-1}. \tag{57}$$

If $\mathbf{U}$ is real and symmetric, χ is symmetric and unitary.

2.2.1. All Channels Closed.   Suppose that the energies are such in all channels that only bound states exist. From (46), (27), and (28), we have that

$$\mathbf{F}(E, \rho) = B^{-1/2}(-1)^l \left\{ \left[\sin \pi\nu(\mathbf{I} - \mathcal{G}\mathbf{J}) - B \cos \pi\nu\mathbf{J} \right] \frac{\xi}{\nu^{1/2}K} \right.$$

$$\left. + \left[\cos \pi\nu(\mathcal{G}\mathbf{J} - \mathbf{I}) - B \sin \pi\nu\mathbf{J} \right] \nu^{3/2}K\theta \right\}. \tag{58}$$

We construct the bound-state solutions by taking linear combinations of the columns of (58) and requiring that the coefficient of the exponentially increasing function ξ should vanish, so that

$$\mathbf{F}(\text{bound}, \rho) = \mathbf{F}(E, \rho)\mathbf{C}. \tag{59}$$

We obtain

$$\left[\sin \pi\nu(\mathbf{I} - \mathcal{G}\mathbf{J}) - B \cos \pi\nu\mathbf{J} \right]\mathbf{C} = 0 \tag{60}$$

which gives, using (50) and (53)

$$|\mathcal{R} + \tan \pi\nu| = 0 \tag{61}$$

as the condition for a bound state. The solutions of (61) give the eigenenergies of the system. Using (60) the normalized eigenfunctions can be obtained in the form

$$F(\text{bound}, \rho) = \theta K Z, \tag{62}$$

where Z is a column vector satisfying

$$(\mathcal{R} + \tan \pi \nu) q Z = 0, \tag{63}$$

$$q = (-1)^l \left(\frac{2}{\pi \nu^3} \right)^{1/2} \cos \pi \nu. \tag{64}$$

The normalization condition (QDT I, Section 7) gives

$$Z^T \zeta Z = 1, \tag{65}$$

where

$$\zeta = 1 + q \left(\frac{\partial \mathcal{R}}{\partial \varepsilon} \right) q. \tag{66}$$

For large ν, $\zeta \simeq 1$.

2.2.2. *Some Channels Open.* Let the channel labeling and energies be such that the first M_o channels are open ($\varepsilon_i > 0$, $i = 1, M_o$) and the rest are closed. We may partition any of the solution matrices into four submatrices containing open–open, closed–open, open–closed, and closed–closed elements, respectively:

$$F = \begin{pmatrix} F_{oo} & F_{oc} \\ F_{co} & F_{cc} \end{pmatrix}, \tag{67}$$

where F_{oo} has dimension $M_o \times M_o$, F_{co} has dimension $(M - M_o) \times M_o$ and so on. The part of F corresponding to a solution of the actual electron–ion scattering problem with the required boundary conditions is the $M \times M_o$ submatrix

$$F_o = \begin{pmatrix} F_{oo} \\ F_{co} \end{pmatrix} \tag{68}$$

with

$$F_{co}(\rho) \to 0 \quad \text{as} \quad \rho \to \infty. \tag{69}$$

If we define $F_o(R_{oo}, \rho)$ and $F_o(S_{oo}, \rho)$ to be such that their open–open parts are given by

$$F_{oo}(R_{oo}, \rho) = s_o + c_o R_{oo}, \qquad \rho > \rho_1 \qquad (70)$$

$$F_{oo}(S_{oo}, \rho) = \phi_{-o} - \phi_{+o} S_{oo}, \qquad \rho > \rho_1, \qquad (71)$$

then the $M_o \times M_o$ matrices R_{oo} and S_{oo} are just the reactance and scattering matrices that would be obtained from a numerical solution of the coupled equations of the scattering problem. In general, R_{oo} and S_{oo} will not be slowly varying but must be expressed in terms of the slowly varying matrices $\mathcal{R}$ and χ. Let

$$F_o(R_{oo}) = F(\mathcal{R})\begin{pmatrix} I_{oo} \\ D_{co} \end{pmatrix} \qquad (72)$$

where I_{oo} is the $M_o \times M_o$ unit matrix. One obtains

$$F_{oo}(R) = F_{oo}(\mathcal{R}) + F_{oc}(\mathcal{R})D_{co}, \qquad (73)$$

$$F_{co}(R) = F_{co}(\mathcal{R}) + F_{cc}(\mathcal{R})D_{co}, \qquad (74)$$

and using (52)

$$F_{co}(R) = c_c \mathcal{R}_{co} + (s_c + c_c \mathcal{R}_{cc})D_{co} \qquad \text{for} \quad \rho > \rho_1. \qquad (75)$$

Using (27), (28), and (36) we obtain

$$F_{co} = 2^{-1/2}(-1)^l \left\{ \left[\cos \pi\nu \mathcal{R}_{co} + (\sin \pi\nu + \cos \pi\nu \mathcal{R}_{cc})D_{co} \right] \nu^{-1/2}\xi/K \right.$$
$$\left. + \left[\sin \pi\nu \mathcal{R}_{co} - (\cos \pi\nu - \sin \pi\nu \mathcal{R}_{cc})D_{co} \right] \nu^{3/2}K\theta \right\}. \qquad (76)$$

If we now choose D_{co} to be such that the coefficient of the exponentially increasing function ξ vanishes in (76) we get

$$D_{co} = -(\tan \pi\nu + \mathcal{R}_{cc})^{-1}\mathcal{R}_{co}, \qquad (77)$$

and substituting back into (73) one obtains an expression for the reactance matrix in terms of the general matrix $\mathcal{R}$, so that

$$R = \mathcal{R}_{oo} - \mathcal{R}_{oc}\left[\mathcal{R}_{cc} + \tan \pi\nu \right]^{-1}\mathcal{R}_{co}. \qquad (78)$$

Similarly, one derives an expression for S in terms of the χ matrix

$$S = \chi_{oo} - \chi_{oc} \left[\chi_{cc} - e^{-2\pi i \nu} \right]^{-1} \chi_{co} . \tag{79}$$

3. Single-Channel Quantum Defect Methods: General Formulas in the Independent-Particle Approximation

The status of QDT around 1958 has been summarized by Seaton[3] in the paper to be referred to as QDM. The formulations apply to the wave function of an electron in a central field that behaves as a Coulomb field for large r. Writing the asymptotic solutions in terms of known analytic functions and convergent power series in the energy, reliable extrapolation techniques were obtained. At that time, detailed calculations of wave functions were a major undertaking, involving several years' work on a mechanical desk calculator. It was an important research task to develop fast methods for the calculation of atomic data that promised to be more reliable than "order of magnitude" or "factor of 2." Methods based on quantum defect theory made the best possible use of experimental data available and the understanding of the physics and mathematics of the problem.

Now, some 25 years later, with the aid of computers, we can calculate the wave functions within a few hours. Given sufficient computer power we can produce results with errors less than 10%. Comparing these results with those obtained applying single-channel quantum defect theory we find agreement to within 10% for cases where the active electron does not couple strongly with the electron cloud of the ion. This remarkable success speaks for the continued use of the general formulas, particularly since their evaluation takes only a few seconds on modern computers.

3.1. Expressions for the Wave Functions

To apply quantum defect methods we seek to express the solutions of the radial equation of an electron in the field of an ion in terms of Coulomb functions that are analytic in the energy plus coefficients that are expressible as power series in the energy. The mathematical background has been summarized in Section 2. The radial equation is approximated by

$$\left[\frac{d^2}{dr^2} - \frac{l(l+1)}{r^2} - V(r) + E \right] P(E,l;r) = 0, \tag{80}$$

where E and V are in Rydberg units, $rV(r) = -2Z$ for $r = 0$, and $rV(r) = -2z$ for $r \to \infty$, with Z the nuclear and z the residual charge on the ion. Asymptotically, Eq. (80) corresponds to the general form of the Coulomb equation (18), where one introduces $\rho = zr$, $\varepsilon = E/z^2 = -1/\kappa^2$, and $l = \lambda - \frac{1}{2}$. If one assumes that $\rho V(\rho) = -2$ for $\rho > \rho_1$, see conditions (43), the solution is given by Eq. (46), where, for the single-channel case considered here, F, I, and J are scalars.

For negative energy states,

$$\varepsilon_n = -(1/\nu_n)^2 \quad \text{and} \quad F(E,\rho) \to 0 \quad \text{as} \quad r \to \infty. \tag{81}$$

We then have, see Eqs. (54a) and (61),

$$\tan \pi\mu(\varepsilon) + \tan \pi\nu_n = 0, \tag{82}$$

which determines the quantum defects of the bound states

$$\mu(\varepsilon_n) = \mu_n \equiv n - \nu_n. \tag{83}$$

Given the quantum defects, the normalized radial functions are

$$P_{nl} = z^{1/2}(-1)^{n-l-1}\theta K Z \tag{84}$$

from Eq. (62). For a single channel, from Eqs. (64)–(66)

$$Z = \left(1 + \frac{2}{\nu^3}\frac{\partial\mu}{\partial\varepsilon}\bigg|_{\varepsilon=\varepsilon_n}\right)^{-1/2}. \tag{85}$$

It is seen that $Z^2 \simeq 1$ for large ν.

For positive energy states we write the solutions in the reactance matrix form (52). In the one-channel case

$$R = \tan \delta, \tag{86}$$

where δ is the phase shift due to the non-Coulomb part of the potential. From Eq. (53) for the generalized reactance matrix, here a scalar,

$$\mathcal{R} = BJ(\mathcal{G}J - I)^{-1} \tag{87}$$

one obtains, using (25), (30), and (31),

$$\tan \delta(\varepsilon) = (1 - e^{-2\pi\gamma})^{-1}\tan \pi\mu(\varepsilon), \tag{88}$$

and, in particular, for the zero-energy phase shift

$$\tan \delta(0) = \tan \pi \mu_\infty \qquad (89)$$

which defines the phase shift uniquely and is the basis for the extrapolation techniques used in this section. We have thus a working formula for finding normalized bound-state wave functions and free-state wave functions with correct asymptotic forms that use observed energy level data.

3.2. General Formulas for Radiative Transition Probabilities

Radiative transition probabilities involve the radial integral

$$R(E'l', E, l) = \int_0^\infty F(E', l'; r) r F(E, l; r) \, dr \qquad (90)$$

For transitions between valence electron states the largest contribution to the radial integral comes from the asymptotic part of the wave function. The relative phases of the two functions determine the amount of overlap or cancellation. One can therefore expect calculations for line strengths between excited states to be quite reliable when expressions (84) and (52) are used at all r to evaluate (90) in place of an "exact" wave function. A cutoff factor must be introduced to remove the divergence in the irregular Coulomb function near the origin. With the Coulomb functions known, R can be tabulated for a range of l and ε. This was first done for transitions between bound states by Bates and Damgaard.[37] With modern computers their formula is easily programmed and evaluated in a few seconds. For strong optically allowed transitions the method gives results within 15% of extensive close-coupling calculations as long as the radiating electron states are fairly "pure," that is, can be represented by a single-channel wave function. Use of experimental quantum defects and functions (84) can indeed produce better results than, for example, the use of single-configuration self-consistent field solutions, particularly for excited states, since (84) gives the exact solution at large distances. In Table 1 we compare close-coupling results for oscillator strengths in O IV[38] with single-configuration calculations[39] and results from the Bates and Damgaard formula. The ion chosen is particularly difficult from the point of view of independent-particle theory, since there is strong interaction between series of configurations $2s^2nl$ and $2s2pml'$. Where the bound states of the interfering series lie close together, the wave functions are completely mixed, and the independent-particle approximations fail utterly. Away from such energies the results compare well with the close-coupling work, with the

TABLE 1

Oscillator Strengths for ns–mp Transitions in O IV

Transition	f		
	Single-configuration Hartree–Fock	Bates and Damgaard	Close-coupling approximation
$2s^23s–2s^23p$	0.689	0.619	0.556
$2s^23s–2s^24p$	0.096	0.085	1.640
$2s^24s–2s^24p$	0.949	1.040	0.870
$2s^24s–2s^25p$	0.099	0.073	0.086
$2s^25s–2s^25p$	1.207	1.419	1.256
$2s^25p–2s^26s$	0.232	0.194	0.179
$2s^26s–2s^26p$	1.463	0.722	0.511

quantum defect method sometimes giving better results than the self-consistent field solutions.

Using free-state functions (52) with phase shifts obtained from extrapolated experimental quantum defects, the method was extended to bound–free transitions by Burgess and Seaton.[40] Peach[41] gave a formulation for free–free transitions. Here, the expression for the radial operator had to be changed to the "acceleration" form to force convergence of the radial integral in the asymptotic region. This formulation also brought improvements in the calculation of photoionization cross sections and oscillator strengths for highly excited states. In 1967, Peach[42] gave a revised general formula for bound–free transitions which remedied some inaccuracies in the Burgess–Seaton approach, which was based on hand calculations. Peach wrote a general computer program (unpublished, but available on request) for the calculation of atomic photoionization cross sections, using her general formula. Extensive results for continuous absorption coefficients were published in 1971.[43] Hoang-Binh and van Regemorter[44–46] give analytical expressions for the radial integrals for bound–bound, bound–free, and free–free transitions. Their method allows an explicit evaluation of the effects of the cutoff parameter. In this analytical treatment they overcome cancellation effects which inhibit the numerical evaluation of the general formula for high Rydberg states, $n > 15$. Oscillator strengths for transitions between high Rydberg states are best obtained by extrapolations from data for free–free and bound–free transitions.

Detailed comparisons between the quantum defect method, using the program of Peach, and the close-coupling method for photoionization have been made for O IV.[38] These show agreement to within 10% for the background cross section for photoionization from states $2s^2nl$, including

$nl = 2p$. For photoionization from $2s2pml$ states, QDT was not generally reliable for $ml = 2p$, but quite good for $m > 2$. It must be emphasized that this single-channel QDT cannot reproduce any resonance structure nor deal correctly with branching at higher thresholds.

3.3. Collision Cross Sections

3.3.1.   Use of Extrapolated Quantum Defects.   For elastic scattering of an electron in the field of an ion the differential cross section can be written in terms of a pure Coulomb part, the Rutherford formula, and a factor due to the non-Coulomb phase. The latter can be estimated from extrapolated quantum defects. This technique was introduced by Seaton.[3]

3.3.2.   Use of Adjusted Calculated Quantum Defects.   Electron impact excitation cross sections are related to the scattering matrix which is obtained from the asymptotic forms of many-channel wave functions. However, in the case of boron like and fluorine like ions, electron scattering off the ^2P ground state is a single-channel problem in LS coupling, if the $l, l + 2$ coupling is neglected. The reactance matrix in intermediate coupling, relevant to the $^2P_{1/2}$–$^2P_{3/2}$ excitation, is obtained by algebraic transformation of the single-channel LS-coupling reactance matrix. By extrapolating the calculated values of $\tan\delta$ to the bound-state region, calculated quantum defects can be compared with experimental results. Moreover, adjustable parameters can be introduced so as to obtain a better fit to the bound-state data. Cross sections are then calculated using the adjusted $\tan\delta$. An application of this technique to electron impact excitation of the fine-structure levels in C^+ and Si^+ is given in reference 3. In a many-channel form, the method is extensively used and discussed by Saraph and Seaton.[47] Seaton[48] derives a variational principle for the scattering-phase matrix, which is required in this work.

3.3.3.   Use of Observed Quantum Defects in the Bethe Approximation.   Excitation cross sections corresponding to an optically allowed transition have a large contribution from high energies and high angular momentum states of the colliding electron. These contributions can be estimated using the van Regemorter formula[49] which involves the oscillator strength for the dipole transition. The oscillator strength can be obtained by the general formulas of Section 3.2. The method is widely used to estimate the vast number of collision rate data required in astrophysics and plasma physics.

3.4.   Summary

The applications of single-channel quantum defect theory cited here demonstrate the potential of the theory. Although the range of reliable extrapolations is restricted by conditions (43) on the non-Coulomb part of the ionic potential, for a wide variety of problems the method gives remarkably good results. It was therefore of great practical consequence when the theory was generalized to the multichannel case by Seaton[4] in 1966.

4. Applications to Simple Multichannel Problems

4.1. The Spectrum of Calcium

The single-channel theory discussed in the previous section is applicable to atomic systems that may be described as consisting of an electron added to an ion whose first excitation energy is large. To a good approximation, the electron in such a system moves in a central field due to the ion core, see Eq. (80), and the energy levels form simple Rydberg series with slowly varying quantum defects, converging onto the same limit, which represents the ion in its ground state. In most systems, however, the above description does not apply: the central field approximation is not adequate, since coupling effects due to the noncentral interactions between external and even core electrons are important. In such systems one observes overlapping series of the same angular momentum and parity converging to the ground and to excited states of the ion. Under these circumstances the quantum defects may not be slowly varying functions of the energy but may exhibit the behavior characteristic of perturbed series. This is illustrated in Fig. 1, where the quantum defects for the Ca $4snp$ $^1P_1$ series, which converges on to the $4s$ level of Ca^+ are plotted against energy. The perturbation is caused by the level $3d4p$ $^1P_1$ which is a member of a series converging on the $3d$ level of the Ca^+ ion. In such a case the assignment of

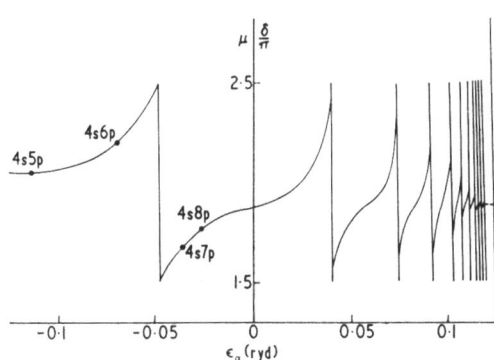

FIGURE 1. Quantum defect and phase shift against energy in rydbergs for $Ca^+ + e^-$ $^1P_1$.

a single configuration such as $3d4p$ to a given energy level is only approximate: the two series form two coupled channels and their wave functions are especially strongly mixed at energies where their respective bound states lie close together. The concepts of *perturbed series* or *configuration mixing* are merely different ways of describing the same physical effect, the failure of the central field approximation to describe the forces experienced by the electron.

4.1.1. Bound States. In order to apply QDT to complex systems it is necessary to use the many-channel formalism. The fundamental equation is the condition for a bound state, as discussed in Section 2, Eq. 61

$$|\mathcal{R} + \tan \pi \nu| = 0,$$

which must be satisfied at all the energy eigenvalues of the system.

Consider a two-channel case with

$$\varepsilon_1 = -1/\nu_1^2 \quad \text{and} \quad \varepsilon_2 = -1/\nu_2^2 .$$

Equation (61) may then be rewritten

$$(\mathcal{R}_{11} + \tan \pi \nu_1)(\mathcal{R}_{22} + \tan \pi \nu_2) - \mathcal{R}_{12}\mathcal{R}_{21} = 0. \tag{91}$$

The elements of $\mathcal{R}$ are slowly varying functions of the energy. If $\mathcal{R}_{12} = 0$, we obtain two noninteracting series, uncoupled, whose energy levels are given by

$$\mathcal{R}_{11} + \tan \pi \nu_1 = 0 \quad \text{and} \quad \mathcal{R}_{22} + \tan \pi \nu_2 = 0.$$

Let us define a continuous variable μ_1 by

$$\tan \pi \mu_1 = \mathcal{R}_{11} - \frac{\mathcal{R}_{12}\mathcal{R}_{21}}{\mathcal{R}_{22} + \tan \pi \nu_2} . \tag{92}$$

At the eigenvalues, $\varepsilon_{1,n} = -1/(n - \mu_{1,n})^2$, determined by the solutions of (91), we have $\mu_1 = \mu_{1,n}$. Equation (92) gives the behavior of the quantum defect as a function of energy. As $\mathcal{R}_{22} + \tan \pi \nu_2$ passes through zero, μ_1 varies rapidly with energy, and Fig. 1 shows a typical case. In quantum defect theory we do not distinguish between perturbed and perturbing levels—we simply treat all energy levels of the same total angular momentum and parity as a set of states each member of which is described as an admixture of (two) channels. Away from a perturbation, each state is predominantly one channel or the other, but near it, strong mixing occurs.

If a sufficient quantity of spectroscopic data is available, Eq. (91) may be used as the starting point of a semiempirical technique for obtaining the matrix $\mathscr{R}$ as a function of energy. Information about unknown energy levels and about the positive energy region, where autoionizing states are found, may then be obtained by interpolating or extrapolating this matrix. The more slowly varying Y matrix and hence the $\mathscr{R}$ matrix have been calculated in QDT II and QDT IV by imposing condition (91) at the known energy levels of the atomic system and solving the resulting nonlinear algebraic equations to obtain the matrix elements. The form of the power series expansion for Y is determined by the number of observed levels available: for an expression that is linear in the energy, six energy levels, preferably in the vicinity of the perturbation and including the perturbing level, are required to determine all the matrix elements. The numerical techniques are described in QDT II and QDT IV. From $Y(E)$ so obtained, the complete set of energy levels can be calculated using Eq. (91). In this way, the calculation can be checked against observed energy levels not included in the fit, and further energy levels can be predicted. Once a well-fitting matrix $Y(E)$ has been obtained, multichannel quantum defect methods can be applied to predict other properties of the system. The asymptotic forms of the configuration mixed wave functions for the bound states may be obtained by computing the coefficients Z_i defined by Eq. (62), which give a measure of the degree of mixing of the channels. Results obtained by Moores in QDT IV for the Ca $^1P_1$ series are shown in Table 2. For this two-channel case the total wave function is written as a linear combination of the form

$$\Psi = Z_1\psi_1 F_1 + Z_2\psi_2 F_2, \qquad (93)$$

where ψ_1 and ψ_2 are wave functions for Ca$^+$ in the $4s$ and $3d$ states, respectively, and

$$F_i = \theta\left(-1/\nu_i^2, l_i; \rho\right) K(\nu_i, l_i). \qquad (94)$$

It is seen from the table that the maximum value of Z_2 occurs for the level

TABLE 2

The $^1P_1$ Series in Calcium

n	T_n (cm^{-1})	Configuration	ν_1	ν_2	Z_1	Z_2
4	23652.32	$4s4p$	2.068	1.670	-0.879	-0.476
5	36731.62	$4s5p$	2.954	2.043	0.925	-0.381
6	41678.94	$4s6p$	3.793	2.268	-0.868	-0.498
7	43933.33	$3d4p$	4.519	2.398	0.849	-0.529
8	45425.39	$4s8p$	5.318	2.498	-0.930	-0.367
9	46479.96	$4s9p$	6.231	2.576	0.970	-0.242
10	47184.43	$4s10p$	7.192	2.633	-0.985	-0.175

$n = 9*$ which has been assigned $3d4p$ by spectroscopists. However, even for this level, we see that $Z_2^2 < Z_1^2$; only the sum of Z_2^2 over levels 4–10 is close to unity. Thus the perturbation may be regarded as smeared out over these levels and no individual level should be assigned configuration $3d4p$.

Wave functions of the form (93) are used in QDT IV to compute oscillator strengths for transitions in Ca in a many-channel extension of the Coulomb approximation of Bates and Damgaard,[37], except that configuration interaction wave functions from the work of Chisholm and Öpik[50] were used for the Ca ^1S ground state. The effect of the perturbation is again seen in the behavior of the f values, which, if plotted against energy and connected by a continuous line, display a resonance profile with a peak in the region of maximum perturbation. This is illustrated in Fig. 2 which is discussed further below.

   4.1.2. Autoionizing States.   The connection between perturbed series and autionizing states becomes apparent when we consider energies for which at least one channel is open. Consider a two-channel case for which $\epsilon_1 > 0$ and $\epsilon_2 < 0$. From Eqs. (78) and (79) we obtain expressions for the scattering matrix elements S_{11} and S_{21} in terms of $\mathcal{R}$ and $\tan \pi \nu_2$, which were determined for all channels closed,

$$S_{11} = \exp 2i(\sigma_1 + \delta_1). \tag{95}$$

We obtain for the non-Coulomb phase, δ_1,

$$\tan \delta_1 = \mathcal{R}_{11} - \mathcal{R}_{12}^2/(\mathcal{R}_{22} + \tan \pi \nu_2). \tag{96}$$

We know from Eq. (89) that the quantum defects $\pi\mu(\epsilon_1 < 0)$ extrapolate to the phase shift δ_1 $(\epsilon_1 > 0)$. Comparing Eqs. (92) and (96) it is seen that the phase shift δ_1 has the same behavior as a function of the energy as the quantum defect μ: whenever $\mathcal{R}_{22} + \tan \pi \nu_2$ passes through zero, δ_1 varies rapidly. This is to be expected, since the zeros of $\mathcal{R}_{22} + \tan \pi \nu_2$ correspond to the eigenvalues of the uncoupled perturbing channel, which gives rise to perturbations of the bound states. In the case of Ca $^1P_1$, the variations in the phase shift occur as a result of the doubly excited series $3dnp\ ^1P_1$, the first member of which lies below threshold. The members located above threshold may autoionize and, due to their coupling with the continuum, are spread over a range of energies. Thus the widths of the resonances due to autoionizing states are a measure of the coupling strength. In QDT IV, positions and widths of the $3dnp\ ^1P_1$ states of Ca have been calculated by extrapolating the Y matrix determined from the energy level data for the

---

*Here n is not a principal quantum number but merely a level index.

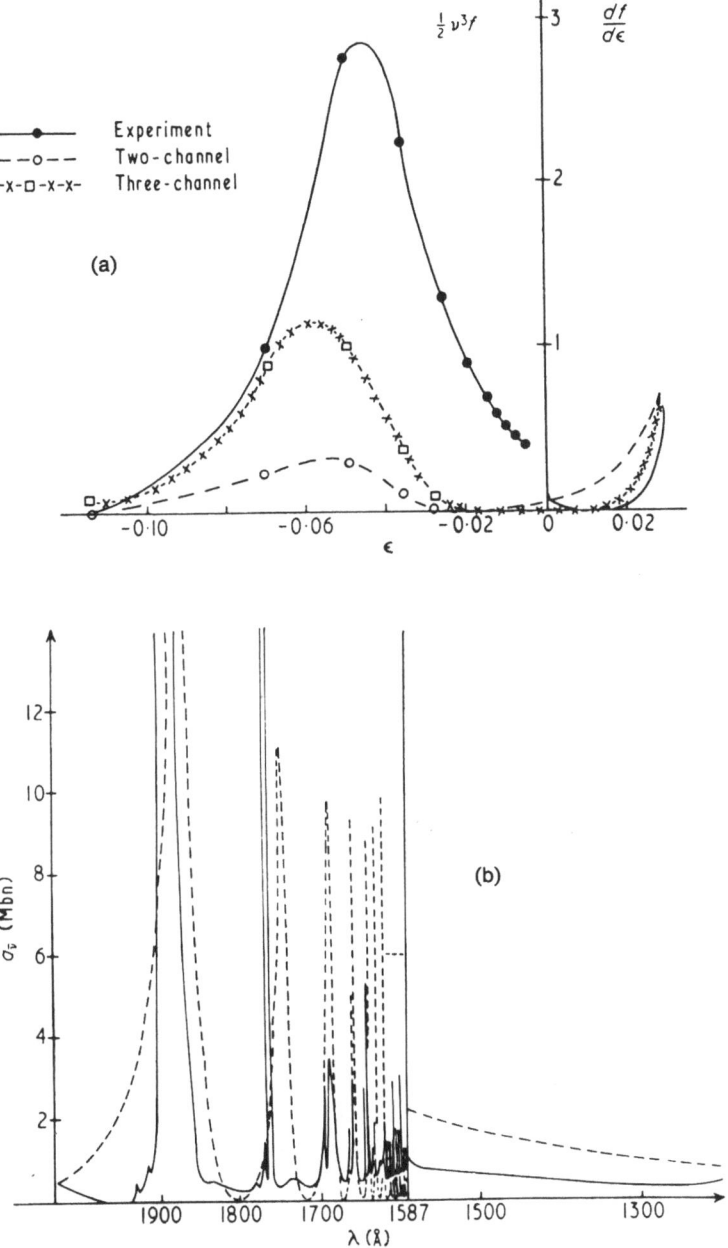

FIGURE 2. Radiative transitions to $^1P_1$ states of calcium. (a) —●—, experiment[52]; --O--, two-channel calculation; ×-□-×, three-channel calculation. (b) ——, experiment[53]; ---, two-channel calculation.

perturbed $^1P_1$ series measured by Garton and Codling[51] to positive values of the energy ε_1. The quantum defect plot and its continuation above threshold are shown in Fig. 1. The closed-channel wave function has asymptotic form

$$F_{21} \sim \theta_2 K_2 Z_{21}, \tag{97}$$

where

$$|Z_{21}|^2 = \frac{\frac{1}{2}\nu_2^3(1 + \tan^2\pi\nu_2)\gamma}{(\tan\pi\nu_2 - \alpha)^2 + \gamma^2} \tag{98}$$

with

$$\alpha = -\mathcal{R}_{22} + \mathcal{R}_{12}\gamma \quad \text{and} \quad \gamma = \mathcal{R}_{12}^2/(1 + \mathcal{R}_{11}^2). \tag{99}$$

Since α and γ are slowly varying functions of the energy, $|Z_{21}|^2$ has peaks with a distorted Lorentz profile at energies where $\alpha = \tan\pi\nu_2$. Away from these maxima, $|Z_{21}|^2$ is small. Taking the total wave function to be normalized and assuming $\gamma \ll 1$, which corresponds to nonoverlapping resonances, it may be shown that

$$\int_0^\infty d\rho \int_{\text{res}} d\varepsilon \, |F_{21}(\rho)|^2 = 1. \tag{100}$$

This result is analogous to that obtained for the perturbed bound series, for which $|Z_2|^2$ summed over a complete perturbation was unity. For small γ, summing over a perturbation or integrating over a resonance is equivalent to including one normalized bound state from channel two. If γ is large this model breaks down since the two channels are coupled so strongly that separation into perturbed and perturbing channels ceases to have much meaning.

The extrapolated **Y** matrix gives the asymptotic forms of the coupled-channel wave functions in the continuum, and hence the generalization of the Burgess–Seaton technique[40] for calculating photoionization cross sections is possible. In QDT IV photoionization cross sections for calcium were obtained from the **Y** matrix for the $^1P_1$ series. As in the case of the bound–bound transitions, the Chisholm-Öpik functions[50] were used for the ground state of Ca. This was the first calculation of a photoionization cross section with the inclusion of autoionizing resonances. It was only made possible at that time (1966) by the use of quantum defect methods because the computer power available then would have made calculations by close-coupling methods, as are used today, quite intractable. In Fig. 2, we show the results obtained in QDT IV for radiative transitions in Ca. The curve obtained by joining the values of $\frac{1}{2}\nu^3 f$ for the bound states extrapolates to $df/d\varepsilon$ for the continuum states. Figure 2 clearly shows the connection between the bound and autoionizing members of the second channel. Plots such as that of Fig. 2 are very useful for assessing the quality of the calculations on either side of a series limit. Extrapolations of $\frac{1}{2}\nu^3 f$ may be used to determine threshold values of photoionization cross sections which may be sensitive to the experimental technique.

Even though the results obtained in QDT IV were a dramatic improvement on previous calculations, which made no allowance at all for resonance effects, agreement with the experimental results of Shabanova[52] in the bound-state region and Ditchburn and Hudson[53] in the autoionizing region was only fair. The calculated oscillator strengths were low, the photoionization minimum near 1960 Å was not given correctly, and positions and widths of resonances in the vicinity of 1740 Å were quite bad. Here, perturbation by another channel occurs. Since the publication of QDT IV, additional experimental work on the absorption cross section has been carried out.[54,55] Carter et al.[56] have achieved some improvement in predicting the minimum using different extrapolation techniques. Recent theoretical work on the Ca spectrum has employed Fano's form of quantum defect theory. An important aspect of this form is that the S matrix is diagonalized by a unitary transformation

$$U_{\alpha\beta}S_{\beta\gamma}U_{\gamma\delta}^{\dagger} = e^{2i\mu_\alpha}\delta_{\alpha\delta}\,. \tag{101}$$

The elements of the unitary matrix U and the eigen quantum defects μ_α are used for extrapolations instead of Y. The condition for a bound state (61) now takes the form

$$\det \mathbf{M} = 0, \quad \text{where} \quad M_{ij} = U_{ij}\sin(\nu_i + \mu_j). \tag{102}$$

In cases for which there are only two independent values of ν, the graphical representation of relation (102) between ν_1 and ν_2 defines a curve known as the Lu–Fano plot, which has proved to be a useful and instructive tool for fitting spectra. If U and μ are constant in energy, the curve repeats itself each time ν_1 increases by one, so that if ν_1 and ν_2 are defined modulo one, a single curve is obtained. Armstrong et al.[24] have carried out a detailed analysis of new data for the $^1S_0, ^{1,3}D_2$ series using this form of the theory. However, multiconfiguration Hartree–Fock calculations[57] reveal discrepancies with the quantum defect calculations, which are attributed to the existence of high-lying perturbers in the continuum above the series being studied. Geiger,[23] working in a limited energy range near the $4\,^2S$ limit, has been able to explain the main features of the ^1P spectrum in terms of a two-channel Fano formalism. By fitting to two experimental oscillator strengths, Geiger was able to determine semiempirical values for $\frac{1}{2}\nu^3 f$ and $df/d\varepsilon$ which did not tend to the same limit at the ionization threshold, indicating that at least one of the experiments should be repeated. The threshold value of $df/d\varepsilon$ is particularly sensitive to the experimental technique used.

4.2. Bound States of Complex Ions by Extrapolation of Calculated Scattering Parameters: Configurations $1s^2 2s^2 2p^q nl$

The preceding example shows how QDT may be used for the analysis of complex spectra and how information for the continuum states may be obtained by extrapolation from negative to positive energies. The process may also be reversed, and scattering data may be extrapolated from positive to negative energies to predict positions of bound states. If all channels are open, the reactance matrix is equal to $\mathcal{R}$ and is a slowly varying function of the energy, suitable for extrapolations.

Calculations for electron scattering by ions with configurations $1s^2 2s^2 2p^q$, $q = 1, \ldots, 5$, were performed by Saraph et al.[58] in the exact resonance and distorted wave approximations. This calculation included only one configuration to describe the target. MQDT was used to assess the error of the calculation, by calculating the positions of bound states $2p^q nl$ and comparing these with experiment. The difference between calculated and measured bound-state energies gave some indication of the error in the scattering calculation. Furthermore, it was found that where this difference was not too large, usually for terms $2p^q np$, adjustments to the calculated scattering parameters could be made, using first-order perturbation theory and a variational principle for the correction of the Y matrix in order to improve the agreement with experiment. Collision strengths calculated from the adjusted matrices could be considered an improvement on the *ab initio* calculation, or, where they differed grossly, as an indication of the inadequacy of the calculation.[47]

The bound-state calculations gave, originally as a by-product, detailed analyses for spectral series $2p^q(S_i L_i)npSL$, where the calculated identification of level quantum numbers was found to be particularly useful. From the reactance matrices of Saraph et al.,[58] calculated at five energies above the highest target term due to configuration $1s^2 2s^2 2p^q$, matrices Y were obtained [Eq. (53)] and these were fitted as functions of the energy:

$$Y = Y_1/Y_2 \tag{103}$$

with

$$Y_1 = \sum_{i=1}^{4} D_i \varepsilon^{i-1} \quad \text{and} \quad Y_2 = D_5 + D_6\varepsilon. \qquad (104)$$

In order to avoid large numbers, the coefficients D were redefined: by diagonalization of the Y matrix at zero energy, one has

$$Y(0)X = X \tan \pi\eta_0,$$

which gives

$$D_1X = X \sin \pi\eta_0 \quad \text{and} \quad D_5X = X \cos \pi\eta_0. \qquad (105)$$

For comparison with spectroscopic data transformation to pair coupling is required, and the angular momenta couple as

$$S_i + L_i = J_i, \quad J_i + l = K, \quad \text{and} \quad K + s = J \qquad (106)$$

and the transformation is performed using Racah algebra as described, for example, by Saraph.[59] Since the scattering problem has many more channels in pair coupling than in LS coupling, the scattering calculation is performed in LS coupling, the fitting of the Y matrices as functions of the energy is done on the Y^{LS}, and the transformation to pair coupling is performed on the coefficient matrices D_i. To find the bound states, the equation

$$\left| \left[(\sin \pi\nu)/B \right] Y_2 + (\cos \pi\nu) Y_1 \right| = 0, \qquad (107)$$

which is equivalent to Eq. (61), was solved by iterative numerical techniques. Note that the diagonal matrix B explicitly accounts for a substantial share of the energy dependence of $\mathcal{R}$. By using calculated term energy differences in B when computing Y from $\mathcal{R}$, but measured energy level differences when computing $\mathcal{R}$ from Y and when solving (107) for bound states, some correction for the effects of the (poor) calculated term differences on $\mathcal{R}$ was possible as well as the introduction of the fine-structure splitting. In particular, the large ^2P term splitting in the rare-gas-like ions could be imposed in this way. As a result, the complex interaction of series $2p^5(^2P_{3/2})nlKJ$ and $2p^5(^2P_{1/2})n'lKJ$ was reproduced by this calculation in a qualitatively correct form, as shown in Figs. 2–4 of Saraph and Seaton.[47]

For each solution of (107), the channel mixing parameters

$$W_\alpha = \frac{A_\alpha^2 Z_\alpha^2}{\sum_{\alpha'} A_{\alpha'}^2 Z_{\alpha'}^2} \qquad (108)$$

were calculated (see extensive tables in reference 47). Here, A_α and Z_α are defined by Eqs. (22) and (62) and α denotes a set of channel quantum numbers. Bound states for which any one W_α is much larger than all the others are defined by that α. Saraph and Seaton[47] show in their tables that often bound states can not be described by any one parent level $(S_i L_i J_i)$, instead, extensive mixing of channels with different parent levels occurs. They also found generally that low bound states are well defined in LS coupling, $\alpha = (L_i S_i) n l L S$, and high bound states in pair coupling, $\alpha = (J_i) n l K J$. This was of course already known from spectroscopy, but the multichannel quantum defect method offered a way of finding the most suitable coupling scheme from calculated data, by seeing which coupling scheme gave the best defined α. The study of the variation of coupling scheme with energy and the finding of suitable sets of identifying level quantum numbers by frame transformations is a vital element in the work of Fano and collaborators,[19–22] and with accurate calculated scattering data made available by up-to-date computing methods it is an important tool in spectral analysis.

4.3. The Spectrum of the H_2 Molecule

As another example of the applications of MQDT to perturbed Rydberg series we consider the $v = 0$ $1s\sigma np\sigma$ $^1\Sigma_u^+$ and $1s\sigma np\pi$ $^1\Pi_u^+$ states of the H_2 molecule. The phenomenon of l uncoupling in these series was the subject of the first application of Fano's form of the theory.[19] The problem is closely analogous to that of the Ne spectrum discussed in the previous section; again, it is necessary to perform a frame transformation to describe the spectrum as the energy changes. One has two Rydberg series whose higher members perturb one another and converge onto different angular momentum states of the ionic core. Two different rotational states of H_2^+ play the same role as the two ^2P levels of Ne^+. Consider an electron in the field of a H_2^+ ion in its lowest electronic and vibrational ($v = 0$) state. In the body-fixed or molecular frame, we expand the total wave function in the eigenfunctions of total angular momentum $X_{JM}^{l\Lambda\eta}(\hat{r}, \hat{R})$ defined by Chang and Fano[60] [their Eq. (3)]. Provided rotational terms in the Hamiltonian are negligible, Λ is a good quantum number (adiabatic nuclei approximation) and can be used to label the energy levels. We shall call this scheme a. In the laboratory frame, we expand in the alternative eigenfunctions $\Phi_{JM}^{lj}(\hat{r}, \hat{R})$, where j is the rotational quantum number of the ion core. The energy levels are now labeled by the value of j, where $\mathbf{J} = \mathbf{j} + \mathbf{l}$, and their parity (scheme b). The first scheme applies to situations in which the electron is at short distances from the core, in contrast to the second which applied to large separations. As one proceeds up a Rydberg series, due to the increasing importance of the external region, one sees a transition from the first scheme to the second. In the case considered by Fano, low-lying members of the series conform very closely to either $\Lambda = 0$ (Σ) or $\Lambda = 1$ (Π) levels. In this case, coupling between different l values is very weak and only the $l = 1$ channel need be considered. For $n > 5$ or 6 unique identification begins to break down due to the coupling of the two Λ values, and the levels go over to two coupled series with ($l = 1$, $j = 0$) and ($l = 1$, $j = 2$), respectively, converging on to the $j = 0$ and $j = 2$ levels of the H_2^+ ion, which are separated by $\Delta E = 12B$.* Between these two thresholds auto-ionizing states converging on the upper level are found. Thus in scheme a we have two uncoupled channels that we label σ and π while in scheme b we have two coupled channels that we label by their j values (0 and 2). Let ε_0 be the energy in Rydbergs relative to the $j = 0$ level of the ion and ε_2 be the energy relative to the $j = 2$ level, and $\varepsilon_j = -1/\nu_j^2$. Then we have $\varepsilon_2 = \varepsilon_0 + 12B$ and $\nu_2^{-2} = \nu_0^{-2} - 12B$.

Let us define matrices $\mathcal{R}^a$ and $\mathcal{R}^b$ pertaining to schemes a and b respectively. They are related by the transformation

$$\mathcal{R}^a = U^\dagger \mathcal{R}^b U \tag{109}$$

where U is unitary. In scheme b, the condition for a bound state, Eq. (61), may be written

$$|\mathcal{R}^b + t^b| = 0 \tag{110}$$

where t^b is a diagonal matrix with elements

$$t_j^b = \tan \pi\nu_j, \quad j = 0, 2. \tag{111}$$

Equation (110) may be written out explicitly

$$\begin{vmatrix} \mathcal{R}_{00}^b + \tan \pi\nu_0 & \mathcal{R}_{02}^b \\ \mathcal{R}_{20}^b & \mathcal{R}_{22}^b + \tan \pi\nu_2 \end{vmatrix} = 0. \tag{112}$$

Using (109) and (110) we obtain

$$|\mathcal{R}^a + t^a| = 0, \tag{113}$$

where

$$t^a = U^\dagger t^b U. \tag{114}$$

*Where B is the rotational constant.

In our example, $\mathfrak{R}^a$ is diagonal, with elements $\mathfrak{R}^a_\sigma$ and $\mathfrak{R}^a_\pi$, and

$$t^a_{\Lambda\Lambda'} = \sum_{j,j'=0,2} U(ljJ\Lambda)\tan(\pi\nu_j)U(l'j'J\Lambda') \qquad (115)$$

with $l = l' = J = 1$ and

$$U(ljJ\Lambda) = (2 - \delta_{\Lambda 0})^{1/2}(-1)^{J+\Lambda}C^{lj}_{-\Lambda\Lambda 0}, \qquad (116)$$

where $C^{l'L}_{mm'M}$ is a Clebsch–Gordan coefficient. One obtains

$$t^a = \frac{1}{3}\begin{bmatrix} t_0 + 2t_2 & \sqrt{2}\,(t_0 - t_2) \\ \sqrt{2}\,(t_0 - t_2) & 2t_0 + t_2 \end{bmatrix}. \qquad (117)$$

Note that in the limit of zero rotational splitting $(B \to 0)$ $t^b_2 \to t^b_0 \to t^b$ and $t^a \to t^b$ as it must do. For low-lying levels one has term energies $E_n \gg 12B$, which gives $\nu_0 \simeq \nu_2 \simeq \nu$ and $t^b_0 \simeq t^b_2 = t^b$. Then t^a becomes diagonal and system splits up into two independent systems with bound states at $\mathfrak{R}^a_\sigma +$ $\tan \pi\nu = 0$ and $\mathfrak{R}^a_\pi + \tan \pi\nu = 0$, respectively. As the threshold is approached, however, $\nu_0 \to \infty$ while $\nu_2 \to (12B)^{-1/2} \simeq 25.0$. Here, t_0 and t_2 are very different and the σ and π channels are strongly coupled: the adiabatic nuclei theory does not describe the system.

Using MQDT it is possible to carry out a quantitative analysis of these perturbing series. In the following, we use the experimental data of Monfils[61] and Herzberg and Jungen.[26] These are shown in Table 3, the levels being labeled by the integer n, starting from $n = 1$. From the data we are able to obtain the elements of the matrix $\mathfrak{R}^a$. Fitting procedures are used which enable the problem to be discussed in a more precise fashion than the more descriptive treatment of Fano.[19] Writing $\mathfrak{R}^a_\sigma = x$, $\mathfrak{R}^a_\pi = y$, Eq. (113) may be rewritten

$$(x + t^a_{\sigma\sigma})(y + t^a_{\pi\pi}) - (t^a_{\sigma\pi})^2 = 0, \qquad (118)$$

and using (117) this equation may be put into the alternative form

$$\tan \pi\nu_0 = \frac{-[(x + 2y)\tan \pi\nu_2 + 3xy]}{3\tan \pi\nu_2 + x + y}. \qquad (119)$$

Equation (119) defines a curve in the ν_0, ν_2 plane (a Lu–Fano plot). If x and y are constant the same curve is obtained for all ν_0, ν_2 (modulo 1). Using a technique to be described elsewhere [Moores[62]], the values of x and y were obtained by fitting levels 10–43 to (119) using an iterative

TABLE 3

The np ($v = 0$) Series in H_2

Label[a]	$\nu_0{}^b$	$\Delta\nu^c$	$Z_\sigma{}^d$	$Z_\pi{}^d$	$I(obs)^e$	$\lvert Z_\sigma + \sqrt{2}\,Z_\pi \rvert^2$
2	1.790	0.000	1.000	0.007	—	5.1
3	2.081	0.000	− 0.006	1.000	—	9.9
4	2.811	0.000	1.000	0.026	—	5.4
5	3.091	0.000	− 0.024	1.000	—	9.7
6	3.828	− 0.002	0.998	0.065	—	6.0
7	4.102	0.000	− 0.061	0.998	—	9.1
8	4.863	+ 0.006	0.991	0.135	7	7.0
9	5.126	+ 0.002	− 0.128	0.992	7	8.1
10	5.884	− 0.007	0.971	0.240	7	8.6
11	6.162	− 0.001	− 0.230	0.974	7	6.6
12	6.935	+ 0.008	0.930	0.368	8	10.5
13	7.234	+ 0.003	− 0.350	0.935	6	4.7
14	7.947	− 0.013	0.877	0.481	8	12.2
15	8.338	− 0.002	− 0.455	0.891	4	3.2
16	8.984	− 0.004	0.839	0.544	8b	13.0
17	9.485	− 0.013	− 0.493	0.870	2	2.7
18	10.013	− 0.004	0.836	0.549	8b	13.0
19	10.726	+ 0.027	− 0.412	0.911	5	3.8
20	11.065	+ 0.003	0.894	0.449	7b	11.7
21	11.883	− 0.003	− 0.058	0.998	7	9.2
22	12.192	+ 0.001	0.996	0.094	7	6.4
23	12.976	+ 0.001	0.238	0.971	7	13.0
24	13.513	+ 0.020	0.991	− 0.140	5	3.2
25	14.032	+ 0.003	0.250	0.968	7	13.1
26	14.804^f	− 0.043	0.979	0.205	8	8.1
27	15.169	+ 0.002	− 0.153	0.988	5	7.8
28	15.984	0.000	0.813	0.583	7	13.4
29	16.615	− 0.010	− 0.403	0.915	1	4.0
30	17.070	+ 0.004	0.871	0.491	6	12.3
31	17.955	0.000	0.233	0.972	6	13.0
32	18.481	− 0.008	0.997	− 0.084	0	3.9
33	19.200^f	− 0.154	0.252	0.968	8ba	13.2
34	19.937	− 0.013	0.839	0.544	6	13.0
35	20.455^f	+ 0.560	− 0.388	0.922	0	4.2
36	21.062	+ 0.003	0.839	0.545	6	13.0
37	21.960	− 0.017	0.308	0.952	6	13.7
38	22.738	+ 0.028	0.997	0.075	0	6.1
39	23.133	0.000	0.044	0.999	5	10.6
40	24.004	− 0.003	0.757	0.653	6	14.2
41	24.945	+ 0.020	0.218	0.976	5	12.8
42	25.456	− 0.038	1.000	0.010	3	5.2
43	26.096	+ 0.018	0.235	0.972	4	13.0

[a] Index of energy level.
[b] Experimental value of effective number ν_0.
[c] $\Delta\nu = \nu_0(obs) - \nu_0(calc)$.
[d] Values of Z_σ and Z_π.
[e] Observed intensity on a scale of 1 to 8.
[f] Observed level badly blended.

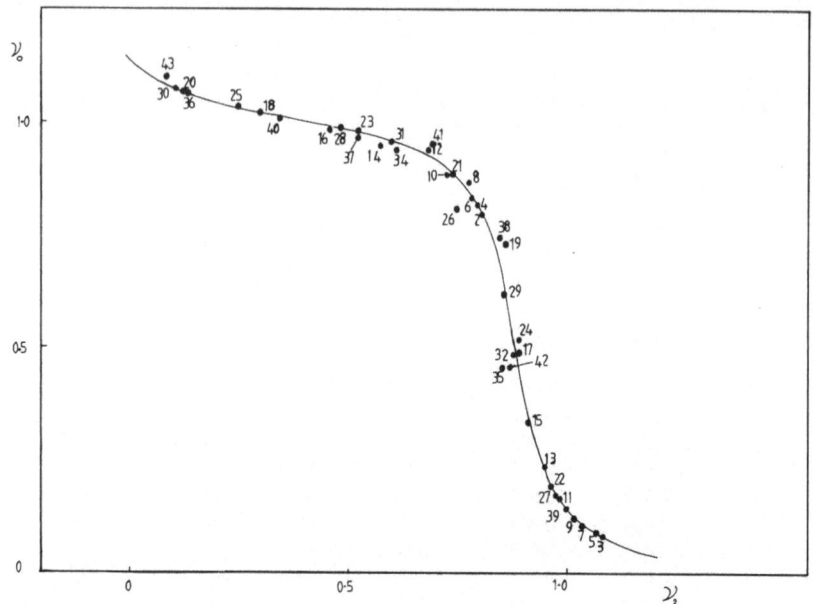

FIGURE 3. Lu–Fano plot for np series in H_2: ●, experimental points; curve, fit to Eq. (119).

nonlinear least-squares method. The lower levels were then also made to satisfy (118) by adding linear and quadratic terms in the energy which for levels 10–43 gave a sufficiently small contribution that the fit was not upset. Figure 3 shows the Lu–Fano plot for the observed points and the calculated curve. The channel mixing is obtained by calculating, from the $\mathcal{R}$-matrix elements, the coefficients Z_σ and Z_π defined in a similar manner to the Z_i of Eq. (93). The total wave function is of the form

$$\Psi = Z_\sigma \Psi_\sigma + Z_\pi \Psi_\pi \tag{120}$$

where the Ψ_Λ are wave functions for each channel normalized to unity. The values obtained for Z_σ and Z_π (Table 3) illustrate the predominantly σ or π character of the low-lying levels, and the effects of channel mixing due to l uncoupling as the energy increases.

From the work of Herzberg and Jungen[26] it may be shown that the intensities of the spectral lines obtained in transitions from the gound state to members of the series depend approximately on the quantity $(Z_\sigma + \sqrt{2} Z_\pi)^2$. This is shown (in arbitrary units) in column 7 of Table 3. In column 6 are given the approximate intensties as observed by Herzberg and

Jungen[26] on a scale of 1–8. It is seen that the variations in the two columns correlate very well.

Above the $j = 0$ series limit but below the $j = 2$ limit, the $j = 2$ states are autoionizing. The quantum defect for $j = 0$ extrapolates to a phase shift for p-electron scattering by H_2^+, $j = 0$, which is given by an expression of the form (96). The rapid variations in the phase shift as $\mathcal{R}_{22}^b + \tan \pi \nu_2$ passes through a zero give rise to autoionizing states for $\nu_2 > 25$ which have been observed[26,63] as apparent emission features, or window-type resonances, in the absorption continuum. Dubau and Seaton[16] show that the photoionization cross section in this region depends on a profile factor

$$P = \frac{(x_2 + q)^2}{(1 + x_2^2)}, \tag{121}$$

where q is the ratio of the strength of the radiative transition to the resonance state to that to the continuum, and $x_2 = \tan[\pi(\nu_2 + \alpha)]/\tanh(\pi\beta)$, where $\alpha + i\beta$ is the complex quantum defect,[12] which varies slowly with energy. Equation (121) applies to the complete series of resonances. Since x_2 is a periodic function of ν_2 the pattern of resonances repeats itself, with minor deviations due to the small energy variation of q, α, and β, each time ν_2 changes by unity. Taking $q = 0$ in (121) and calculating x_2 from the extrapolated reactance matrix elements, one obtains a series of window resonances whose minima occur at $\nu_2 = n + 0.89$, $n \geqslant 25$. This compares reasonably well with experiment[26,63] below $n = 31$, but for larger ν_2 the resonances are perturbed strongly by coupling to vibrational channels. This has been taken into account by Jungen and Dill[32] in an elegant application of Fano's adaptation of the theory to molecules. Excellent agreement with experiment is obtained.

5. Extrapolation of the Generalized Reactance Matrix

The basis for this work is Eq. (54)

$$\mathcal{R}(E) = \mathcal{Q}(E)\mathcal{P}^{-1}(E),$$

which expresses the generalized reactance matrix as a ratio of two power series in the energy. Calculated reactance matrices **R** are identical with $\mathcal{R}$ at energies where all channels are open and can then be fitted in the same way. At energies where some channels are closed, **R** has poles due to quasibound states in these channels and is obtained by direct solution of the close-coupling equations or by extrapolation of $\mathcal{R}$ and use of Eq. (78)

$$\mathbf{R} = \mathcal{R}_{oo} - \mathcal{R}_{oc}(\mathcal{R}_{cc} + \tan \pi \nu)^{-1} \mathcal{R}_{co}.$$

At energies just below a new threshold, direct solution of the close-coupling equations becomes numerically difficult, but the quantum defect method (78) remains unaffected and provides an easy means to obtain $\mathbf{R}$ at these energies.

If $\mathbf{R}$ has been obtained from solutions of the close-coupling equations at energies where some channels are closed, it can still be fitted to expressions of the form (54), where the denominator has zeros at the energies of the quasibound states due to the closed channels. If all closed channels have the same energy, $\mathbf{R}$ can be fitted to the form (78), giving $\mathcal{R}$.

5.1. Discussion of Extrapolation Methods

5.1.1. Restrictions on the Validity of Extrapolation Methods. In principle, expression (54), having been obtained from calculated reactance matrices, should contain all the scattering and bound-state data for the electron plus ion system over the entire range of positive and negative energies. In practice, the range of validity is restricted by (i) the long-range nature of the potential, (ii) certain intrinsic errors in the scattering calculations, and (iii) the truncation of the power series expansions in (54).

Error source (i) concerns the breakdown of condition (iii) of (43): the actual potential for electron–ion interactions is not purely Coulomb-type, even at large distances. As a result, QDT expressions for the scattering data as functions of energy are only approximately correct. In practice, it is found that these expressions are good working formulas for cases where the quantum defects or phase shifts are mainly due to the short-range interactions.

Error source (ii) warrants some detailed discussion. The scattering calculations suffer from three intrinsic error sources that affect the viability of extrapolation techniques.

Firstly, the close-coupling expansion of the total wave function for an ion with $N + 1$ electrons

$$\Psi(SL\pi, E; \mathbf{r}) = \mathcal{C}\left[\sum_i \phi_i(S_i L_i \pi_i, E_i; \bar{\mathbf{r}}_{N+1}) \left(\frac{1}{r_{N+1}} \right) F(SL\pi, l_i, \varepsilon_i; r_{N+1}) \right.$$

$$\left. + \sum_j \Phi_j(SL\pi; \mathbf{r}) c_j(E) \right] \tag{122}$$

is for practical reasons trunctated, so that the sum over i, the number of free channels, is smaller than 20, and the sum over bound and correlated ion channels, j, is less than 100. In (122), S, L, π, and E are the angular momenta, parity, and energy of the total ($N + 1$ electron) system, S_i, L_i, π_i, and E_i those of the ion state belonging to channel i, ϕ_i represents a target state i and the angular part of the ($N + 1$)th electron's wave function and depends on all coordinates except r_{N+1}. The target states are kept "frozen." The Φ_j are introduced to compensate for the orthogonality conditions imposed on $\mathbf{F}$, and they are used to compensate for effects due to higher target terms omitted in the first sum. Expansion (122) is discussed in detail in Chapter 1. Here, we are only concerned with the fact that reactance matrices calculated from the asymptotic form of $\mathbf{F}$ have certain shortcomings. The truncation of the sum over i restricts the validity of the scattering data to energies below those where the omitted channels would be open or would cause perturbations due to quasibound states. Obviously, the same restrictions limit the range over which reactance matrices obtained from these calculations can be extrapolated.

Secondly, the additional states Φ_j, made up of frozen target and pseudo-orbitals, can introduce pseudoresonances, unphysical states that look like real perturbers. When such pseudoresonances are well isolated and narrow, it is possible to avoid energies where they occur when obtaining $\mathbf{R}(E)$ from the calculated reactance matrices. It is then possible to use $\mathbf{R}(E)$ to interpolate at energies where the close-coupling solutions give spurious results.

Thirdly, numerical errors in the calculation for the radial function $\mathbf{F}$ are strongly energy dependent and can introduce spurious energy dependence in $\mathbf{R}(E)$. This numerical error can be controlled, in practice, by careful computing techniques and checking of results, so that it is kept smaller than other errors.

Finally, we must consider that the close-coupling solutions are also in error due to the use of inexact target-state functions. In order to obtain quantum defects and interaction patterns that are comparable with experiment, it is essential to use experimental energy differences for the target terms. For example, consider two interacting series converging to terms i and j, respectively. The effective quantum numbers in channel i are $\nu_i = z/(-\varepsilon_i)^{1/2}$. The mth bound state of channel j has energy $\varepsilon_{jm} = -z^2/\nu_{jm}^2$ with respect to term j but has energy $\varepsilon_{im} = \varepsilon_{jm} + E_j - E_i$ with respect to term i. The perturbing channel j has an effective quantum number $\nu_{im} = z/(-\varepsilon_{im})^{1/2}$ in channel i. If the energy difference $E_j - E_i$ is wrong, the perturbing level will be wrongly positioned within the sequence that converges to term i, and the interference pattern will be quite wrong. If the fine-structure splitting of terms is important, the B matrix can be used to replace term energies by level energies as described in detail in Section 4.2. Therefore, the use of inexact target wave functions can be compensated for to some extent by the use of experimental energies for the target states. The residual error $\delta\mu$ will then be nearly energy independent. But since $\delta\mu$ is still strongly l-dependent, calculations of perturbation effects between series with different valence electron angular momenta l are a severe test for the approximation used.

The third error source, truncation of the power series expansions for $\mathbf{R}(E)$, should be assessed in the light of the foregoing discussion. Clearly, only the first few terms will be dominated by the physical effects that we want to describe. Effects due to error sources (i) and (ii) will play a major role in the coefficent of higher order in E, so the expansions should be kept short and extrapolations over long distances should be avoided. Close-coupling solutions can be obtained at appropriate intervals so that the $\mathbf{R}(E)$ can be checked or newly calculated.

5.1.2. Fitting Techniques. A computer program, RANAL, has been developed by Seaton and many collaborators and is widely used for the fitting of calculated reactance matrices as functions of energy and for the extrapolation and interpolation of scattering data. One picks, as input, reactance matrices $\mathbf{R}$ obtained as solutions of the scattering problem at three or four energies that are free from resonance effects and, to account for any resonances, $\mathbf{R}$ at about five energies near to any one resonance due to a closed or bound channel. Instead of solving the matrix equation (54) directly for the expansion coefficients, as was indeed done in the work described in Section 4.2, it is computationally more economical to first use the method of least-squares fitting on individual matrix elements and to find the pole positions:

$$R_{ij}(E) = \left(\sum_{r=1}^{q} D_{ij}^{(r)} E^{r-1} \right) \left(1 - \sum_{r=q+1}^{m} D_{ij}^{(r)} E^{r-q+1} \right)^{-1}, \qquad (123)$$

where the number of input $\mathbf{R}$ matrices is larger than m, and the number of poles in the energy range is smaller than $m - q$. The quality of the fit can be examined and, if necessary, improved by varying q or m or adding more input data. Next, the pole positions are determined, for individual matrix elements, from the zeros of the denominators. Due to numerical inaccuracies these pole energies vary from element to element, and, if the spread is too large, the fitting procedure must be repeated with a more suitable range of input $\mathbf{R}$. Once fair agreement has been obtained, these pole positions are used as the starting points for the iterative scheme

$$R_{ij}(E) = \sum_{r=1}^{s} A_{ij}^{(r)} E^{r-1} - \sum_{p=1}^{NP} B_{ij}^{(r)} (E - E_p)^{-1}, \qquad s + NP < m \quad (124)$$

where the differences between the input $\mathbf{R}$ and $\mathbf{R}(E)$ are minimized by varying E_p. Expression (124) gives $\mathbf{R}$ in terms of a slowly varying part (the background value) and parts with simple poles at the energies E_p. We know that the poles arise from closed–open and closed–closed elements of the symmetric $\mathfrak{R}$ matrx. The residuals, the coefficients $B_{ij}^{(p)}$, may therefore be factorized into matrix elements connecting the open channels to the closed channel that causes the pole at E_p:

$$R_{ij}(E) = \sum_{r=1}^{s} A_{ij}^{(r)} E^{r-1} - \sum_{p=1}^{NP} C_{ip}(E - E_p)^{-1} C_{jp}, \qquad C_{ip} = C_{pi}. \quad (125)$$

If the parent levels of the quasi-bound states causing the poles are known, $\mathbf{R}$ may be fitted to an expression resembling the quantum defect formula (78). At the poles, the eigenvalues $\mathfrak{R}_p$ of $\mathfrak{R}_{cc}$ satisfy

$$\mathfrak{R}_p + \tan \pi\nu(E_p) = 0. \quad (126)$$

So instead of expression (5.4) one uses

$$R_{ij}(E) = \sum_{r=1}^{s} A_{ij}^{(r)} E^{r-1} - \sum_{p=1}^{NP} Q_{ip}[\mathfrak{R}_p + \tan \pi\nu(E)]^{-1} Q_{pj}, \quad (127)$$

which represents the repetitive character of the perturbation pattern over a range of energies. In (126) and (127) the continuous effective quantum number $\nu(E)$ refers to the target terms onto which the closed-channel bound states converge. For large ν extrapolations over a short range of energies using (127) will faithfully reproduce several resonances, since the pattern is repeated every $\delta\nu = 1$. The formula is therefore particularly useful at energies just below a threshold, where close coupling calculations are numerically difficult. But quite generally, Eq. (127) provides a practical alternative to c.c. calculations at a very fine energy mesh.

5.2.   Applications

5.2.1.   *Collision Strengths in LS Coupling.*   Expression (125) may be used to give a detailed description of resonances over a short range of energies. Such data are necessary for the evaluation of collision rates when the electron velocity distribution has a maximum overlapping with these resonances.

At energies just below a new threshold, resonances due to quasibound states converging onto that threshold become so crowded that only their average effect is of practical interest. Using the analytical properties of the χ matrix, Gailitis[18] and Seaton[10] have derived expressions for $|S_{ij}|^2$

averaged over resonances

$$\langle |S_{ij}|^2 \rangle = |\chi_{ij}|^2 + \sum_{pq} \frac{\chi'_{ip}\chi'_{pj}\chi'^*_{iq}\chi'^*_{qj}}{1 - \chi'_p\chi'^*_q} \tag{128}$$

The χ-matrix elements in (128) are obtained from the extrapolated generalized reactance matrix $\mathscr{R}$. The relation between χ and χ' involves the eigenvectors of the closed–closed part of χ:

$$\chi_{cc}X = X\chi'_c, \quad \chi'_{oc} = \chi_{oc}X, \quad \text{and} \quad \chi'_{co} = X^T\chi_{co}. \tag{129}$$

The collision strength is given by

$$\Omega(S_iL_i - S_jL_j) = \tfrac{1}{2} \sum_{SL\pi} (2S+1)(2L+1)\sum_{l_il_j} |S^{SL\pi}(S_iL_il_i - S_jL_jl_j)|^2 \tag{130}$$

5.2.2. Collision Strengths for Excitations between Fine-Structure Levels.
For light ions where relativistic effects are not important, reactance matrices in pair coupling can be obtained from reactance matrices calculated in *LS* coupling by an algebraic transformation and these can be used to compute collision strengths for excitations between fine-structure levels.[59]

In order to explore resonance effects or obtain cross sections for excitations between two levels at energies just below a threshold one may obtain the generalized $\mathscr{R}$ matrices at the energies of interest, transform these to pair coupling, and then proceed as in Section 5.2.1. For each $SL\pi$ the resonances occur at different energies, and in order to obtain the very complex resonance structure for the total collision strengths in intermediate coupling the $\mathscr{R}(E)$ are required at a very fine mesh. For optically forbidden transitions, only the first few partial waves contribute significantly. The method has been used to obtain detailed collision strengths for $^2D_{5/2}-^2D_{3/2}$ transitions in ions with outer configuration p^3. In order to use the quantum defect formula (78) here, one has to neglect the small fine-structure splitting of the term $p^3\,^2P$. To carry out the calculations, the program JAJOM[59] was adapted by Pradhan (JAJOM II, to be submitted) to obtain the resonance profiles, rate coefficients, and Gailitis average (128) using $\mathscr{R}^{LS}$ obtained at energies below the ^2P threshold by the program RANAL as input. Results for O II, Ne V, and S II have been published.[11,64-66]

5.2.3. Photoionization Cross Sections. The use of close-coupling wave functions for the initial and final states in the calculation of photoionization cross sections makes it possible to calculate and identify the observed resonance pattern. Here, as in the calculation for electron impact excitation, quantum defect methods can be used to obtain detailed resonance analyses from solutions of the close-coupling equations at relatively few energies. In 1973, Luke[67] produced the first results combining the two methods. He dealt with autoionization states in the photoionization spectrum of neon. In the following years, computational techniques were improved and generalized while results for many complex ions were obtained. In QDT XIII, Dabau and Seaton[16] summarize the theoretical formulation required in this application of quantum defect methods.

The differential oscillator strength for photoionization to an open channel i' is given by

$$\frac{d}{dE} f_{i',a} = \tfrac{2}{3}(E - E_a)|(\Psi_{i'}(\text{out}, E)|\vec{P}|\Psi_a)|^2, \tag{131}$$

where $\vec{P}$ is the dipole operator

$$\vec{P} = \sum_{n=1}^{N+1} \vec{r}_n . \tag{132}$$

We use close coupling wave functions as given by (122). We shall write $\bar{r}_{N+1}$ to denote all coordinates except r_{N+1} and we shall use r (without subscript) to denote r_{N+1}. Ψ_a is the wave function for the initial bound state and $\Psi_{i'}$ has an outgoing wave in channel i', so the normalization conditions for the F are

$$\sum_i \int_0^\infty F_i^*(E_a) F_i(E_a') dr + \sum_j c_j(E_a) c_j(E_a') = \delta(E_a, E_a'), \tag{133}$$

$$\sum_i \int_0^\infty F_{ii'}^*(\text{out}, E) F_{ii''}(\text{out}, E') dr + \sum_j c_{ji'}(E) c_{ji''}(E') = \delta(i', i'')\delta(E - E').$$

$$\tag{134}$$

The outgoing wave has asymptotic form

$$F(\text{out}, zr) = \phi^+ - \phi^- S^* \tag{135}$$

which differs from the form of a scattered wave (48). Equation (131) may be rewritten to separate the part that depends on the photoelectron energy, $\varepsilon_i = E - E_i$. Writing

$$\frac{d}{dE} f_{i',a} = \tfrac{2}{3}(E - E_a)|\vec{G}_{i'a}|^2, \tag{136}$$

one has

$$\vec{G}_{i'a} = \sum_i \int_0^\infty F_{ii'}^*(\text{out}, E)\vec{g}_{ia}\, dr + \sum_j c_{ji'}(E)\int_0^\infty \Phi_j \vec{P}\Psi_a\, d\mathbf{r}_1 \cdots d\mathbf{r}_{N+1} \quad (137)$$

and

$$\vec{g}_{ia} = r\int_0^\infty \phi_i^* \vec{P}\Psi_a\, d\vec{r}_{N+1}. \quad (138)$$

Expression (137) shows that the amplitude of the differential oscillator strength will have poles at the same energies as the **S** matrix.

In practical calculations one works with real functions that have **R** matrix asymptotic forms. The computed reactance matrices may be fitted to quantum defect expressions (125) or (127) using the program RANAL, for example. The amplitudes $\vec{G}'$, from real functions **F**, are computed separately by the program PHOTUC, which was designed to calculate photoionization cross sections from the solutions of the close-coupling equations.[68] An extension to RANAL was written (by A. Pradhan, unpublished) that fits the $\vec{G}'$ to quantum defect expressions

$$\vec{G}_{i'}(E) = X_{i'} - \sum Y_p(E - E_p)^{-1}C_{i'p} \quad (139)$$

or

$$\vec{G}_{i'}'(E) = X_{i'} - \sum Z_p\left[\mathfrak{R}_p + \tan \pi\nu(E)\right]^{-1}Q_{i'p} \quad (140)$$

using E_p and **C** or $\mathfrak{R}_p$ and **Q** obtained from the RANAL fit (125) or (127). The amplitudes can thus be interpolated or extrapolated to the desired energies and then normalized to **S*** asymptotic form

$$\vec{G}(\text{out}(S^*), E) = \vec{G}'(\mathbf{R}, E)\{(\text{Im}\,\mathbf{S}) + i(1 + \text{Re}\,\mathbf{S})\} \quad (141)$$

and the photoionization cross section obtained. The **S** matrices in (141) are also computed from the fitted **R** matrices.

The method used here is quite general: it is not restricted to a small number of open or closed channels and there are no restrictions on the magnitude of the interaction between the autoionizing state and the continuum. As early as 1961, Fano[69] derived a formula that allowed accurate fitting of the resonance profile due to the interference of one discrete state with the photoionizing continuum. The more general expressions derived here reduce to the Fano formula for the case of one closed channel giving rise to a narrow resonance, i.e., weak interaction, as demonstrated by Dubau and Seaton.[16]

5.2.4. *Electron Impact Ionization Cross Sections.* The Coulomb–Born exchange approximation to the cross section for electron impact ionization of a positive ion with $N + 1$ electrons is given by[70]

$$Q(k_0^2) = \frac{1}{2} \int_0^{E/2} \sigma(k_0^2, \chi^2) \, d\chi^2, \tag{142}$$

where $\sigma(k_0^2, \chi^2)$ is the cross section for ejected electron energy χ^2, and $E = (1/2)k_0^2 - I$, I being the ionization energy.

$$\sigma(k_0^2, \chi^2) = \frac{k\chi}{4\pi k_0 \omega_0} \sum_{\gamma M} \int q_\gamma(\boldsymbol{\chi}, \mathbf{k}) \, d\hat{k}_0 \, d\hat{k} \, d\hat{\chi}, \tag{143}$$

where

$$q_\gamma(\boldsymbol{\chi}, \mathbf{k}) = |f_\gamma(\boldsymbol{\chi}, \mathbf{k})|^2 + |f_\gamma(\mathbf{k}, \boldsymbol{\chi})|^2 - \mathrm{Re}\, f_\gamma^*(\boldsymbol{\chi}, \mathbf{k}) f_\gamma(\mathbf{k}, \boldsymbol{\chi}), \tag{144}$$

and $f_\gamma(\boldsymbol{\chi}, \mathbf{k})$ is the Coulomb–Born ionization amplitude

$$f_\gamma(\boldsymbol{\chi}, \mathbf{k}) = -(2\pi)^{-5/2} \big(\Psi_\gamma(\mathrm{out}, \chi^2) |\vec{V}_{N+1}| \Psi_0\big), \tag{145}$$

where

$$\vec{V}_{N+1} = \int \phi^*(z, -\mathbf{k}_0, \mathbf{x}_{N+2}) \sum_{i=1}^{N+1} \frac{1}{r_{i,N+2}} \phi(z, -\mathbf{k}, \mathbf{x}_{N+2}) \, d\mathbf{x}_{N+2}. \tag{146}$$

Here, $\mathbf{k}_0, \boldsymbol{\chi}$, and $\mathbf{k}$ are the wave vectors for the incident, ejected, and scattered electron, respectively, ω_0 is the statistical weight of the initial state, and z the residual charge on the ion. The symbol $\mathbf{x}_i$ denotes space and spin coordinates of electron i, and $\phi(z, -\mathbf{k}, \mathbf{x})$ are Coulomb wave functions [reference 70, their Eq. (3.1)]. Expressing $f_\gamma(\boldsymbol{\chi}, \mathbf{k})$ in this form emphasizes the similarity between the theory of photoionization and the Coulomb–Born treatment of electron impact ionization, as the form of the matrix element in (145) is the same as that of the transition matrix element for photoionization in (131), but with the dipole operator $\vec{P}$ replaced by the $(N + 1)$-electron multipolar operator $\vec{V}_{N+1}$. Jakubowicz and Moores[70] have written a computer program similar to PHOTUC[68] for the calculation of Coulomb–Born ionization cross sections from Eqs. (142)–(146). Since close-coupling wave functions calculated by the program IMPACT[71] are used for Ψ_0 and Ψ_γ, resonances will appear in the cross section (143) in the same photoionization. Jakubowicz[72] has adapted the program RANAL to treat the more general case of electron impact ionization in a similar fashion to the extension for photoionization. Several additional complications occur: The operator $\vec{V}_{N+1}$ is energy dependent, but this dependence is so slow that, provided the resonance is not too broad, it can be neglected over these small energy intervals. The ionization cross section involves a summation over a large number of final states $\Psi_f(SL\pi)$. Each partial wave $SL\pi$ contributing will contain resonances, but all at different energies. So each case must be treated separately, for the purpose of fitting as a function of the energy. Moreover, the integration in (142) must be performed over an integrand that varies rapidly with χ^2. Techniques for handling these problems have been developed by Jakubowicz.[72]

6. Conclusions

We have indicated a wide range of applications where the use of quantum defect methods supplements and extends the results of direct computation or experiment. Even so, we have only given a mathematical background that applies to electrons in the field of a nonrelativistic positive ion, and we have restricted the examples of applications accordingly. The usefulness of the method has led to its extension to a wider range of atomic fields. We have already mentioned developments for the treatment of heavy ions, where relativistic effects are important.[34-36] However, the use of quantum defect methods is not restricted to cases where the asymptotic form of the potential is Coulombic. Ross and Shaw,[73] Green et al.,[74] and Doughty et al.[9] have developed forms of the theory applicable to potentials that are zero for $r > r_0$. The M-matrix method of Ross and Shaw[73] has been applied by Damburg and Peterkop[75] to the location of resonances in $e^- + H$ scattering. When Coulomb potentials are absent, the effects of potentials behaving asymptotically as r^{-p}, $p \geqslant 2$, are more important. The strong energy dependence of the M matrix for $e^- + K$ scattering in the vicinity of the excitation threshold[76-78] may be attributed to the effects of the long-range dipole-type polarization potential. A method for taking this into account has been developed by Watanabe and Green.[79] Other work on long-range potentials of the form r^{-p}, $p \geqslant 2$, including the r^{-2} dipole potential has been carried out by a number of authors.[6,15,74,80]

Both in the extension of its applications and in the development of the theory for different potentials, quantum defect theory continues to offer a very rewarding challenge.

References

1. D. R. HARTREE, *The Calculation of Atomic Structure*, Wiley, New York, 1957.
2. F. B. HAM, in *Solid State Physics*, Vol. 1, F. Seitz and D. Turnbull, Eds., Academic, New York, 1955, pp. 127–192.
3. M. J. SEATON, The quantum defect method, *Mon. Not. R. Astron. Soc.* **118**, 504–518 (1958).
4. M. J. SEATON, Quantum defect theory I. General formulation, *Proc. Phys. Soc.* **88**, 801–814 (1966).
5. M. J. SEATON, Quantum defect theory II. Illustrative one-channel and two-channel problems, *Proc. Phys. Soc.* **88**, 815–832 (1966).
6. O. BELY, Quantum defect theory III. Electron scattering by He^+, *Proc. Phys. Soc.* **88**, 833–842 (1966).
7. D. L. MOORES, Quantum defect theory IV. The absorption of radiation by calcium atoms, *Proc. Phys. Soc.* **88**, 843–859 (1966).
8. D. L. MOORES, Quantum defect theory V. Autonionizing and bound states of the neutral beryllium atom, *Proc. Phys. Soc.* **91**, 830–841 (1967).
9. N. A. DOUGHTY, M. J. SEATON, and V. B. SHEOREY, Quantum defect theory VI. Extrapolations along isoelectronic sequences, *J. Phys. B* **1**, 802–812 (1968).
10. M. J. SEATON, Quantum defect theory VII. Analysis of resonance structures, *J. Phys. B* **2**, 5–11 (1969).
11. P. DE A. P. MARTINS and M. J. SEATON, Quantum defect theory VIII. Resonances in the collision strengths for O^+ $2p^3$ $^2D_{3/2}-^2D_{5/2}$, *J. Phys. B* **2**, 333–340 (1969).
12. D. W. NORCROSS and M. J. SEATON, Quantum defect theory IX. Complex quantum defects for the e–Be^+ system, *J. Phys. B* **3**, 579–584 (1970).
13. J. DUBAU and J. WELLS, Quantum defect theory X. Photoionization, *J. Phys. B* **6**, 1452–1460 (1973).
14. M. J. SEATON, Quantum defect theory XI. Clarification of some aspects of the theory, *J. Phys. B* **11**, 4067–4093 (1978).
15. J. DUBAU, Quantum defect theory XII. Complex quantum defects for the He^+ $+e^-$ system, *J. Phys. B* **11**, 4095–4107 (1978).
16. J. DUBAU and M. J. SEATON, Quantum defect theory XIII. Further discussion of photoionization, *J. Phys. B*, submitted.
17. O. BELY, D. L. MOORES, and M. J. SEATON, "Many-channel quantum defect theory in atomic collision processes," in *Proceedings of the Third International Conference on the Physics of Electronic and Atomic Collisions London 1963*. M. R. C. McDowell, Ed., North Holland, Amsterdam, 1964, pp. 304–311.
18. M. GAILITIS, Behavior of cross sections near threshold of a new reaction in the case of a Coulomb attraction field, *Sov. Phys.-JETP* **17**, 1328–133 (1963).
19. U. FANO, Quantum defect theory of *l*-uncoupling in H_2 as an example of channel interaction treatment, *Phys. Rev. A* **2** 353–365 (1970).
20. U. FANO, Unified treatment of perturbed series, continuous spectra and collisions, *J. Opt. Soc. Am.* **65**, 979–987 (1975).
21. K. T. LU, Spectroscopy and collision theory. The Xe absorption spectrum, *Phys. Rev. A* **4**, 579–596 (1971).
22. M. LEE, Spectroscopy and collision theory III. Atomic eigenchannel calculation by a Hartree–Fock–Roothaan method, *Phys. Rev. A* **10**, 584–600 (1974).
23. J. GEIGER, Energy loss spectra of xenon and krypton and their analysis by energy-dependent multichannel quantum defect theory, *Z. Phys. A* **282**, 129–141 (1977).
24. J. A. ARMSTRONG, P. ESHERICK, and J. J. WYNNE, Bound even-parity $J = 0$ and 2 spectra of Ca: A multichannel quantum defect theory analysis, *Phys. Rev. A* **15**, 180–196 (1977).
25. C. W. CLARK and K. T. TAYLOR, Eigenchannel analysis of neon negative ion resonances, *J. Phys. B* **15**, L213–L219 (1982).
26. G. HERZBERG and CH. JUNGEN, The absorption spectra of the molecules H_2, HD, and D_2. Part IV, *J. Mol. Spectrosc.* **41**, 425–486 (1972).

27. O. ATABEK, D. DILL, and CH. JUNGEN, Quantum defect theory of excited $^1\Pi_u^-$ levels of H_2, *Phys. Rev. Lett.* **33**, 123–126 (1974).

28. O. ATABEK and CH. JUNGEN, Quantum defect theory of excited $^1\Sigma_u^+$ levels of H_2, in *Electron and Photon Interactions with Atoms*, H. Kleinpoppen and M. R. C. McDowell, Eds., Plenum, New York, 1976, pp. 613–620.

29. CH. JUGEN and O. ATABEK, Rovibrational interactions in the photoabsorption spectrum of molecular hydrogen and deuterium: An application of multichannel quantum defect methods, *J. Chem. Phys.* **66**, 5584–5609 (1977).

30. C. M. LEE, Multichannel dissociative recombination theory, *Phys. Rev. A* **16**, 109–122 (1977).

31. A. GIUSTI, A multichannel quantum defect approach to dissociative recombination, *J. Phys. B* **13**, 3867–3894 (1980).

32. CH. JUNGEN and D. DILL, Calculation of rotational-vibrational preionization in H_2 by multichannel quantum defect theory, *J. Chem. Phys.* **73**, 3338–3345 (1980).

33. A. GIUSTI and CH. JUNGEN, "A Multichannel Quantum Defect Treatment of the Competition Between Preionization and Predissociation in the NO Molecule," in *Abstracts of Contributed Papers, XII International Conference on the Physics of Electronic and Atomic Collisions, Gatlinburg, 1981*, S. Datz, Ed., pp. 71–72 (1981).

34. W. R. JOHNSON and K. T. CHENG, Quantum defects for highly stripped ions, *J. Phys. B* **12**, 863–879 (1979).

35. C. M. LEE and W. R. JOHNSON, Scattering and spectroscopy: Relativistic multichannel quantum defect theory, *Phys. Rev. A* **22**, 979–988 (1980).

36. W. R. JOHNSON, K. T. CHENG, K.-N. HUANG, and M. LE DOURNEUF, Analysis of Beutler–Fano autoionizing resonances in the rare-gas atoms using the relativistic multichannel quantum defect theory, *Phys. Rev. A* **22**, 989–997 (1980).

37. D. R. BATES and A. DAMGAARD, The calcualtion of the absolute strengths of spectral lines, *Phil Trans. R. Soc. Lond. Ser. A* **242**, 101–122 (1949).

38. H. E. SARAPH, O IV: Bound states, oscillator strengths, and photoionization cross sections, *J. Phys. B* **13**, 3129–3148 (1980).

39. R. P. MCEACHRAN and M. COHEN, Theoretical oscillator strengths for the boron isoelectronic sequence, *J. Quant. Spectrosc. Radiat. Transfer* **11**, 1819–1826 (1971).

40. D. BURGESS and M. J. SEATON, A general formula for the calculation of atomic photoionization cross sections, *Mon. Not. R. Astron. Soc.* **120**, 121–151 (1960).

41. G. PEACH, A general formula for the calculation of absorption cross sections for free–free transitions in the field of positive ions, *Mon. Not. R. Astron. Soc.* **130**, 361–377 (1965).

42. G. PEACH, A revised general formula for the calculation of atomic photoionization cross sections, *Mem. R. Astron. Soc.* **71**, 13–27 (1967).

43. G. PEACH, Continuous absorption coefficients for nonhydrogenic atoms, *Mem. R. Astron. Soc.* **73**, 1–123 (1970).

44. H. VAN REGEMORTER, D. HOANG-BINH, and M. PRUD'HOMME, Radial transition integrals involving low or high effective quantum numbers in the Coulomb approximation, *J. Phys. B* **12**, 1053–1061 (1979).

45. D. HOANG-BINH and H. VAN REGEMORTER, The calculation of radial integrals for bound–free and free–free transitions, *J. Phys. B* **14**, L329–L335 (1981).

46. D. HOANG-BINH and H. VAN REGEMORTER, Photoionisation from low or high excited states, *J. Phys. B* **12**, L715–L718 (1979).

47. H. E. SARAPH and M. J. SEATON, The calculation of energy levels for atoms in configurations $1s^2 2s^2 2p^q nl$, *Phil. Trans. R. Soc. Lond. Ser. A* **271**, 1–39 (1971).

48. M. J. SEATON, A variational principle for the scattering phase matrix, *Proc. Phys. Soc.* **89**, 469–470 (1966).

49. H. VAN REGEMORTER, Rate of collisional excitation in stellar atmospheres, *Astrophys. J.* **136**, 906–915 (1962).

50. C. D. H. CHISHOLM and U. ÖPIK, A simplified Hartree–Fock procedure for atoms with two electrons outside closed shells, *Proc. Phys. Soc.* **83**, 541–547 (1964); and corrigendum in *Proc. Phys. Soc.* **84**, 1041 (1964).

51. W. R. S. GARTON and K. CODLING, Ultraviolet extensions of the arc spectra of the alkaline earths: the absorption spectrum of calcium vapour, *Proc. Phys. Soc.* **86**, 1067–1075 (1965).

52. L. N. SHABANOVA, Oscillator strengths of the spectral lines of Ca I, *Opt. Spectrosc.* **15**, 450–451 (1963).

53. P. W. DITCHBURN and R. D. HUDSON, The absorption of light by calcium vapour (2100 to 1080Å), *Proc. R. Soc. Lond. Ser. A* **256**, 53–61 (1960).

54. G. H. NEWSOM, The absorption spectrum of calcium vapour: 1660–2028Å, *Proc. Phys. Soc.* **87**, 975–982 (1966).

55. T.J. MCILRATH and R. J. SANDEMAN, Revised absolute absorption cross sections of Ca I at 1886.5 and 1765.1Å, *J. Phys. B* **5**, L217–L219 (1972).

56. V. L. CARTER, R. D. HUDSON, and E. L. BREIG, Autoionization in the uv photoabsorption of atomic calcium, *Phys. Rev. A* **4**, 821–825 (1971).

57. C. FROESE-FISCHER and J. E. HANSEN, Comparison of multichannel quantum defect theory and multiconfiguration–Hartree–Fock wave functions for alkaline earth atoms, *Phys. Rev. A* **24**, 631–634 (1981).

58. H. E. SARAPH, M. J. SEATON, and J. SHEMMING, Excitation of forbidden lines in gaseous nebulae I. Formulation and calculations for $2p^q$ ions, *Phil. Trans. R. Soc. Lond. Ser. A* **264**, 77–105 (1969).

59. H. E. SARAPH, Fine structure cross sections from reactance matrices, *Comput. Phys. Commun.* **3**, 256–268 (1972).

60. E. S. CHANG and U. FANO, Theory of electron–molecule collisions by frame transformations, *Phys. Rev. A* **6**, 173–185 (1972).

61. A. MONFILS, The absorption spectra of the molecules H_2, HD, and D_2; Part VII, *J. Mol. Spectrosc.* **25**, 513–517 (1968).

62. D. L. MOORES, to be submitted.

63. D. M. DEHMER and W. A. CHUPKA, Very high resolution study of photoabsorption, photoionization, and predissociation in H_2, *J. Chem. Phys.* **65**, 2243–2273 (1976).

64. A. PRADHAN, Collision strengths for [O II] and [S II], *Mon. Not. R. Astron. Soc.* **177**, 31–38 (1976).

65. K. GILES, Collision strengths for [Ne V], *Mon. Not. R. Astron. Soc.* **187**, 49p–51p (1979).

66. C. MENDOZA, Compilation of transition probabilities, electron excitation rate coefficients, and photoionization cross sections, in *Proceedings of IAU Symposium No. 103: Planetary nebulae*, D. R. Flower, Ed., D. Reidel, Dordrecht, Holland (1983).

67. T. M. LUKE, Autoionization states in the photoionization of neon, *J. Phys. B* **6**, 30–41 (1973).

68. H. E. SARAPH, Photoionization cross sections from solutions of the close-coupling equations, *Comput. Phys. Commun.*, to be published.

69. U. FANO, Effects of configuration interaction on intensities and phase shifts, *Phys. Rev.* **124**, 1866–1878 (1961).

70. H. JAKUBOWICZ and D. L. MOORES, Electron impact ionization of Li-like and Be-like ions, *J. Phys. B* **14**, 3733–3760 (1981).

71. M. A. CREES, M. J. SEATON, and P. M. H. WILSON, IMPACT, a program for the solution of the coupled integro-differential equations of electron–atom collision theory, *Comput. Phys. Commun.* **15**, 23–83 (1978).

72. H. JAKUBOWICZ, Theoretical studies of the electron impact ionization of positive ions, Ph.D. thesis, University of London, 1980.

73. M. H. ROSS and G. L. SHAW, Multichannel effective range theory, *Ann. Phys.* **13**, 147–186 (1961).

74. C. GREENE, U. FANO, and G. STRINATI, General form of the quantum defect theory, *Phys. Rev. A* **19**, 1485–1509 (1979).

75. R. DAMBURG and R. PETERKOP, Application of the multichannel effective range theory to electron–hydrogen scattering, *Proc. Phys. Soc.* **80**, 1073–1077 (1962).

76. D. L. MOORES and D. W. NORCROSS, Alkali-metal negative ions. I. Photodetachment of Li^-, Na^- and K^-, *Phys. Rev. A* **10**, 1598–1604 (1974).

77. C. M. LEE, Spin polarization and angular distribution of photoelectrons in the Jacobi–Wick helicity formalism. Application to autoionization resonances, *Phys. Rev. A* **10**, 1598–1604 (1974).
78. D. W. NORCROSS and K. TAYLOR, private communication (1974).
79. S. WATANABE and C. GREENE, Atomic polarizability in negative-ion photo-detachment, *Phys. Rev. A* **22**, 158–169 (1980).
80. C. GREENE, Dependence of the photoabsorption spectra on long range fields, *Phys. Rev. A* **22**, 149–157 (1980).

7

ELECTRON–ION PROCESSES IN HOT PLASMAS

Jacques Dubau and Henri van Regemorter

1. Introduction

The spacecraft observations of astrophysical objects in the UV–X-ray wavelength range and the production of hot plasmas in new devices developed for nuclear fusion have given a new impetus to the physics of highly charged ions.

Since the important discoveries of Grotrian[1] and Edlén[2] in 1939 and 1942—the identification of a few solar coronal lines with forbidden fine-structure transtions in highly ionized atoms like Fe xiv—it has been known that the temperature of the solar corona is of the order of 10^6 K or higher and that the electron density is of the order of 10^8–10^{10} cm^{-3}.

Areas of enhanced magnetic flux within the corona, which are the sites of solar flares, were later identified. In these regions, the temperature varies from 10^6 to 4×10^7 K and the density is less than 10^{13} cm^{-3}. Spectral analysis of the corona was limited until spacecraft astronomical observatories made it possible to reach the UV–X ray wavelength range where hot plasmas emit most strongly. Instrumental improvements of the earth-satellite spectrometer detectors have also been very helpful in analyzing transient plasmas, and it is now possible to obtain both good theoretical and observed line intensities that may be analyzed with the aid of theoretical intensities to obtain reliable diagnostics.

Simultaneously with these developments in astronomical plasma diagnostics, experimental hot plasmas were produced in the magnetic field confinement Tokamak or in laser-produced plasmas (for fusion purpose):

JACQUES DUBAU and HENRI VAN REGEMORTER • Observatoire de Paris–Meudon, 92190 Meudon, France.

1. In Tokamak, the plasma temperature can be of the order of 10^7 K and the density, although higher than in the solar corona, is smaller than 10^{15} cm^{-3}. The magnetic field is also higher than in the solar corona and the continous injection of neutral particles into the plasma gives an important role to charge exchange processes that modify the ionization equilibrium. Nevertheless, in many aspects Tokamak plasmas are very similar to that of the solar corona and many techniques initially developed for the study of astrophysical plasmas can be applied to Tokamak plasmas.

2. In laser plasmas, the physical situation is very different: the temperatures are still comparable to Tokamaks but densities are so high, $N_e > 10^{20}$ cm^{-3}, and the plasma life is so short (10^{-8} s) that their study requires new atomic concepts and much more work to go beyond the present unsatisfactory stage of the theory. Such refinements are not considered here.

In the present chapter, we shall be concerned with plasmas such as coronal and Tokamak plasmas. The temperature will be above 10^6 K and the electron density will not be too high, $N_e \ll 10^{15}$ cm^{-3}. For such plasmas, ion–electron processes are the most important mechanisms for populating the emitting levels.

It will be seen that although resonances in the collisional excitation process play an important role, radiative transitions can perturb them to cause dielectronic recombination. Emphasis will be put on the use of a distorted-wave approximation which gives a reliable representation of the electron–ion system. We will also discuss the validity of approximate methods derived from perturbation theory for highly charged ions.

Section 2 will be concerned with a survey of the connection between atomic calculations and spectroscopic diagnostics, i.e., allowed and forbidden line intensity ratios and X-ray satellite lines diagnostics.

Section 3 is organized as follows: In Section 3.1 the role of the radiation field, usually neglected in collision theory, is emphasized for highly ionized system. In Section 3.2 the formal treatment of the $A^{+z} + e$ system is given in the framework of perturbation theory applied to continuum states. In Section 3.3 and 3.4 conditions for the validity of the distorted-wave approximation for high-Z systems are discussed with particular reference to the calculation of resonance contributions to collisional excitation. Section 3.5 is devoted to very approximate methods giving reliable results in the case of strong allowed transitions. Sections 3.6 and 3.7 give a survey of relativistic effects for collisional fine-structure transitions and autoionization probabilities. The latter atomic data are required for satellte line calculations.

It is not our ambition to provide an exhaustive survey of electron–ion processes in hot plasmas but to point out a few important features that we hope the reader might find valuable to understanding the peculiarities of hot plasmas compared to other plasmas. For advanced research on the subject, very good reviews on atomic problems in fusion research have appeared recently, in particular, two reports on "Atomic and Molecular Data for Fusion"[3] presented at International Atomic Energy Agency meetings held at Culham in 1976 and at Fontenay-aux-Roses in 1980. Reference must also be made of the book *Atomic and Molecular Physics in Controlled Thermonuclear Research*.[4]

Review papers on X-ray spectroscopic diagnostics of laboratory plasmas include the articles by Presnyakov[5] and de Michelis and Mattioli.[6] Useful compilations of data are regularly published by the International Atomic Energy Agency,[7] the Oak Ridge Laboratory, and the National Bureau of Standards,[8] as well as by the Nagoya Institute of Plasma Physics.[9]

A number of review articles on diagnostics in astrophysical objects have been published recently, including papers by Gabriel and Jordan,[10] Dupree,[11] Jordan,[12] and Feldman.[13] The last includes much information on recent rocket and satellite observations.

Ion–electron collision theory has been reviewed by Bely and Van Regemorter[14] and by Seaton.[15] More recently, Henry[16] has given an exhaustive comparison of available calculations for electron excitation cross sections of positive ions.

2. Line Intensities

The emission line spectra observed in high-temperature plasmas can be interpreted to give both energy loss and diagnostics for electron temperature and density. These lines originate from levels that are populated mainly by electron collisional excitation, but radiative cascades from upper levels as well as atomic processes involving other stages of ionization can also be very effective in transient plasmas. Comparison of allowed and forbidden lines gives good electron density diagnostics, and from the ratios of satellite to resonance line intensities in the X-ray range a reliable electron temperature determination has been obtained.

2.1. Level Populations

Various atomic data must be known before starting an analysis of the line radiation. A very precise wavelength identification is necessary in the

UV–X ray range where hundreds of lines can be clearly resolved in a very short wavelength range. However, wavelengths are not the only atomic data required; for diagnostics many electron cross sections and radiative probabilities have to be calculated with good precision.

For a transition between levels α and α' in A^{+z}, the intensity of the emitted line in ergs s^1 sr^{-1} is

$$I_{\alpha\alpha'} = \frac{h\nu}{4\pi} \int_V N_\alpha A^r_{\alpha\alpha'} dV. \tag{1}$$

Here ν is the frequency of the photon (s^{-1}), N_α the level population (cm^{-3}), $A^r_{\alpha\alpha'}$ the radiative probability (s^{-1}) of the transition α to α', and V the emitting volume (cm^3). To derive Eq. (1) two assumptions were made. The first is to consider the plasma as optically thin; this is usually the case for low-density hot plasmas. The second is to separate the different atomic processes one from the other; this is possible as long as the processes do not interfere. We shall examine the problem in detail in Section 3. If these two assumptions are valid, the population equations can be written as follows:

$$\frac{dN_\alpha}{dt} = \sum_{\alpha'' \neq \alpha} (C_{\alpha''\alpha}N_{\alpha''} - C_{\alpha\alpha''}N_\alpha), \tag{2}$$

where $C_{\alpha\alpha''}$ denotes the transition rates from α to α''. They include different processes such as electron collisional excitation and ionization, radiative cascades, electron–ion recombination and sometimes ion–ion collisions.

At low density and especially for very strong resonance lines the main excitation process is usually electron–ion collisions, and we have

$$C_{\alpha''\alpha} = N_e \mathcal{C}_{\alpha''\alpha}(T_e), \qquad \alpha'' < \alpha \tag{3}$$

where N_e is the electron density and $\mathcal{C}_{\alpha''\alpha}(T_e)$ the electron excitation rate coefficient corresponding to a thermal distribution of electrons for temperature T_e. Now if the main depopulation process is a radiative transition, we have

$$C_{\alpha\alpha''} = A^r_{\alpha\alpha''}, \qquad \alpha'' < \alpha. \tag{4}$$

For a stationary plasma $(dN_\alpha/dt = 0)$ in which Eqs. (3) and (4) are applicable for a transition between the gound level α' and an excited level α, Eq. (2) gives the so-called *coronal model* population equation:

$$N_e N_{\alpha'} \mathcal{C}_{\alpha'\alpha}(T_e) = N_\alpha A^r_{\alpha\alpha'}. \tag{5}$$

Absolute line intensities described by Eqs. (1) and (5) are selected for the *emission measure analysis*.[17] We shall not consider such an analysis here but shall concentrate on relative line intensity diagnostics.

For nonstationary plasmas, such as solar flares or laboratory plasmas, it is impossible to assume $dN_\alpha/dt = 0$ but it is usually possible to restrict the number of differential equations (2) by classifying the levels into two categories according to their relaxation time in the following way: Except for metastable levels, which are rare, and for the ground levels of different stages of ionization of the element A, the lifetimes of excited levels are many orders of magnitude shorter than those of ground levels. Therefore, it is possible to assume that as the ground and metastable levels evolve, the excited levels will stay in equilibrium with them, which means that the population of the excited levels is known when the ground-level evolution has been determined.[18]

By means of this classification Eq. (2) can be reduced to a few differential equations that are solved simultaneously, with a larger set of linear equations taking the populations of the ground and metastable levels as known quantities. When there are no metastable states these linear equations need to be solved only once, after separating the contribution of the different stages of ionization which behave independently. This approximation is also valid for sufficiently low electron densities because then the metastables behave as ordinary excited levels. The population of the level i of A^{+z} is given by

$$N_\alpha(t) = N_e(t)\left[N_{z-1}(t)\bar{\mathcal{C}}_\alpha^-(t) + N_z(t)\bar{\mathcal{C}}_\alpha^0 + N_{z+1}(t)\bar{\mathcal{C}}_\alpha^+(t)\right], \qquad (6)$$

where $N_{z-1}(t)$, $N_z(t)$, and $N_{z+1}(t)$ are the number densities of A^{+z-1}, A^{+z}, and A^{+z+1} (we have neglected the atomic processes involving nonadjacent stages of ionization). $\bar{\mathcal{C}}_\alpha^-$, $\bar{\mathcal{C}}_\alpha^0$, and $\bar{\mathcal{C}}_\alpha^+$ are the contributions from the different stages of ionization. Such a study has been made for Fe xxv resonance lines[19,20] observed in solar flares and Tokamak.

These studies include electron excitation and ionization, radiative and dielectronic recombination, and radiative cascades. Using these results it has been possible to analyze the evolution of Fe xxv lines formed in solar flares.[21]

For stationary plasmas the method is simpler, involving only the solution to a set of linear equations, which are usually called *statistical equilibrium equations*. They are used simultaneously with results of the ionization equilibrium calculations.[22–24]

2.2. Forbidden Lines

The presence of forbidden lines arising from metastable levels is very important for electron density diagnostics. The calculation of their intensi-

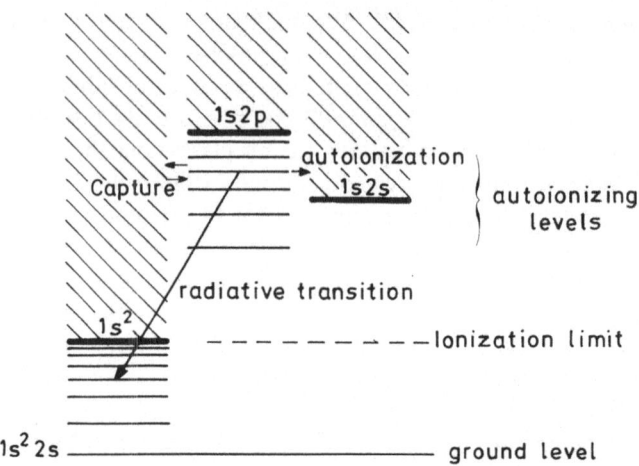

FIGURE 1. Energy diagram of the lithiumlike and heliumlike ions.

ties requires great accuracy for the rates of the processes populating the levels as well as for the radiative transition probabilities. The latter are usually small and are therefore very sensitive to the approximation used in their calculation.

The radiative probabilities being small, configuration interaction or relativistic effects, such as spin–orbit interaction, can play a role. For example, the $1s2s\,^3S_1$–$1s^2\,^1S_0$ transition probability is naught if relativistic effects are not included in the theory of the magnetic dipole transition operator.[25] The same relativistic effects have to be introduced in O^+ to explain the line intensity of $1s^2 2s^2 2p^3\,^2D_{5/2}$–$^4S_{3/2}$.[26] As for the $1s2p\,^3P_1$–$1s^2\,^1S_0$ transition, it is an electric dipole transition but only through the spin–orbit interaction between $^1P_1$ and $^3P_1$. All these forbidden transitions become large for very highly ionized elements becaues they scale as z^8 and z^{10}, whereas for allowed transitions the z dependence is at most z^4. Thus for very highly ionized elements the metastable levels behave like normal excited levels even for very large electron densities.

Since the various population processes are also small for metastable levels, radiative cascades can be as important as direct excitations. When the metastable level is close to the ground level, which is usually the case for levels belonging to the same configuration, proton excitation must not be neglected. The numerous calculations of proton collision rates have been based principally on semiclassical methods. A quantum treament was formulated by Faucher[27] and excellent agreement with semiclassical results was obtained.[28] Finally, indirect electron collisional processes, i.e., the decay of collisional resonances, are often as important as direct excita-

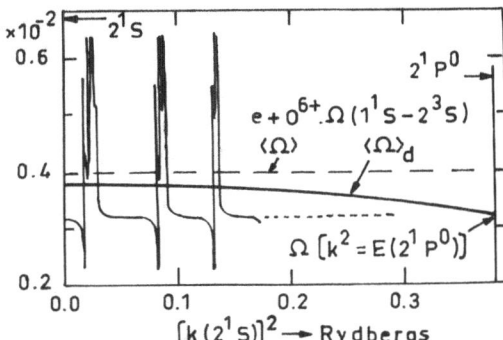

FIGURE 2. $1s^2\,{}^1S_0-1s2s\,{}^3S_1$ collision strength for O VII energy range between the $1s2s\,{}^1S_0$ and $1s2p\,{}^1P_1^{\circ}$ thresholds.

tions for metastable levels. To evaluate the contribution of these resonances one can use a close-coupling or distorted-wave approximation (see Section 3.4). Comparisons have been made for O III and C V and show a good agreement between the different methods.[29,30] Unfortunately, these calculations do not consider the radiative decay of collisional resonances which may be an important effect decreasing the resonance effect by a factor 2 for moderately charged ions and by a much larger factor for highly ionized elements.

To illustrate these resonance effects, we shall consider the $1s2p\,nl$ autoionizing states, which have energies situated between the $1s2s$ and $1s2p$ ionization limits. They can be populated by resonance capture ($1s^2 + e$) and decay into $1s2s + e$. The $1s2s$ configuration can therefore be excited directly from $1s^2$ or through the decay of a $1s2p\,nl$ resonance which was populated by $1s^2 + e$ capture. But the $1s2p\,nl$ autoionizing state can also decay by radiative transitions to $1s^2\,nl$. The efficency of the indirect excitation of $1s2s$ through resonance is therefore reduced by radiative decay. Figure 1 shows the position of the different levels. In Figs. 2 and 3

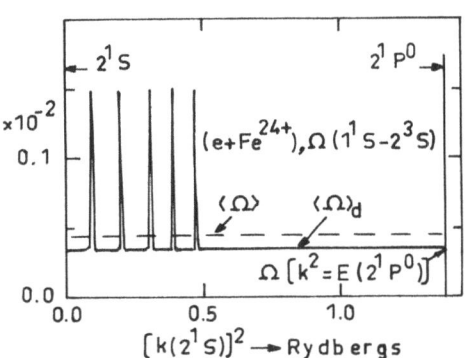

FIGURE 3. $1s^2\,{}^1S_0-1s2s\,{}^3S_1$ collision strength for Fe XXV energy range between the $1s2s\,{}^1S_0$ and $1s2p\,{}^1P_1^{\circ}$.

are reproduced the results of numerical calculations by Pradhan[31] for the $1s^2\,{}^1S_0-1s2s\,{}^3S_1$ collision strengths of O^{6+} and Fe^{24+}. The resonances appear on a background caused by the direct excitation process alone. $\langle\Omega\rangle$ is the average resonance effect, in a theory excluding radiative decay, and $\langle\Omega_d\rangle$ is the same average resonance effect but in a theory including radiative decay. Of course, $\langle\Omega_d\rangle$ is between the background and $\langle\Omega\rangle$. It is seen that for Fe^{24+} radiative decay annihilates the resonance contribution.

2.3. Satellite Lines

When the $1s2p\,nl$ autoionizing levels decay by radiative transitions, they emit a photon at wavelengths close to the $1s2p-1s^2$ He-like resonance lines. The emission may appear either in the profile of the resonance lines or well separated from them.

All of these unresolved or resolved lines are called *satellite lines*. Indeed they correspond to $1s2p\,nl\to1s^2\,nl$ lines which are similar to $1s2p\to1s^2$ lines in the presence of a "nl" spectator electron which perturbs the core electrons only for small nl. In Fig. 4 are plotted two curves

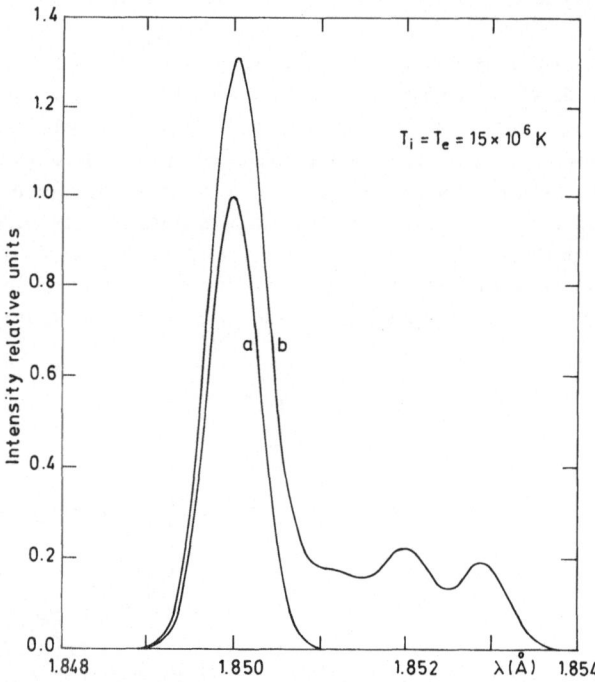

FIGURE 4. Fe xxv resonance line intensity: (a) pure resonance line; (b) apparent resonance line (including unresolved satellites).

corresponding to the "pure" resonance line $1s2p\,{}^1P_1 - 1s^2\,{}^1S_0$ for Fe^{24+} and to the "apparent" resonance line after inclusion of the unresolved satellite lines, $n > 3$, at an electron and ion temperature $T_e = T_i = 15 \times 10^6$ K. One sees the importance of the unresolved satellite intensities compared to the direct excitation process. In the calculation it was assumed that these satellite lines were populated only by the resonance capture process (also called *dielectronic capture*) in which dielectronic recombination proceeds in two steps: resonance capture followed by satellite emission.

The $1s2p\,nl$ (configuration) autoionizing levels are not only populated by resonance capture from $1s^2 + e$ continuum, but also by inner-shell excitation. Inner-shell excitation followed by autoionization[32] can be efficient in ionizing the $A^{+(z-1)}$ system and it is now included in all the ionization equilibrium calculations.[22–24] But for very highly ionized elements, inner-shell processes are not followed by autoionization but by radiative transitions. The corresponding satellite lines are called *inner-shell satellites*. For $n = 2$ and $1s2p\,nl$, the resolved satellites are used for diagnostic purposes. The intensity of a satellite line relative to the corresponding resonance line gives a reliable electron temperature diagnostic if the satellite line arises from a pure "resonance capture" level and allows ion abundance ratios to be determined in the case of a pure inner-shell satellite line.[33–35]

To illustrate the satellite line diagnostics, we give in Fig. 5 a helium-like Ca^{18+} spectrum which contains $1s2p$, $1s2s-1s^2$ resonance and forbidden lines, as well as their satellites $1s2p\,nl$. The spectrum was recorded by

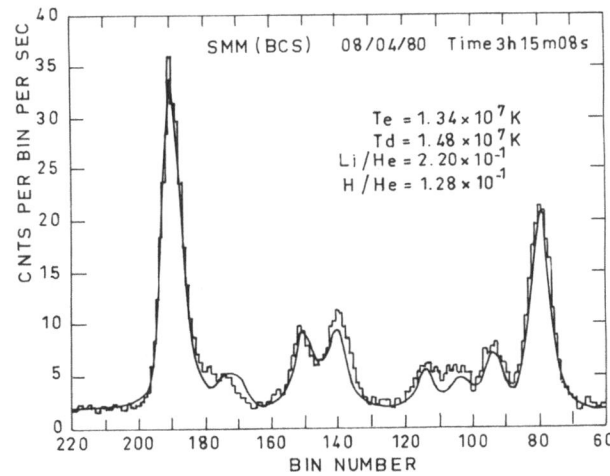

FIGURE 5. Fitting of a Ca spectrum recorded by XRP in the range 3.17–3.22 Å by a synthetic spectrum.

the X-ray polychromator aboard the Solar Maximum Mission Space Observatory. The wavelength range is 3.17–3.22 Å. The best fit obtained includes both Ca^{18+} lines and their Ca^{17+} satellites. It corresponds to an electron temperature $T_e = 13.4 \times 10^6$ K, a Doppler profile temperature $T_D = 14.8 \times 10^6$ K, and relative ion abundances $N_{Ca^{17+}}/N_{Ca^{18+}} = 0.22$ and $N_{Ca^{19+}}/N_{Ca^{18+}} = 0.128$. To obtain such satellite line intensities it is necessary to calculate all the possible autoionization probabilities possible for $1s2p\ nl$ in all the possible $A^{+z} + e$ systems. This is not difficult for three-electron systems but becomes cumbersome for more complex systems such as $1s2s^2\ 2p^3$ autoionizing levels. Such calculations have nevertheless been made for Fe XXI by Doschek et al.[36] and for Cr XIX by the Tokamak at Fontenay-aux-Roses (T.F.R.–C.E.A.).[37] Lines of Fe XXI are observed in solar flares and Cr XIX in the Fontenay-aux-Roses Tokamak.

3. Electron–Ion Processes

3.1. Introduction

Collision physicists usually pay no attention to the lifetimes of atomic levels. The collision time τ_c is usually much shorter than atomic relaxation times, the radiative lifetime $(A^r)^{-1}$, and the collisional lifetime γ_c^{-1}. Radiative and collisional widths of the levels can then be neglected.

Theoreticians are concerned with an ideal situation in which the ion–electron system $A^{+z} + e$ is completely isolated from the plasma and the external radiation fields surrounding the system. The implicit approximation must be checked for collisions of an electron with highly charged ions, remembering that the spontaneous radiative probability scales with z^4.

$$A^r_{\alpha\alpha'} = \frac{4}{3} \frac{e^2}{\hbar^4 c^3} \Delta E^3_{\alpha\alpha'} \sum_s |\langle\alpha|r_s|\alpha'\rangle|^2 \qquad (7)$$

where $\Delta E_{\alpha\alpha'}$ is the transition energy from α to α'.

The collisional width $\gamma_c = N_e \bar{v} q$, where N_e is the electron density, $\bar{v}$ the mean electron velocity, and q a sum of excitation cross-sections decreases with z^{-1}. N_e can be very large, in particular, for laser plasmas.

Two conditions must be fulfilled to justify the usual treament of the collision problem for an isolated quantum system $A^{+z} + e$:

$$\gamma_c \tau_c \ll 1 \quad \text{and} \quad A^r \tau_c \ll 1. \qquad (8)$$

The first condition is always satisfied for low-lying excited states of ions in low-density astrophysical and laboratory plasmas: The collision time is shorter than the average time between two collisions and the collision occurs singly; this is the impact approximation.

The second condition must be checked. By introducing the concept of collisional probability, it can be replaced by another condition that is more precise and more physical.

Within the semiclassical approach, the cross section is

$$Q_{\alpha\alpha'} = \int 2\pi\rho P_{\alpha\alpha'}(\rho)\, d\rho, \tag{9}$$

where $P_{\alpha\alpha'}$ is the dimensionless probability of transition $\alpha \to \alpha'$ induced by a perturber with impact parameter ρ.

We define the collisional probability per unit of time as

$$A^c = \frac{P_{\alpha\alpha'}(\rho)}{\tau_c}, \tag{10}$$

where τ_c is the collision time.

The usual treatment of the collisional problem, neglecting the influence of the radiation field, will be valid if

$$A^c \gg A^r, \tag{11}$$

a condition that is more restrictive than $\tau_c^{-1} \gg A^r$, where A^r is the radiative transition probability.

In terms of the partial collision strength $\Omega_{\alpha\alpha'}^{l_i}$ the quantum expression of the cross section is

$$Q_{\alpha\alpha'} = \frac{\pi}{k_i^2} \frac{1}{\omega_\alpha} \sum_{l_i} \Omega_{\alpha\alpha'}^{l_i} \tag{12}$$

where ω_α is the statistical weight of the initial ionic level α. The quantity $\Omega_{\alpha\alpha'}$ is symmetric and dimensionless. Since the impact area is $2\pi\rho\, d\rho = (\pi/k_i^2)(2l_i + 1)$, the quantum analogue to the transition probability becomes

$$P_{\alpha\alpha'}(\rho) = \frac{1}{\omega_\alpha} \frac{1}{2l_i + 1} \Omega_{\alpha\alpha'}^{l_i}. \tag{13}$$

We are interested in low values of l_i in the case of ionized systems.

From the classical theory of Coulomb excitation [see Alder et al.[38]] it is known that the collisional time τ_c is of the order of

$$\tau_c = \frac{2\pi a}{v_i} = 2\pi \frac{ze^2}{mv_i^3} \tag{14}$$

where a is the half-axis of the hyperbolic trajectory, when $l_i K_i \ll 1$.

The probability that one perturber with a small velocity v_i and momentum l_i collides with the target state α is

$$A^c = \tau_0^{-1} K_i^3 z^2 \frac{1}{\omega_\alpha} \frac{\Omega_{\alpha\alpha'}^{l_i}}{2l_i + 1} , \qquad (15)$$

where τ_0 is the time of revolution in the first Bohr orbit,

$$\tau_0^{-1} = 6.58 \times 10^{+15}\ \mathrm{s}^{-1} = \frac{1}{2\pi} \frac{\hbar}{m} a_0^{-2} ,$$

and K_i^2 is defined by $k_i^2 = K_i^2 z^2 a_0^{-2}$ with $k_i = m v_i / \hbar$.

In the case of autoionization, the initial channel corresponds to an electron $n_i l_i$ bounded to the ion in state α, and the autoionization probability, which can be obtained immediately from Eq. (15), is

$$A^a = \tau_0^{-1} \frac{z^2}{n_i^3} \frac{1}{\omega_\alpha} \frac{\Omega_{\alpha\alpha'}^{l_i}}{2l_i + 1} . \qquad (16)$$

As in Eq. (15), $\Omega_{\alpha\alpha'}^{l_i}$ is the partial collision strength at threshold. In both formulas, and in the following, the statistical weight of the initial channel is $\omega_\alpha(2l_i + 1)$. This gives the A^a or the A^c averaged over the states $A^{+z} + e(n_i l_i)$.

We shall see in Sections 3.4 and 3.7 how A^a is defined in LS and intermediate coupling [see also Seaton and Storey[39]].

A^a and A^c are collisional probabilities in both cases and must be compared to A^r, the radiative transition probability of the ion in state α which can be written in the form

$$A^r = \tau_0^{-1} \frac{8\pi}{3} \alpha^3 \Delta E_{\alpha\alpha'}^3 \frac{S_{\alpha\alpha'}}{\omega_\alpha} , \qquad (17)$$

where α is the fine-structure constant and $S_{\alpha\alpha'}$ the line strength in atomic units.

The collision strength $\Omega_{\alpha\alpha'}^{l_i}$ does not vary with energy near threshold and scales as z^{-2} when z increases. Therefore A^a and A^c are independent of z whereas A^r varies as z^4. For highly ionized systems radiative decay can occur long before autoionization even for low-n resonances which completely disappear.

This coupling with the radiation field reduces the resonance contribution to excitation. When $n_i \rightarrow \infty$, $A^a \rightarrow 0$ and there is no real discontinuity at the energy of a new threshold. In highly ionized systems the Gailitis jump disappears and the average resonance contribution decreases smoothly as n increases. This decrease is given by the branching ratio $A^a/(A^a + A^r)^{-1}$ when autoionization competes with radiative decay.[40,41] Usually one assumes that capture is always followed by autoionization

$$A_{\alpha'}^{+z} + e(k_i'l_i') \rightleftarrows A_\alpha^{+z} + e(n_il_i). \tag{18}$$

If α' is the ion ground state, radiative decay gives rise to recombination

$$A_{\alpha'}^{+z} + e(k_i'l_i') \rightarrow A_\alpha^{+z} + e(n_il_i) \rightarrow A_{\alpha'}^{+z} + e(n_il_i) + h\nu, \tag{19}$$

where $A_{\alpha'}^{+z} + e(n_il_i)$ is a true bound state.

The reduction of the resonance contribution to excitation due to (19) is equivalent to a loss of electron flux which is compensated by a gain of photon flux due to radiative decay.

In low-density plasma, for lower excited ionic states, collisional excitation is usually followed by radiative decay (see Section 2.2). This gives rise to the normal emission lines.

For high-z systems, most of the resonance states behave as true bound states; an electron capture is followed by radiative decay. This gives rise to the satellite emission lines or to the broadening of the corresponding parent ion resonance line when the satellite lines for high-n states cannot be resolved.

Assuming a Maxwellian distribution of electron velocities, the capture rate can be written in the form of excitation coefficient rate [see Eq. (5)]:

$$\mathcal{C}_{\alpha'\alpha} = 4\pi\left(\frac{m}{2\pi kT}\right)^{3/2} Q_{\alpha'\alpha}^{l_i} v_{i'}^2 \frac{dv_{i'}^2}{2} \exp\left(\frac{-E_c}{kT}\right), \tag{20}$$

where E_c is the energy of the doubly excited state c measured from the ion ground state α', and $dv_{i'}^2$ corresponds to the energy difference between two successive doubly excited states, that is,

$$\frac{dv_{i'}^2}{2} = \frac{z^2}{n_i^3} \frac{1}{a_0^2} \frac{\hbar^2}{m^2}. \tag{21}$$

As the capture cross section is

$$Q_{\alpha'\alpha}^{l_i} = \frac{\pi}{K_{i'}^2 z^2} \frac{1}{\omega_{\alpha'}} \Omega_{\alpha'\alpha}^{l_{i'}} a_0^2 \tag{22}$$

by using expression (16) for the autoionization probability A^a the capture rate may be expressed in the known form

$$\mathcal{C}_{\alpha'\alpha} = \left(\frac{2\pi\hbar^2}{mkT}\right)^{3/2} \frac{\omega_\alpha(2l_i + 1)}{\omega_{\alpha'}} A^{\rm a} \exp\left(\frac{-E_c}{kT}\right) \tag{23}$$

which derives [see Eq. (A31) in Appendix] from the symmetry of the S matrix or of the collision strength $\Omega_{\alpha\alpha'} = \Omega_{\alpha'\alpha}$.

The intensity of a satellite line formed by capture–collisional excitation of the resonance state followed by radiative decay will be proportional to

$$I_s = \left(\frac{2\pi\hbar^2}{mkT}\right)^{3/2} \frac{\omega_c}{\omega_{\alpha'}} \exp\left(\frac{-E_c}{kT}\right) A^{\rm a} \frac{A^{\rm r}}{A^{\rm a} + A^{\rm r}}, \tag{24}$$

where $\omega_c = \omega_\alpha(2l_i + 1)$. In high-$z$ systems, when $A^{\rm r} \gg A^{\rm a}$

$$I_s = \left(\frac{2\pi\hbar^2}{mkT}\right)^{3/2} \frac{\omega_c}{\omega_{\alpha'}} \exp\left(\frac{-E_c}{kT}\right) A^{\rm a}. \tag{25}$$

For high-n unresolved satellites, the corresponding photon flux is emitted within the resonance transition of the parent ion. The loss of electron flux as the resonance contribution to excitation decreases is compensated by an increase in photon flux within the *satellite* lines. This has been discussed by Presnyakov and Urnov[41] and in many papers on the dielectronic recombination process.

On the other hand, the photon flux emitted in the *resonance* transition of the ion after collisional excitation according to Eq. (5) is

$$I_r = \int_{v_0}^{\infty} Q_{\alpha'\alpha} f(v) v \, dv \qquad \text{with} \quad \tfrac{1}{2}mv_0^2 = \Delta E_{\alpha\alpha'}. \tag{26}$$

Measurements of satellite lines and of the apparent profile of resonance transitions in Tokamak do give an experimental test of the validity of the theory given above for the competition between collisional and radiative processes and, in particular, of the validity of the branching ratios.

In fact, for highly ionized systems the interactions with the colliding electron and with the radiation field are both weak and can be treated by perturbation theory. There is no interference between them. The branching ratio between $A^{\rm a}$ and $A^{\rm r}$ have a clear meaning even when these probabilities are of the same order of magnitude.[42]

It is more difficult to estimate the reduction of the cross section due to radiative decay. One usually assumes that coupling between open channels implies

$$A_{\alpha'}^{+z} + e(k_i'l_i') \rightleftarrows A_\alpha^{+z} + e(k_i l_i); \tag{27}$$

neglecting the radiative decay of the ion

$$A_\alpha^{+z} + e(k_i l_i) \rightarrow A_{\alpha'}^{+z} + e(k_i l_i) + h\nu. \tag{28}$$

Near the excitation threshold, the collision time can be of the same order as the orbiting time of a resonance state, during which the ion excited state α may disappear through radiative decay. This collision time τ_c is given by (14) in the case of elastic scattering (28) by the ion Coulomb field. In the inelastic case, an average collision time must be defined to estimate the collisional probability A^c given by (10). When proper account is taken of the average electron velocity, the collision time τ_c corresponding to the angular momentum $l = mv\rho/\hbar$ is of the order of

$$\tau_c = \frac{2\pi}{\bar{v}} \left(a^2 + \rho^2\right)^{1/2}; \tag{29}$$

a symmetrized expression in which a and ρ are given in terms of the average electron velocity $\bar{v} = \sqrt{v_i v_{i'}}$.[38,69]

In an estimate of the corresponding collision probability $A_{\alpha\alpha'}^c = P_{\alpha\alpha'}\tau_c^{-1}$, which is symmetric because $\omega_\alpha A_{\alpha\alpha'}^c = \omega_{\alpha'} A_{\alpha'\alpha}^c$, can be deduced and compared to the transition probability A^r. The process of radiative decay, usually neglected in the treatment of a collision problem, destroys the excitation cross section at threshold and reduces it near threshold.

Neglecting radiative decay, the excitation cross section or the collision strength Ω are finite at threshold where many values of the angular momentum l give contributions to Ω in the special case of positive ions. Because of the Coulomb field, the distance of closest approach $r_c = l^2/2me^2z$ of the perturbing electron when its velocity tends to zero remains finite for all values of l. But in fact radiative decay occurs long before excitation when the electron velocity tends to zero and the cross section is zero at threshold.

Usually, excitation rates are not very likely to be reduced by radiative decay, because at a given temperature of the order of $kT = z^2 I_H$, where the ions actually exist, the average collision times involved are much shorter. More work remains to be done to solve the complete *colliding particles + electromagnetic field system*, a radiative collision study that takes proper account of the coupling with the radiation field.

3.2. The $A^{+z} + e$ System

All electron–ion processes require a good quantal representation of the system. For atomic bound-state wave functions, there are two approaches.

The first one is derived from the Hartree–Fock concept and is called the close-coupling approximation. The second one is derived from the central potential model and is known as the distorted-wave method. Furthermore, to represent collisional excitation processes of an N-electron ion, a multiconfiguration expansion has to be used with the $(N + 1)$-electron configurations containing a free electron.

In the following we assume the representation of the N-electron target is "exact," recalling that for complex ions the cross sections can be more sensitive to the representation of the target than to the approximation used for the collisional problem.

In the close-coupling approximation, the wave function of the $(N + 1)$-electron system is expanded in the complete set of the eigenfunctions of the N-electron ionic system

$$\Psi(E) = \sum_i \mathcal{Q}\Theta_i \frac{F_i(r_{N+1})}{r_{N+1}}, \tag{30}$$

where $\mathcal{Q}$ is the antisymmetrization operator and E is the energy of the total system. Θ_i is a vector coupled function containing the ion system wave function χ_i as well as the spin and angular part of the colliding electron. For a nonrelativistic Hamiltonian H, the coupling corresponds to fixed values of S^T and L^T, the total spin and angular momentum (each of which commutes with the Hamiltonian H).

$$\Theta_i \equiv \Theta(\Gamma_i S_i L_i l_i S^T L^T M_S^T M_L^T) = \sum_{M_{S_i} M_{L_i}} \sum_{m_{s_i} m_{l_i}} C^{S_i (1/2) S^T}_{M_{S_i} m_{s_i} M_S^T} C^{L_i l_i L^T}_{M_{L_i} m_{l_i} M_L^T}$$

$$\times \chi(\Gamma_i S_i L_i M_{S_i} M_{L_i} | x_1 x_2 \cdots x_N) Y_{l_i}^{m_{l_i}} (\hat{\mathbf{r}}_{N+1}) \delta(m_{s_i}, \sigma_{N+1}) \tag{31}$$

where $S_i L_i$ and $\frac{1}{2} l_i$ are the spin and angular momenta of the ion and colliding electron, respectively. (The ionic state is $\alpha = \Gamma_i S_i L_i; M_{S_i} M_{L_i}$.)

The radial function $F_i(r_{N+i})$ of the colliding electron is defined by

$$F_i(r) \equiv F_{k_i l_i}(r), \quad \varepsilon_i = k_i^2/2 \quad \text{and} \quad F_i(0) = 0. \tag{32}$$

where $E = E_i + \varepsilon_i$, E_i and ε_i being the energies of the ion and of the colliding electron, respectively (in atomic units).

The Schrödinger equation (see Ref. 43) is replaced by

$$\langle \Theta_i | H - E | \Psi(E) \rangle = 0. \tag{33}$$

Equations (30) and (33) lead to a set of integro-differential equations to be solved for the $F_{k_i l_i}(r)$. In practice, this set is restricted to a few equations that include the ionic levels considered in the excitation processes plus the adjacent states which interfere most strongly with the transition considered. Improved methods have also been applied where pseudostates replace some adjacent states to simulate the effect of excited ionic levels not included in Eq. (30).

For ionized atoms, the nucleus interaction Z/r is relatively more important than the two-electron interactions that are responsible for the coupling in the integro-differential equations. It is logical when Z is large to replace the sophisticated close-coupling approach by a "perturbation in the continuum" approximation [see Dirac,[44] Fano,[45] and Mies[46]].

The total Hamiltonian is separated as follows:

$$H = H^0 + V, \tag{34}$$

where

$$H^0 = H_N + h_{N+1}; \qquad h_{N+1} = -\frac{\nabla^2_{N+1}}{2} - U(r);$$

$$V = \sum_{i=1}^{N} \frac{1}{r_{i\,N+1}} - \frac{Z}{r} + U(r). \tag{35}$$

Here H_N is the ion Hamiltonian and $U(r)$ a potential which takes a different form according to the approximation adopted (distorted-wave or Coulomb–Born).

The wave function for the $(N + 1)$-electron system is developed on all the bound and free eigenfunctions of the Hamiltonian H^0,

$$\Psi(E) = \sum_m a_m(E)\phi_m + \sum_i \int_0^\infty d\varepsilon\, b_\varepsilon^i(E)\phi_\varepsilon^i, \tag{36}$$

where ϕ_m is a normalized bound function and ϕ_ε^i is a free function normalized to the Dirac function. We have

$$\langle \phi_m | \phi_{m'} \rangle = \delta_{mm'}$$
$$\langle \phi_\varepsilon^i | \phi_{\varepsilon'}^{i'} \rangle = \delta_{ii'}\delta(\varepsilon - \varepsilon'). \tag{37}$$
$$\langle \phi_m | \phi_\varepsilon^i \rangle = 0$$

Similarly to (30), ϕ_ε^i is written

$$\phi_\varepsilon^i = \mathscr{C}\Theta_i \frac{G_i(r_{N+1})}{r_{N+1}} \tag{38}$$

From (35) one finds that $G_i(r)$ satisfies the radial equation

$$\left[\frac{d^2}{dr^2} - \frac{l_i(l_i + 1)}{r^2} + 2U(r) + k_i^2 \right] G_i(r) = 0. \tag{39}$$

For large r, $G_{k_i l_i}(r)$ converges to the Coulombic solutions and can be expanded on the two linearly independent solutions in sine and cosine for $\varepsilon_i > 0$ to give

$$G_i(r) \underset{r \to \infty}{\sim} M \frac{1}{\sqrt{k_i}} \sin[\eta_i(r) + \tau_i], \qquad G_i(0) = 0 \tag{40}$$

with

$$\eta_i(r) = k_i r - \frac{l_i \pi}{2} + \frac{z}{k_i} \ln(2k_i r) + \arg \Gamma\left(l_i + 1 - i \frac{z}{k_i} \right).$$

The constant phase shift τ_i is a function of ε_i. It comes from the potential $U(r)$ being different from z/r for small r, where $z = Z - N$ is the charge of the ion. M is an arbitrary constant which should be chosen to satisfy the normalization condition (37) on ϕ_ε^i. From Green's theorem one obtains

$$M = \left(\frac{2}{\pi} \right)^{1/2}. \tag{41}$$

To simplify the formalism it is convenient to diagonalize first the total Hamiltonian with respect to the bound wave function ϕ_m. The bound eigensolutions are

$$|\bar{\phi}_{m'}\rangle = \sum_m |\phi_m\rangle\langle\phi_m | \bar{\phi}_{m'}\rangle \tag{42}$$

such that

$$\langle\bar{\phi}_{m'}|H|\bar{\phi}_{m''}\rangle = \delta_{m'm''}\bar{E}_{m'}. \tag{43}$$

$\langle\phi_m | \bar{\phi}_{m'}\rangle$ are the matrix components of the $\bar{\phi}_{m'}$ on the ϕ_m basis. Equation (36) is now written in the form

$$\Psi(E) = \sum_m \bar{a}_m(E)\bar{\phi}_m + \sum_i \int_0^\infty d\varepsilon\, b_\varepsilon^i(E)\phi_\varepsilon^i \tag{44}$$

which defines the coefficients $\bar{a}_m(E)$.

As usual in perturbation theory, the coefficient $\bar{a}_m$ and b^i are obtained by projecting the Schrödinger equation on the eigenfunctions ϕ_m and ϕ_ε^i of H^0, i.e.,

$$\langle \bar{\phi}_m | H - E | \Psi(E) \rangle = 0$$
$$\langle \phi_\varepsilon^i | H - E | \Psi(E) \rangle = 0. \tag{45}$$

In the Appendix it is shown how one can obtain the coefficients $\bar{a}_m$ and b_ε^i which satisfy (45) as well as the boundary condition that derives from (38) and can be written as follows, providing a definition of $F_{ii'}(r)$: $[r_{N+1} \to \infty$ and r_i finite $(i = 1, \ldots, N)]$

$$\Psi_{i'}(E) \sim \sum_i \int d\varepsilon\, b_\varepsilon^{ii'}(E)\phi_\varepsilon^i \sim \sum_i \mathcal{Q}\Theta_i \frac{F_{ii'}(r_{N+1})}{r_{N+1}}. \tag{46}$$

There are n degenerate solutions to this problem. They are characterized by the index i' in (46). One set of solutions is defined by

$$F_{ii'}^\rho(r) \underset{r \to \infty}{\sim} \frac{1}{\sqrt{k_i}} \left[\sin(\eta_i + \tau_i)\delta_{ii'} + \cos(\eta_i + \tau_i)\rho_{ii'} \right]. \tag{47}$$

In (46), the summation runs only over open channels, the n channels for which $\varepsilon_i > 0$. Each solution corresponds to one entrance channel i' (the exist channel is i).

The value of $\rho_{ii'}$ obtained in Appendix A is

$$\rho_{ii'} = -\pi \left[V_{ii'}(E) + \sum_m \frac{\bar{V}_{im}(E)\bar{V}_{mi'}(E)}{E - \bar{E}_m} \right]. \tag{48}$$

The same formula was obtained by Eissner and Seaton[47] applying the variation principle.

Depending on the problem, it can be convenient to choose boundary conditions such as

$$F_{ii'}^R(r) \underset{r \to \infty}{\sim} \frac{1}{\sqrt{k_i}} \left[\sin(\eta_i)\delta_{ii'} + \cos(\eta_i)R_{ii'} \right]$$

$$F_{ii'}^S(r) \underset{r \to \infty}{\sim} \frac{1}{\sqrt{k_i}} \left[\exp(-i\eta_i)\delta_{ii'} - \exp(i\eta_i)S_{ii'} \right]. \tag{49}$$

All these solutions are related to one another by matrix transformations:

$$\mathbf{R} = (\sin\tau + \cos\tau \cdot \boldsymbol{\rho})(\cos\tau - \sin\tau \cdot \boldsymbol{\rho})^{-1}, \tag{50}$$

$$\mathbf{S} = (1 + i\mathbf{R})(1 - i\mathbf{R})^{-1}, \tag{51}$$

and

$$\mathbf{S} = \exp(i\tau)(1 + i\boldsymbol{\rho})(1 - i\boldsymbol{\rho})^{-1}\exp(i\tau). \tag{52}$$

Choosing the usual definition of the spherical harmonics and using real radial functions for the target, one obtains real and symmetric $\boldsymbol{\rho}$ matrices. In this case, the matrix $\mathbf{R}$ is also real and symmetric.

From (50) and (51) one finds that if $\mathbf{R}$ is a real matrix (see above) the scattering matrix S is unitary. Consequently there is no loss or gain of electron flux during the collison. The matrix S is also symmetric, in accordance with the microreversibility principle. The perturbation method could have been applied directly to the S matrix but in that case S would not have been unitary. The perturbation techniques giving ρ or R directly and S using (51) are called unitarized approximations. They are particularly valuable when some resonances are present.

The excitation cross sections, deduced from the asymptotic expression of the $A^{+z} + e$ system, are

$$Q_{\alpha'\to\alpha}(\varepsilon_i) = \frac{\pi}{k_i^2\omega_{\alpha'}}\,\Omega_{\alpha\alpha'} \tag{53}$$

with

$$\Omega_{\alpha\alpha'} = \tfrac{1}{2}\sum_{L^T S^T l_i l_{i'}} (2L^T + 1)(2S^T + 1)|T_{ii'}|^2 \tag{54}$$

where $\alpha = \Gamma_i S_i L_i$ are the quantum numbers for the target, $T = 1 - S$ is the transition matrix, and ω_α is the statistical weight of the ion in the state α. The collisions strength $\Omega_{\alpha\alpha'}$ like $S_{ii'}$ is symmetric.

3.3. Coupling between Open Channels

Our main concern is to justify the use of approximate methods in the particular case of highly ionized systems. Consider the total Hamiltonian for the $(N + 1)$-electron system

$$H[Z, N+1] = H[Z, N] - \frac{\nabla_{N+1}^2}{2} - \frac{z}{r_{N+1}} - \left(\frac{N}{r_{N+1}} - \sum_{i=1}^{N}\frac{1}{r_{iN+1}}\right) \tag{55}$$

which, using the scaled electronic coordinate $\rho_{N+1} = zr_{N+1}$, $\rho_{i\,N+1} = zr_{i\,N+1}$ becomes

$$H[Z, N+1] = H[Z, N] - z^2 \left[\frac{\nabla^2_{\rho_{N+1}}}{2} + \frac{1}{\rho} + \frac{1}{z} \left(\frac{N}{\rho_{N+1}} - \sum_{i=1}^{N} \frac{1}{\rho_{i\,N+1}} \right) \right].$$

(56)

Since all electron interaction terms decrease as z^{-1}, the Coulomb–Born approximation apparently should give good results when $z \to \infty$. In Eq. (35) H^0 can be taken such that $U(r) = z/r$.

But when $z = Z - N \to \infty$ for a fixed number of electrons, i.e., along a given isoelectronic sequence, the results must converge toward the results for hydrogenic ions with $z = \infty$.

When $z \to \infty$ not only the electron interactions involving the $N + 1$ electrons are weak but all the electrostatic interactions between the N target electrons become infinitesimal relative to that between the electron and the nucleus. Using $\rho'_i = Zr_i$ and dividing the Hamiltonian by Z^2

$$\frac{1}{Z^2} H[Z, N+1] \equiv H_\rho[Z, N+1] = - \sum_{i=1}^{N+1} \left(\frac{\nabla_i^2}{2} + \frac{1}{\rho_{i'}} \right) + \sum_{i<j} \frac{1}{Z\rho'_{ij}}.$$

(57)

H^0 in (35) can be written as a sum of one-electron Hamiltonians

$$H_\rho^0 \equiv - \sum_{i=1}^{N+1} \left(\frac{\nabla_i^2}{2} + \frac{1}{\rho_{i'}} \right) \quad \text{and} \quad V_\rho \equiv \sum_{i<j} \frac{1}{Z\rho'_{ij}}. \qquad (58)$$

Consequently, the interaction V is a sum of two-body interactions. The scaled cross section $Z^4 Q$ and the scaled R matrix $Z \cdot R$ can be expressed in terms of the R-matrix element for the corresponding transitions in a hydrogenic ion with $Z = \infty$.

Nonrelativistic Coulomb–Born calculations including exchange have been carried out for all transitions between the $1s$, $2s$, $2p$ states in He^+ and in hydrogenlike ions with a large nuclear charge. The $Z = \infty$ results may also be used to obtain estimates of cross sections in nonhydrogenic highly ionized atoms when $Z \gg N$. This has been shown by Burgess et al.[48]

For complex ions, this method of interpolation is accurate to within about 40% when $Z/N \geqslant 3$ and improves when Z increases. Sampson et al.[49] have included the dominant relativistic corrections for heavy atoms using perturbation theory (see Section 3.6).

The potential V in formula (35) must be weak to justify a perturbation treatment. In the Coulomb–Born approximation, V is given by Eq. (35) with $U(r) = z/r$, and all the $R_{ii'}$ matrix elements in (50) must be much smaller than one. The S matrix can be calculated using (51) to take account of the coupling entirely neglected at the first stage.

In the distorted-wave approximation, V is given by Eq. (35) with $U(r) = Z_e(r)/r$ which tends to z/r for $r > r_0$ of the order of the size of the atom. The additional phase shifts due to the short-range potential $Z_e(r)/r - z/r$ are the τ_i appearing in the asymptotic form given in (40). This approximation is valid when all the matrix elements $\rho_{ii'}$ are much smaller than one. The S matrix can be calculated using formula (51).

The distorted-wave approximation reduces to the Born approximation only when the τ_i are very small. In the former, a model potential such as the Thomas–Fermi potential is used for estimating $Z_e(r)/r^{-1}$ and only the weak differences between the true direct and exchange potentials and the potential from this model are treated as perturbations.

The phase shifts τ_i are far from being small compared to unity for partial waves l_i in (39) corresponding to penetrating orbits ($l_i = 0, 1, 2$). The phase shifts decrease as z^{-1}, as does the ion radius r_0, but from quantum defect theory one knows that for small values of l_i the phase shifts at low energies can be very large in low stages of ionization. In practice, the values for z of interest in hot plasmas are still too low to give phase shifts small enough to justify the Coulomb approximation. The particular case of excitation of strong allowed transitions for which low values of l_i play a minor role in the total cross section will be discussed below in Section 3.5 together with the validity conditions of more approximate methods.

In the distorted-wave approximation for excitation all coupling effects between open channels are taken into account only at the stage when the S matrix is calculated from Eq. (52) in terms of the approximate ρ. We are concerned here only with the variation with z of these coupling terms.

Equation (52) shows that for inelastic cross sections the off-diagonal elements $S_{ii'}$ are functions only of ρ. If ρ, which varies as z^{-1}, is small, we obtain the expansion

$$|T_{ii'}|^2 = \left| 2i\rho_{ii'} - 2\sum_{i''} \rho_{ii''}\rho_{i''i'} + \cdots \right|^2 \qquad (59)$$

from which it is clear that the contribution to T of the coupling terms varies as z^{-2} as long as open channels are present.

On the other hand, the coupling with a closed channel c gives rise to resonances and it is known from the work of Gailitis[50] that the average contribution of the resonances to excitation of the transition $i' \to i$ is of the order of the excitation cross section of that channel c. Therefore, this contribution to S as well as the nonresonance contribution vary as z^{-1} and have to be taken into account properly in highly charged ions.

On the contrary, away from resonances, the contribution of a closed channel varies as z^{-2}. This coupling is equivalent to a polarization potential αr^{-4} where the polarizability α scales with z^{-4}. The corresponding contribution to the S matrix scales with z^{-2} like the coupling of an open channel. This polarization term can be approximately included in the unperturbed Hamiltonian H^0 in Eq. (35). This has been done by McDowell et al.[51,52] in their distorted-wave polarized orbital approximation.

Resonance contributions to excitation in high-z systems will be considered in Section 3.4 in much more detail.

Considering now the coupling effects between open channels, we must try to answer the question: Is it possible to say that for a large enough value of z all coupling effects can be included properly using the distorted-wave approximation?

It has been shown, in particular, by Saraph et al.,[53] Eissner,[54] and Flower and Launay,[55] that when all channels are open, the distorted-wave approximation gives good results, compared to the more elaborate close-coupling calculations, for ions with $z > 3$.

Practically, for $z \geqslant 3$ most matrix elements $\rho_{ii'}$ are smaller than 0.5 but exceptions may be found for dipole transitions with small $\Delta E_{ii'}$ at small kinetic energies $k_{i'}^2$.

Because the strongest coupling effects are found in hydrogenic ions, Hayes and Seaton[56] have made a systematic comparison of distorted-wave and three-state close-coupling calculations of the $1s$–$2s$ and $2p$ excitation cross sections which are strongly perturbed by the coupling between the $2s$ and $2p$ degenerate states. They have shown that for the weak $1s$–$2s$ transition in low-z ions the cross sections in the two approximations are very different near threshold. Using the distorted-wave approximation, one can expect good accuracy (uncertainty less than 10%) only for $z > 10$. The accuracy is better for $1s$–$2p$ transitions but there are still significant errors near threshold for small values of z. This can be seen in Fig. 6 which shows, for C^{5+} and Ne^{9+}, the values of $Z^2 \Omega_{\alpha\alpha'}$ as a function of $X = k_{i'}^2 \Delta E_{ii'}^{-1}$ obtained by Hayes and Seaton.[56]

This result could have been predicted from the circumstances that the $2s$–$2p$ coupling R-matrix element must be smaller than 1. The $2s$–$2p$ element of R can in fact be estimated using the Bethe approximation (see Section 3.5) in which this R-matrix element is simply proportional to the square root of the $2s$–$2p$ line strength.

Both the close-coupling and the distorted-wave results converge to the Coulomb–Born exchange results of Burgess et al.[48] when $z \to \infty$.

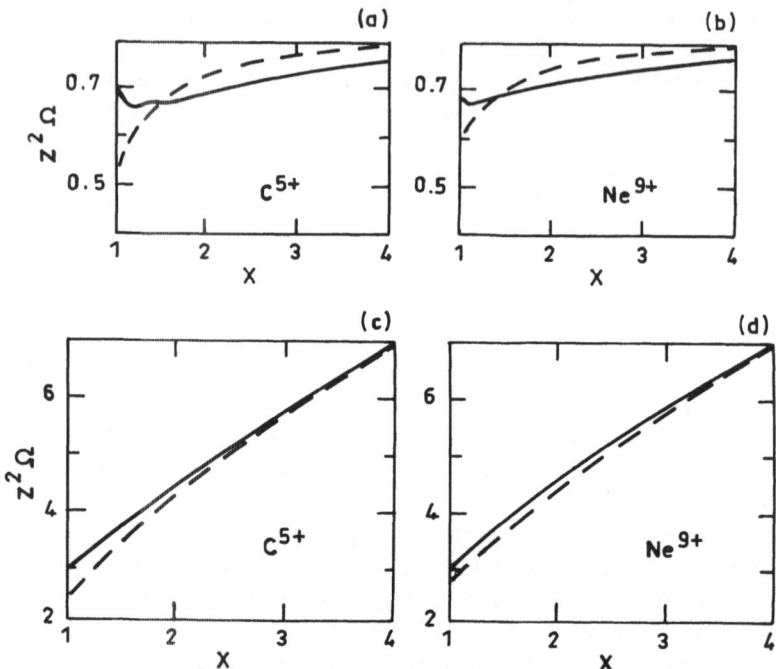

FIGURE 6. Reduced collision strengths for C^{5+} and Ne^{9+} in function of reduced energy: (a) and (b) correspond to $1s-2s$ transition; (c) and (d) correspond to $1s-2p$ transition.

On the other hand, it can be argued whether or not the close-coupling results are physically meaningful near threshold. At small energies above nearly degenerate thresholds the numerical calculations involve enormous values of the radial distance r, the $2s-2p$ coupling becoming *asymptotic*. As a by-product, this coupling affects the collision strengths $1s-2s$ and $1s-2p$ and gives the $a + b\varepsilon_i^{+1/2}$ behavior of the collision strength [see Dubau[57]], which is apparent in Fig. 6 very near threshold, ε_i being the final energy of the colliding electron and a and b being constants.

In fact, this anomaly disappears when proper account is taken of the strong radiative coupling between the $2p$ and the $1s$ levels. For highly charged ions, the lifetime of $2p$ is related to the radiative probability and not to the collisional probability; therefore, the asymptotic coupling between $2s$ and $2p$ does not exist.

3.4. Resonance Contribution to Collisional Excitation

Resonance excitation can be accounted for by either the close-coupling approach or a perturbation treatment, both described in Section

3.3. It is also possible to use hybrid methods based on various combinations of these two methods.

In the close-coupling approximation, the resonances are produced by the *closed channels*. A closed-channel c corresponds to a negative energy radial equation which has radial solutions $F_{ci'}(r)$ satisfying the condition

$$F_{ci'}(r) \xrightarrow[r \to \infty]{} 0, \tag{60}$$

where i' is the (open) entrance channel and c the closed channel. Each closed channel gives a Rydberg resonance series converging to the excitation threshold of the associated target state γ.

The close-coupling method is costly in computer time, and the inclusion in the calculation of resonances converging to several thresholds can require huge amounts of computer storage. To avoid these inconveniences, a hybrid method has been proposed in which the result (30) for $\Psi(E)$ is replaced by

$$\Psi(E) = \sum_i \mathcal{C} \Theta_i \frac{F_i(r_{N+1})}{r_{N+1}} + \sum_j c_j \Phi_j, \tag{61}$$

where Φ_j are bound functions of the $(N+1)$-electron system. The Φ_j are responsible for additional resonance effects at an energy of the total system $E \simeq E_j$; c_j plays a similar role to a_m in Eq. (44) and will have an expression similar to (A4)—see Eissner and Seaton.[47]

The superposition of configuration (SOC) expansions $\sum_j c_j \Phi_j$ or $\sum_m \bar{a}_m \bar{\Phi}_m$ appearing both in (61) and (44) is well suited for representing short-range electron correlations due to the coupling with low-lying resonance states.

Many comparisons of the full close-coupling method with the hybrid (CC + SOC) method have been done. We shall only mention the results of Hayes and Seaton[58] in the case of $1s$–$2s$ and $1s$–$2p$ excitation cross section in hydrogenic ions for energies below the $3s$, $3p$, and $3d$ excitation thresholds.

The six-state closed-coupling (6CC) calculations are compared to the three-state ($1s$, $2s$, $2p$) calculations that include with the most important correlation functions corresponding to the $3ln'l'$ configurations (3CC + SOC of $3ln'l'$).

For excitation of the $2s$ and $2p$ levels, the excitation energy from $1s$ is $E = \frac{3}{4}Z^2$ Ry. The $3ln'l'$ resonances appear at energies $E_j(n')$ (rydbergs) given by

$$E_j(n') = \frac{4}{3} \Delta E \left[\frac{8}{9} - \left(\frac{Z-1}{Z} \right)^2 \frac{1}{n'^2} \right] \tag{62}$$

When $Z \to \infty$, $E_j \to 1.037 \, \Delta E$ for $n' = 3$. These low-lying resonances appear just above the $n = 2$ excitation threshold. They will influence the excitation rates when the radiative decay rate from these resonances is small.

Hayes and Seaton[58] have found good agreement between the two methods in the case of C^{5+} and Ne^{9+}, the agreement improving as Z increases. The use of correlation functions is therefore an economical way to take account of resonance effects without increasing the size of the close-coupling integro-differential system of equations.

We have seen in Section 3.3 that for highly charged ions the distorted-wave method gives reliable results. Although the distorted-wave method usually does not include any $\overline{\Phi}_m$ in the perturbation treatment presented in Section 3.2, some versions, such as the one developed at University College London, includes term of this type. As for the close-coupling method, the accuracy depends on the number of terms included in summation (36). In using the distorted-wave approach in the resonance region, it is necessary to include additional bound states $\overline{\Phi}_m$ to represent the resonances.

For a comparison between close-coupling calculations and the distorted-wave approximation in the resonance regions, there is only one published investigation. This is the work of Hershkowitz and Seaton[59] on the resonances in the $2s^2 \, {}^1S - 2s2p \, {}^3P^\circ$ transitions of C^{2+} and O^{4+}. Figure 7 shows the partial collision strength corresponding to the total $(N + 1)$-electron system with angular momenta $L^T = 1$, $S^T = \tfrac{1}{2}$ below the $2s2p \, {}^1P^\circ$ threshold, especially for the $l_{i'} = 1$, $l_i = 0$ transitions. The agreement is very good both for positions and widths of resonances. If the agreement is not

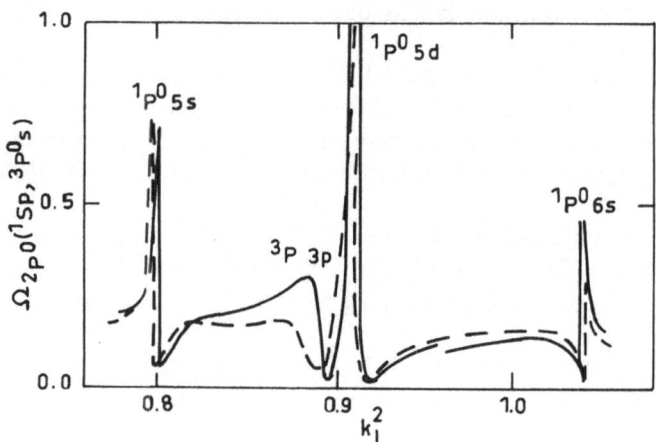

FIGURE 7. The partial collision strength $\Omega^2_{2p^0}({}^1Sp, {}^3P^\circ s)$ for O V: ——, DW calculations; ---, close-coupling calculations.

good in the vicinity of the $2p^2\,{}^3P\,3p$ and $2s2p\,{}^1P\,5d$, it is because there are complicated interference effects between these resonances. The distorted-wave method used by Hershkowitz and Seaton[59] is somewhat different from the method explained before because it uses some ideas from quantum defect theory to build the resonances, but we do not think that the results would be very different if the method was strictly the same as the one described before. We must mention that the discontinuity at 0.95 Ry is peculiar to the method of Hershkowitz and Seaton which uses a different bound function $\bar{\Phi}_m$ on either side of the discontinuity.

The distorted-wave approximation has been applied recently by Pradhan et al.,[60] to electron impact excitation of a number of helium like ions, for all transitions between the $1s^2$, $1s2s$, and $1s2p$ levels. Full allowances for the resonance contribution is made by using three different techniques:

1. Resonances between the $n = 2$ states are obtained by extrapolating the R-matrix elements below the excitation thresholds of excitation using quantum defect theory.

2. Low-lying resonances above the $n = 2$ states are accounted for by explicitly adding bound functions $\bar{\Phi}_m(1s3l3l')$ within the distorted-wave approach, in the same way as expansion (61) can be used within the close-coupling approach.

3. The higher-lying resonance of the type $1s3ln'l'$ for $n' > 3$ were estimated by the Gailitis[50] averaging procedure described in Section A.3 of the Appendix.

In this paper all couplings with the radiation field are neglected. Resonances are calculated without taking account of the competition between autoionization and radiative decay. Within this approximation, the resonance contribution remains significant with increasing z and, as is well known, becomes very important in the case of forbidden or intercombination transitions.

For the $1\,{}^1S$–$2\,{}^3S$ transitions in C^{4+}, the resonances dominate the near-threshold region. The contribution to the collisional rate is such that $\mathcal{C}_r/\mathcal{C}_b$ (where $\mathcal{C}_r$ is the result including the resonances and $\mathcal{C}_b$ includes only the background DW) is of the order of 6 at $T = 10^4$ K and still 1.5 at $T = 5 \times 10^6$ K. For the $2\,{}^3S$–$2\,{}^1P$ transitions in Fe^{24+} this ratio is larger than 5 at $T = 5 \times 10^7$ K.

This systematic study of excitation in helium like ions by Pradhan et al.[60,61] would have important consequences for the analysis of line intensities in laboratory or astrophysical plasmas. But, before reaching any conclusion in this respect, it is necessary to come back to the importance of radiative processes in highly ionized systems.

The Gailitis formula (A35) gives the average resonance contribution to excitation $\alpha' \to \alpha$ below the excitation threshold of level γ in terms of matrix elements of the transition T matrix extrapolated below that threshold (A35)

$$\langle |T_{ii'}|^2 \rangle = \sum_c \frac{|T_{ic}^B|^2 |T_{ci'}^B|^2}{\sum_{i''} |T_{ci''}^B|^2} + |T_{ii'}^B|^2, \tag{63}$$

where i, i', and i'' are open channels corresponding to α, α', and α'' target states and c is one of the closed channels associated with γ. In Eq. (63), it is assumed that the T^B elements can be extrapolated from above to energies below the γ excitation thresholds.

The relation between the autoionization probabilities A_{mi}^a and $|T_{ci}|^2$ is given in the Appendix, (A33) and (A36), where m denotes a resonance in the closed-channel c. The result is

$$A_{mi}^a = \frac{z^2}{2\pi n^3} |T_{ci}^B|^2. \tag{64}$$

Therefore expression (63) becomes

$$\langle |T_{ii'}|^2 \rangle = \sum_c \frac{A_{mi}^a}{\sum_{i''} A_{mi''}^a} |T_{ci'}^B|^2 + |T_{ii'}^B|^2. \tag{65}$$

When radiative decay competes with autoionization, the probability that breakup of the resonance leads to i' must be replaced by

$$\langle |T_{ii'}|^2 \rangle = \sum_c \frac{A_{mi}^a}{\sum_{i'} A_{mi''}^a + \sum_{m'} A_{mm'}^r} |T_{ci'}^B|^2 + |T_{ii'}^B|^2, \tag{66}$$

where m' denotes another bound state of the $(N + 1)$-electron system. The difference between expression (65) and (66) corresponds to radiative losses.

As emphasized in Section 3.1, the spontaneous radiative probabilities $A_{mm'}^r$ vary as z^4, whereas the autoionization probabilities $A_{mi''}^a$ do not depend on z. Radiative decay reduces the contribution to excitation quite significantly in highly charged ions. This is clear from the results for resonance excitation of $1s^2\,{}^1S_0-1s2s\,{}^3S_1$ in O^{6+} and Fe^{24+}, given in Figs. 2 and 3, which have been obtained in using Eqs. (65) and (66) and have already been described in Section 2.2.

3.5. Approximate Methods for Strong Allowed Transitions

In the particular case of positive ions, the distance of closest approach r_c of the perturbing électron when its velocity tends to zero remains finite for all finite values of its angular momentum l, i.e., $r_c = l^2/2z$. Consequently, the collision strength includes a finite contribution for all values of l even at threshold. Apparently Ω is finite at threshold.

For strong allowed transitions, the long-range character of the dipole field gives significant weight to intermediate or large values of l even at low energies, that is, $l \neq 0$, and this means that short-range interactions play a minor role in the cross section.

If the perturbing electron remains outside the target most of the time, many more approximations can be made: exchange can be neglected and the Coulomb–Born approximation can replace the distorted-wave approximation. Even the Coulomb–Bethe approximation becomes reliable.

In the Coulomb–Bethe approximation which was used extensively in the early 1960s,[46,62,63] the Coulomb–Born R-matrix element

$$R_{ii'} = -\sum_N \langle i | \frac{2}{r_{N,N+1}} | i' \rangle. \tag{67}$$

Here i and i' are wave functions of the form $\Theta_i(N)F_i(r_{N+1})/r_{N+1}$, where the $F_i(r_{N+1})$ are the "undistorted" Coulomb functions. In Eq. (67) it is assumed that

$$\frac{1}{r_{N,N+1}} = \sum_\lambda P_\lambda(\hat{r}_N, \hat{r}_{N+1}) \frac{r_N^\lambda}{r_{N+1}^{\lambda+1}} \tag{68}$$

which implies that the colliding electron of coordinate r_{N+1} remains outside the target, i.e., $r_{N+1} > r_N$ for all N; P_λ is the Legendre polynomial.

When the S matrix is calculated from these Coulomb–Bethe R-matrix elements, according to Eq. (51), allowance is made for coupling effects. This is the unitarized form of the approximation. When all of these coupling effects are small, which is the case when the allowed transition under study is strong, the collision strength can be expressed in a very simple form:

$$\Omega_{\alpha\alpha'} = \frac{8\pi}{3\sqrt{3}} S_{\alpha\alpha'} g(k_i k_{i'}), \tag{69}$$

where S is the line strength in atomic units, and $g(k_i k_{i'})$ is the Gaunt factor which is of the order of unity and independent of z. With $l_>$ the larger of $(l_i l_{i'})$ we have

$$g(k_i k_{i'}) = \frac{2\sqrt{3}}{\pi} \sum_{l_i l_{i'}} l_> \left| \int F_{k_i l_i} F_{k_{i'} l_{i'}} \frac{dr}{r^2} \right|^2. \tag{70}$$

The validity of the Coulomb–Bethe approximation justifies the use of expression (70) with an effective $\bar{g}$ Gaunt factor. This approximation is used extensively by plasma physicists and astrophysicists for extrapolating and interpolating excitation cross sections along a given isoelectronic sequence, in particular, for the strong allowed $n = 0$ transitions. But it must be clear also that, for weak allowed transitions, coupling effects are important and reduce g considerably when S is correctly calculated in terms of R. On the other hand, it is hopeless to extend such a procedure of extrapolation to forbidden transitions which are sensitive to short-range interactions and to resonance effects.

The effective $\bar{g}$ is usually determined by comparison with accurate theoretical cross-section data. An assessment of the validity of this approximation has been made by Younger and Wiese,[64] who also discuss the electron correlation effects within the target.

Important resonance contributions are usually given by strong dipole coupling with the initial and final levels of a given transition. Therefore, the Coulomb–Bethe approximation may offer a simple and reliable way to calculate the average contribution of resonances.

In many situations, the resonance contribution to excitation of a forbidden transition 1–2 below the threshold of a strong allowed transition 1–3 is simply given by the extrapolation of Q_{13} and Q_{23} below the threshold of $n = 3$. Approximate methods can be applied to the calculation of Q_{13} and to Q_{23} when they are strongly allowed.

The Coulomb–Born technique has been used in this way by Bely and Petrini[65] for the $4s$–$3d$ transiton in Ca^+ below the $4p$ level, and by Petrini[66] for calculating the resonance contribution to the fine-structure transition $3p(\frac{1}{2}-\frac{3}{2})$ in Fe^{13+} which is of important astrophysical interest.

Even the semiclassical analogue of the Coulomb–Bethe approximation has been applied successfully by Malinovsky[30] for the resonance contribution to the forbidden $2s^2\,^1S$–$2s2p\,^3P$ transition in O^{4+} below the threshold of $2s2p\,^1P$.

All of these simple calculations compare very well with close-coupling computations when the low-l contributions are correctly estimated and do not play an important role. The validity of the semiclassical approach given by Burgess[67] in the case of hyperbolic trajectories, or of the Bethe approximation[62] when properly applied, has been generally underestimated. Because of their simplicity, they make it possible to pay more attention to the representation of the target. Experience shows that for complex highly charged ions the cross sections for strong allowed transitions are in fact more sensitive to correlation effects within the target than to short-range interactions with the colliding electron.

In particular, this explains why reliable results have been obtained for a long time in applying these approximate methods to dielectronic recombination processes[68] and to satellite line intensities[10] or to line broadening calculations.[69]

The Coulomb–Bethe approximation is appropriate for estimating the branching ratio $A^a/(A^a + A^r)$ appearing when radiative decay competes with autoionization. A^a, given by Eq. (16), can be calculated using for the partial collision strength

$$\Omega^{l_i}_{\alpha\alpha'} = \frac{8\pi}{3\sqrt{3}} S_{\alpha\alpha'} g_{l_i}(k_i = 0, k_{i'}), \tag{71}$$

where the partial Gaunt factor g_{l_i} is derived from Eq. (70).[62]

When autoionization occurs only in the continuum corresponding to the ion ground state α', the ratio A^a/A^r takes a very simple form

$$\frac{A^a}{A^r} = \frac{4}{\sqrt{3}} \frac{z^2}{n_i^3} \alpha^{-3} \Delta E_{\alpha\alpha'}^{-3}(2l_i + 1)^{-1} g_{l_i}(k_i = 0, k_{i'}) \tag{72}$$

which is independent of the line strength, that is, of all correlation effects within the target.

Formulas of this type can also be used to estimate the reduction of an excitation cross section near threshold due to radiative decay, which has been discussed in Section 3.1. Eventual overestimation of low-l contributions can be avoided in using the unitarized form of the Coulomb–Bethe approximation, each **R**-matrix element being proportional to the square root of the line strength and to the radial integral appearing in the Gaunt factor (70).

Numerical methods for calculating these radial integrals have been discussed in a few papers.[48,62,70] Useful sum rules have been given by Burgess[71] for evaluating sums over large values of l_i and $l_{i'}$ using the Coulomb–Bethe approximation, which are used in most of the calculations of optically allowed transitions.

3.6. Relativistic Effects

For highly ionized atoms it is necessary to include relativistic effects in the atomic structure as well as in the electron–ion process calculations. The best approach is to use a fully relativistic wave equation constructed from the Dirac monoelectronic equation and the Breit two-electron interaction. This approach is currently used for bound-state wave functions in two types of approximation:

1. The first is an extension of the multiconfiguration Hartree–Fock method to the relativistic case by Desclaux[72] and Grant et al.[73]

2. The second one, which uses a perturbation treatment in which the zero-order Hamiltonian is the monoelectronic Dirac Hamiltonian including a central statistical potential model, was used by Klapisch and Luc-Koenig.[74]

Generalization to the collisional problem is still in progress. There have been some attempts to extend the close-coupling approximation to the Breit–Dirac Hamiltonian, but to our knowledge the only one relativistic collision calculation that has been done so far is an extension of the Coulomb–Born method to the case of electron impact excitation of hydrogenic ions by Walker.[75]

Nevertheless, for moderately charged ions, one expects that nonrelativistic methods can be used to obtain zero-order wave functions, the relativistic Breit–Pauli corrections being included as perturbations.

The Breit–Pauli corrections are the one-body terms (mass, Darwin, and spin–orbit corrections) and the two-body terms (fine-structure: mutual spin–orbit, spin–other orbit, and spin–spin interactions; non-fine-structure: two-body Darwin, contact spin–spin, and orbit–orbit interactions). For bound states, various computer programs have been developed, based either on the Hartree–Fock approach (Froese-Fisher[76] and Cowan and Griffin[77] or on a central statistical potential model [Eissner et al.[78]]. Very few extensions to the collision problem have been done. The code most utilized for astrophysical plasmas is the U.C.L. package, which is composed for atomic structure calculations of SUPERSTRUCTURE[78] and for collisional fine-structure calculations of DW-JJOM.[47,79]

First, the DW program[48] calculates nonrelativistic collisional parameters in the distorted-wave approximation using a nonrelativistic target (as in Section 3.2). Afterwards, the JJOM program[79] uses results obtained in SUPERSTRUCTURE and DW to calculate fine-structure collision strengths between intermediate coupling levels of the target. In the first part of SUPERSTRUCTURE, multiconfigurational wave functions are obtained by diagonalization of the nonrelativistic Hamiltonian: they are the same as the ones used in DW: $\chi(\Gamma_i S_i L_i M_{L_i} M_{S_i})$, see Eq. (31). In the second part of SUPERSTRUCTURE, multiconfiguration wave functions are obtained by diagonalizing the Breit–Pauli Hamiltonian: $\Phi(\Delta_i J_i, M_{J_i})$, where Δ_i is the eigenvalue level while Γ_i is the eigenvalue term index. The $\Phi(\Delta_i J_i, M_{J_i})$ are expanded in the nonrelativistic $\chi(\Gamma_i S_i L_i, M_{S_i} M_{L_i})$ wave functions:

$$\Phi(\Delta_i J_i, M_{J_i}) = \sum_{S_i L_i} f_{J_i}(\Delta_i \Gamma_i S_i L_i)\bar{\chi}(\Gamma_i S_i L_i J_i, M_{J_i})\bar{\chi}(\Gamma_i S_i L_i J_i, M_{J_i}) \quad (73)$$

$$\chi(r_i s_i L_i J_i M_{J_i}) = \sum_{M_{S_i} M_{L_i}} C^{L_i S_i J_i}_{M_{L_i} M_{S_i} M_{J_i}} \chi(\Gamma_i S_i L_i, M_{S_i} M_{L_i}) \quad (74)$$

where $C^{L_i S_i J_i}_{M_{L_i} M_{S_i} M_{J_i}}$ is a Clebsch–Gordan coefficient.

These expansion coefficients are used in JJOM to transform the LS target into the intermediate-coupling target and consequently the transition matrix $T^{S^T L^T}$ is transformed into T^{J^T}, using the pair coupling scheme:

$$L_i + S_i = J_i; \qquad J_i + l_i = K; \qquad K + s = J^T; \qquad (75)$$

J^T being the total angular momentum of the total system. The transformation is performed using Racah recoupling coefficients W:

$$T^{J^T}(\Gamma_i S_i L_i J_i l_i K, \Gamma_{i'} S_{i'} L_{i'} J_{i'} l_{i'} K')$$

$$= \sum_{L^T S^T} X(S^T L^T J^T, S_i L_i J_i; l_i K)$$

$$\times T^{S^T L^T}(\Gamma_i S_i L_i l_i, \Gamma_{i'} S_{i'} L_{i'} l_{i'}) X(S^T L^T J^T, S_{i'} L_{i'} J_{i'}; l_{i'} K'); \qquad (76)$$

$$X(S^T L^T J^T, S_i L_i J_i; l_i K) = W(L^T l_i S_i J_i; L_i K) W(L^T J^T S_i \tfrac{1}{2}; S^T K)$$

$$\times \left[(2S^T + 1)(2L^T + 1)(2K + 1)(2J_i + 1) \right]^{1/2}. \qquad (77)$$

The final T is obtained from Eq. (73):

$$T^{J^T}(\Delta_i J_i l_i K, \Delta_{i'} J_{i'} l_{i'} K') = \sum_{S_i L_i} \sum_{S_{i'} L_{i'}} f_{J_i}(\Delta_i \Gamma_i S_i L_i)$$

$$\times T^{J^T}(\Gamma_i S_i L_i J_i l_i K, \Gamma_{i'} S_{i'} L_{i'} J_{i'} l_{i'} K') f_{J_i}(\Delta_{i'} \Gamma_{i'} S_{i'} L_{i'}). \qquad (78)$$

The operator K commutes with the Hamiltonian of the total system as long as the colliding electron is nonrelativistic, which is one of the two hypotheses of the approximation, the second being that the levels corresponding to the dominant term $\Gamma_i S_i L_i$ have practically the same energy.

To illustrate the importance of the relativistic effects, Fig. 8 shows the reduced collision strength $(z^2\Omega)$ for $1s^2 \, ^1S_0$ to $1s2p \, ^3P_1$ and $^3P_2$ in function of reduced energy (E/E_0) in the case of O VII and Fe XXV. It is seen that the reduced collision strength for $^3P_2$ does not change from O VII to Fe XXV as expected, while for $^3P_1$ the reduced collision strength becomes larger for Fe XXV than for O VII at high energy. This effect is explained by the relativistic interaction between $^3P_1$ and $^1P_1$ in the target. For elements in low stages of ionization such as O VII this effect is negligible and the

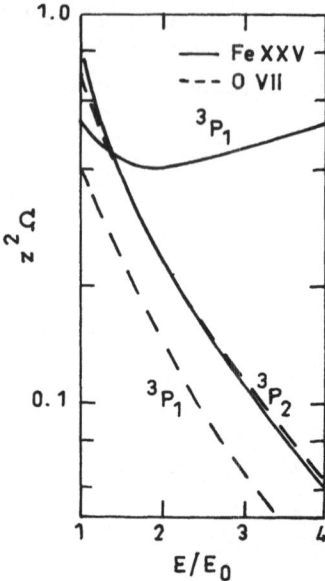

FIGURE 8. Reduced collision strength for the $1s^2\,{}^1S_0-1s2p\,{}^3P_1^o,\,{}^3P_2^o$ transitions in O VII (---) and Fe XXV (—).

collision strength has the normal behavior for a forbidden transition; it decreases with increasing energy, while for highly ionized elements such as Fe XXV, this effect becomes important and the collision strength has the usual behavior for an allowed transition, that is, it increases as $\log E$.

The U.C.L. package has been used extensively for astrophysical plasmas. There exist other codes based on similar approximations, for example, the one by Sampson and Golden.[80] Where a comparison is possible, the results obtained from these codes are very similar.

Finally, we mention that a very interesting discussion on relativistic collisional theory is presented by Jones.[81] This paper also gives a possible treatment of the distorted-wave approximation when the colliding electron is relativistic.

3.7. Relativistic Effects in Autoionization

In Section 2.3 the importance of satellite lines in high-temperature plasmas was shown. These lines are mostly emitted from autoionizing states which produce resonance effects in the electron–ion processes. In Section 3.4 we have seen that it is necessary to take account of the radiative decay of resonance states to calculate collisional excitation rates. For such ions, relativistic effects have to be included to represent the autoionizing states.

In this section we use an approach for autoionizing states similar to the one used in Section 3.6 for target states. We assume that the colliding electron is nonrelativistic and therefore we can use the pair coupling defined by Eq. (76). The transformation between LS-coupling interactions matrix elements (A2),

$$\overline{V}_{im}^{S^{T}L^{T}} = \langle \Gamma_i S_i L_i l_i S^{T}L^{T} | V | \Gamma_m S^{T}L^{T} \rangle, \tag{79}$$

and the intermediate-coupling interactions matrix element,

$$\overline{V}_{im}^{J^{T}} = \langle \Delta_i J_i l_i KJ^{T} | V | \Delta_m J^{T} \rangle, \tag{80}$$

can be obtained by (76) and (78):

$$\overline{V}_{im}^{J^{T}} = \sum_{S_i L_i} \sum_{S^{T}L^{T}} f_{J_i}(\Delta_i \Gamma_i S_i L_i) X(S^{T}L^{T}J^{T}, S_i L_i J_i; l_i K) \overline{V}_{im}^{S^{T}L^{T}} f_{J^{T}}(\Delta_m \Gamma_m S^{T}L^{T}).$$
$$\tag{81}$$

To obtain the f_J for the target and autoionizing states, the SUPERSTRUCTURE[78] program is used twice. Calculations have been done for astrophysical and laboratory data by Bely–Dubau et al.[82] and Dubau et al.[82] Similar treatment has also been used by Cowan[36] and Vainstein and Sofronova.[84] When comparison is possible between the different methods, the results are in good agreement (e.g., for the satellite lines of hydrogenlike and heliumlike resonance lines).

4. Conclusion

From this short study we can conclude that if the understanding of physical conditions in hot plasmas of low densities requires accurate ionic structure calculations with allowance for the main relativistic effects for high-Z systems, an approximate method like the distorted-wave approximation is very reliable for solving the collision problem, provided the coupling with the radiation field is taken into account each time radiative decay completes with autoionization or excitation.

The role of radiative decay as the stabilization process in dielectronic recombination in highly charged ions was recognized a long time ago. Surprisingly, the logical consequences of these phenomena in reducing the cross section for resonance excitation has been ignored until recently. Moreover, the reduction of the excitation cross section just above threshold due to the finite lifetime of an excited ion state has been neglected up to now, although it may affect the excitation collisional rates for highly ionized systems.

All these conclusions only apply to highly charged ions present in hot plasmas ($T > 10^6$ K) and cannot be generalized to low-z ions whose study may require more elaborate *methods* like the close-coupling approximation. However, the study of high-z ions may require more elaborate *models* in which allowance is made for the radiative decay of the $A^{+z} + e$ system.

ACKNOWLEDGMENTS

We wish to express our thanks to Drs. W. Eissner, D. G. Hummer, M. Loulergue, and C. Zeippen for useful suggestions and critical readings of the manuscript.

Appendix

A.1.　Derivation of $a_m(E)$ and $b_\varepsilon^i(E)$

We insert Eq. (44) in (45):

$$\bar{a}_m(E)\left(\bar{E}_m - E\right) + \sum_{i'}\int d\varepsilon'\, b_{\varepsilon'}^{i'}(E)\left\langle\bar{\phi}_m|V|\phi_{\varepsilon'}^{i'}\right\rangle = 0,$$

$$b_\varepsilon^i(E)(E_i + \varepsilon - E) + \sum_m \bar{a}_m(E)\left\langle\phi_\varepsilon^i|V|\bar{\phi}_m\right\rangle$$

$$+ \sum_{i'}\int d\varepsilon'\, b_{\varepsilon'}^{i'}(E)\left\langle\phi_\varepsilon^i|V|\phi_{\varepsilon'}^{i'}\right\rangle = 0.$$

$$(A1)$$

The matrix elements $\left\langle\bar{\phi}_m|V|\phi_{\varepsilon'}^{i'}\right\rangle$ and $\left\langle\phi_\varepsilon^i|V|\phi_{\varepsilon'}^{i'}\right\rangle$ vary slowly with ε and ε', and $b_\varepsilon^i(E)$ has a pole for $\varepsilon = \varepsilon_i \equiv E - E_i$. We can extract these matrix elements from the integrals and we define

$$\alpha_{i'}(E) = \int_0^\infty d\varepsilon'\, b_{\varepsilon'}^{i'}(E),$$

$$A_i(E) = \sum_m \bar{a}_m(E)\bar{V}_{im}(E) + \sum_{i'} V_{ii'}(E)\alpha_{i'}(E),$$

$$\bar{V}_{im}(E) = \left\langle\phi_{\varepsilon_i}^i|V|\bar{\phi}_m\right\rangle,$$

$$V_{ii'}(E) = \left\langle\phi_{\varepsilon_i}^i|V|\phi_{\varepsilon_{i'}}^{i'}\right\rangle.$$

$$(A2)$$

The matrix elements $\bar{V}_{im}(E)$ and $V_{ii'}(E)$ are calculated for $\varepsilon_{i'} = E - E_i$ and $\epsilon_{i'} = E - E_{i'}$. Equation (A1) becomes

$$\bar{a}_m(E)(E - \bar{E}_m) = \sum_{i'} \bar{V}_{mi'}(E)\alpha_{i'}(E)$$

$$b_\varepsilon^i(E)(\varepsilon_i - \varepsilon) = A_i(E)$$

(A3)

which gives [see Dirac[44]]

$$\bar{a}_m(E) = \sum_{i'} \frac{\bar{V}_{mi'}(E)\alpha_{i'}(E)}{E - \bar{E}_m},$$

$$b_\varepsilon^i(E) = A_i(E)\left[\gamma_i(E)\delta(\varepsilon_i - \varepsilon) + \mathrm{PP}\frac{1}{\varepsilon_i - \varepsilon}\right].$$

(A4)

The symbol PP represents the principal part.

If we substitute (A4) into (A2), we obtain

$$\alpha_i(E) = \gamma_i(E)A_i(E)$$

(A5)

Here $\gamma_i(E)$ is arbitrary; it can be chosen from the asymptotic expression for $F_{ii'}^\rho(r)$ in (47).

From Eqs. (38) and (46) follows

$$F_{ii'}^\rho(r) = \int_0^\infty d\varepsilon\, b_\varepsilon^{ii'}(E)G_i(r),$$

(A6)

where i denotes an open channel ($\varepsilon_i > 0$) and i' one of the n independent solutions. From (47) we have

$$A_{ii'}(E)\gamma_{ii'}(E) = \left(\frac{\pi}{2}\right)^{1/2}\delta_{ii'}$$

$$A_{ii'}(E) = -\frac{1}{\pi}\left(\frac{\pi}{2}\right)^{1/2}\rho_{ii'}(E)$$

(A7)

The combination of (A2), (A4), (A5), and (A7) gives

$$\rho_{ii'}(E) = -\pi\left[V_{ii'}(E) + \sum_m \frac{\bar{V}_{im}(E)\bar{V}_{mi'}(E)}{E - \bar{E}_m}\right].$$

(A8)

A.2. Resonances

In order to simplify the formalism, we define the matrix elements

$$a_{ii'} = \pi V_{ii'} \quad \text{and} \quad b_{im} = \sqrt{\pi} \, \bar{V}_{im}. \tag{A9}$$

Then (A8) becomes

$$\rho_{ii'} = -a_{ii'} - \sum_{mm'} b_{im} \left(E - \bar{E}_m \delta_{mm'} \right)^{-1} b_{mi}. \tag{A10}$$

In matrix notation (A10) has the form

$$\rho = -\left[\mathbf{a} + \mathbf{b}(E - \bar{E})^{-1} \mathbf{b}^t \right], \tag{A11}$$

where E is a diagonal matrix $(E_{mm'} = \delta_{mm'} E_m)$ and $\mathbf{b}^t$ is the transpose matrix of $\mathbf{b}$.

Also we define $\mathbf{S}_\rho$ as

$$\mathbf{S}_\rho = (1 + i\rho)(1 - i\rho)^{-1}. \tag{A12}$$

The relation between S and S_ρ is given by Eq. (52)

$$\mathbf{S} = \exp(i\tau) \mathbf{S}_\rho \exp(i\tau). \tag{A13}$$

and therefore for $i \neq i'$ we have

$$|T_{ii'}|^2 = |S_{ii'}|^2 = |(S_\rho)_{ii'}|^2. \tag{A14}$$

Inserting (A11) into (A12) we obtain

$$\mathbf{S}_\rho = \left[1 - i\mathbf{a} - i\mathbf{b}(E - \bar{E})^{-1} \mathbf{b}^t \right] \left[1 + i\mathbf{a} + i\mathbf{b}(E - \bar{E})^{-1} \mathbf{b}^t \right]^{-1}. \tag{A15}$$

If we neglect $\mathbf{a} \cdot \mathbf{b}$, $\mathbf{b}^t \cdot \mathbf{a}$ in (A15) we obtain

$$\mathbf{S}_\rho \simeq (1 - i\mathbf{a}) \left[1 - i\mathbf{b}(E - \bar{E})^{-1} \mathbf{b}^t \right] \left[1 + i\mathbf{b}(E - \bar{E})^{-1} \mathbf{b}^t \right]^{-1} (1 + i\mathbf{a})^{-1}. \tag{A16}$$

Now it is easy to verify that

$$\mathbf{b}^t \left[1 + i\mathbf{b}(E - \bar{E})^{-1} \mathbf{b}^t \right]^{-1} \equiv \left[1 + i\mathbf{b}^t \mathbf{b}(E - \bar{E})^{-1} \right]^{-1} \mathbf{b}^t. \tag{A17}$$

Therefore,

$$\mathbf{S}_p \simeq (1 - \mathrm{i}a)\left\{1 - 2\mathrm{i}\mathbf{b}\big[(E - \bar{E}) + \mathrm{i}\mathbf{b}^t\mathbf{b}\big]^{-1}\mathbf{b}^t\right\}(1 + \mathrm{i}a)^{-1}. \quad \text{(A18)}$$

And if we neglect a^2 we obtain

$$\mathbf{S}_p \simeq 1 - 2\mathrm{i}a - 2\mathrm{i}\mathbf{b}\big[(E - \bar{E}) + \mathrm{i}\mathbf{b}^t\mathbf{b}\big]^{-1}\mathbf{b}^t. \quad \text{(A19)}$$

From (A14) we can write

$$|T_{ii'}|^2 \simeq 4\left|a_{ii'} + \sum_{mm'} b_{im}\big[(E - \bar{E}) + \mathrm{i}\mathbf{b}^t\mathbf{b}\big]^{-1}_{mm'} b_{m'l'}\right|^2. \quad \text{(A20)}$$

In order to analyze overlapping effects one has to diagonalize the complex matrix

$$\mathbf{W} = \bar{E} + \mathrm{i}\mathbf{b}^t\mathbf{b}. \quad \text{(A21)}$$

According to matrix theory, it is always possible to perform such a diagonalization because W is symmetric. If we write the eigenvalues of (A21) as

$$W_m = \mathcal{E}_m - \frac{\mathrm{i}}{2}\gamma_m \quad \text{(A22)}$$

$|T_{ii'}|^2$ can be rewritten as

$$|T_{ii'}|^2 = 4\left|a_{ii'} + \sum_m \frac{\tilde{b}_{im}\tilde{b}_{mi'}}{(E - \mathcal{E}_m) + (\mathrm{i}/2)\gamma_m}\right|^2. \quad \text{(A23)}$$

In the case of nonoverlapping resonances

$$\gamma_m \simeq 2\sum_{i''} b_{i''m} b_{mi''}; \qquad \mathcal{E}_m \simeq E_m \qquad (\tilde{b}_{im} \simeq b_{im}) \quad \text{(A24)}$$

we shall consider only the nonoverlapping case below.

Replacing a and b by their values on (A9), (A23) becomes

$$|T_{ii'}|^2 = 4\pi^2\left|V_{ii'} + \sum_m \frac{\bar{V}_{im}\bar{V}_{mi'}}{(E - \bar{E}_m) + (\mathrm{i}/2)\gamma_m}\right|^2, \quad \text{(A25)}$$

where

$$\gamma_m = \sum_{i''} A^{\mathrm{a}}_{mi''}. \tag{A26}$$

Here γ_m is the total autoionization probability from "bound state" m, the autoionization probability from m to the free channel i being equal to

$$A^{\mathrm{a}}_{mi} = 2\pi |\overline{V}_{mi}|^2 \tag{A27}$$

with $\overline{V}_{im} = \overline{V}_{mi}$. In the case of negligible $V_{ii'}$, $|T_{ii'}|^2$ inserted into (53) gives a resonance cross section with a Lorentz profile

$$|T_{ii'}|^2 = 2\pi \sum_m \frac{A^{\mathrm{a}}_{mi}}{\gamma_m} A^{\mathrm{a}}_{mi'} F_m(E); \quad F_m(E) = \frac{1}{2\pi} \frac{\gamma_m}{\left(E - \overline{E}_m\right)^2 + \gamma_m^2/4}. \tag{A28}$$

Here $F_m(E)$ is normalized to 1, i.e.,

$$\int_{-\infty}^{+\infty} F_m(E)\, dE = 1. \tag{A29}$$

The physical interpretation of (A28) is that an electron e is captured by $A^{+z}_{\alpha'}$ (entrance channel i') forming the $A^{+z} + e$ "bound state" m which decays, via autoionization, to A^{+z}_{α} (exit channel i).

For hot plasma applications, we consider a Maxwellian distribution of electron velocity $v_{i'}$ incoming on $A^{+z}_{\alpha'}$. In atomic units

$$v_{i'} f(v_{i'})\, dv_{i'} = 4\pi \left(\frac{1}{2\pi kT} \right)^{3/2} \exp\left(-\frac{\varepsilon_{i'}}{kT} \right) 2\varepsilon_{i'}\, d\varepsilon_{i'},$$

$$\varepsilon_{i'} = \frac{k_{i'}^2}{2} = \frac{v_{i'}^2}{2}, \tag{A30}$$

and using (53), (54), and (A28) we have

$$\langle v_{i'} Q_{i' \to i} \rangle = \frac{1}{2\omega_{\alpha'}} \left(\frac{2\pi}{kT} \right)^{3/2} \sum_m \omega_m \frac{A^{\mathrm{a}}_{mi}}{\gamma_m} A^{\mathrm{a}}_{mi'} \exp\left[-\frac{\left(\overline{E}_m - E_{i'}\right)}{kT} \right], \tag{A31}$$

where $\omega_m = (2L^{\mathrm{T}} + 1)(2S^{\mathrm{T}} + 1)$ is the statistical weight of m. The expression (A31) is very similar to the dielectronic recombination coefficient rate formula [see, e.g., Seaton and Storey[39]]. For dielectronic recombination the branching ratio is different because

$$A_{mi}^{a} \to A_{mm'}^{r} \quad \text{and} \quad \gamma_m = \sum A_m^{a} + \sum A_m^{r}, \tag{A32}$$

where $A_{mm'}^{r}$ is the radiative transition probability from m to m'.

A.3. Gailitis Formula

If we consider a Rydberg series of bound levels of m of the $A^{+z} + e$ system, they are described by a target level $A_{\alpha''}^{z}$ and the quantum numbers of a bound electron $nl_{i''}$. The $A_{\alpha''}^{z}$ excited level energy must be larger than the colliding electron energy. The Rydberg series converges to the $A_{\alpha''}^{z}$ excitation threshold and the matrix element V_{im} is replaced by $V_{ii''}$ above this threshold. Using the analytical properties of the solutions of the radial equations it is possible to obtain an expression for the continuation between V_{im} and $V_{ii''}$ ($m = i'' nl_{i''}$) (see Chapter 6).

$$\lim_{\varepsilon_{i''} \to 0} |V_{ii''}|^2 = \lim_{n \to \infty} \frac{n^3}{z^2} |V_{im}|^2 = \lim_{n \to \infty} \frac{n^3}{2\pi z^2} A_{mi}^{a} \quad \text{if} \quad \overline{V}_{mi} \simeq V_{mi}. \tag{A33}$$

Therefore the average of $|T|^2$ over the resonance m can be written

$$\frac{1}{\Delta E} \int |T_{ii'}|^2 dE \frac{n^3}{z^2} 2\pi \sum_m \frac{A_{mi}^{a}}{\gamma_m} A_{mi'}^{a} + 4\pi^2 |V_{ii'}|^2, \tag{A34}$$

where

$$\overline{E}_m \simeq E_{i''} - \frac{z^2}{2n^2} \quad \text{and} \quad \Delta E = \frac{z^2}{n^3} \Delta n \quad (\Delta n = 1).$$

Inserting (A33) into (A34) we obtain

$$\lim_{n \to \infty} \frac{1}{\Delta E} \int |T_{ii'}|^2 dE = \lim_{\varepsilon_{i''} \to 0} \frac{|T_{ii''}^{B}|^2 |T_{i''i'}^{B}|^2}{\sum_{i'''} |T_{i'''i'''}|^2} + |T_{ii'}^{B}|^2, \tag{A35}$$

where

$$|T_{ii'}^{B}|^2 = 4\pi^2 |V_{ii'}|^2 \tag{A36}$$

is the background of the T matrix element as a function of energy. It corresponds to the "direct" transition matrix element.

References

1. W. Grotrian, *Naturwissenschaften* **27**, 214 (1939).
2. B. Edlén, *Z. Astrophys.* **22**, 30 (1942).
3. *Atomic and Molecular Data for Fusion*. Reports of the I.A.E.A. in *Phys. Rep.* **37**, 125 (1978) and *Phys. Scr.* **23**, 69 (1981).
4. *Atomic and Molecular Physics in Controlled Thermonuclear Research*, M. R. C. McDowell and A. M. Ferendeci, Eds., Plenum, New York, 1980.
5. L. P. Presnyakov, *Sov. Phys. Usp.* **19**, 387 (1976).
6. C. de Michelis and M. Mattioli, *Nucl. Fusion* **21**, 677 (1981).
7. *International Bulletin on Atomic and Molecular Data for Fusion*, K. Katsonis and R. A. Langley, Eds., I.A.E.A. Agency, Vienna.
8. *Atomic Data for Fusion*, D. H. Crandall, C. F. Barnett, and W. L. Wiese, Eds., Oak Ridge Laboratory and National Bureau of Standards.
9. *Reports of the Institute of Plasma Physics*, Nagoya University, Japan.
10. A. H. Gabriel and C. Jordan, in *Case Studies in Atomic and Molecular Collision Physics*, Vol. 2, E. W. McDaniel and M. R. C. McDowell, Eds., North Holland, Amsterdam, 1972, p. 209.
11. A. K. Dupree, *Adv. At. Mol. Phys.* **14**, 393 (1978).
12. C. Jordan, in *Progress in Atomic Spectroscopy*, W. Hanle and H. Kleinpoppen, Eds., Plenum, New York, 1979.
13. U. Feldman, *Phys. Scr.* **24**, 681 (1981).
14. O. Bely and H. Van Regemorter, *Ann. Rev. Astron. Astrophys.* **8**, 329 (1970).
15. M. J. Seaton, *Adv. At. Mol. Phys.* **11**, 83 (1975).
16. J. W. Henry, *Phys. Rep.* **68**, 1 (1981).
17. A. H. Gabriel and H. E. Mason, in *Applied Atomic Collision*, Vol. 1, H. S. W. Massey, B. Bederson, and E. W. McDaniel, Eds., Academic Press, New York, 1982, to be published.
18. D. R. Bates, A. E. Kingston, and R. W. P. McWhirter, *Proc. R. Soc. Lond. Ser. A* **267**, 297 (1962).
19. R. Mewe and J. Schrijver, *Astron. Astrophys.* **65**, 115 (1978).
20. F. Bely-Dubau, J. Dubau, P. Faucher, and A. H. Gabriel, *Mon. Not. R. Astron. Soc.*, **198**, 239 (1982).
21. R. Mewe, J. Schrijver, and J. Sylvester, *Astron. Astrophys. Suppl.* **40**, 323 (1980).
22. C. Jordan, *Mon. Not. R. Astron. Soc.* **148**, 17 (1970).
23. H. P. Summers, *Mon. Not. R. Astron. Soc.* **158**, 255 (1972).
24. V. L. Jacobs, J. Davis, P. C. Keffe, and M. Blaha, *Astrophys. J.* **211**, 605 (1977).
25. G. W. F. Drake, *Phys. Rev. A* **3**, 908 (1971).
26. W. Eissner and C. J. Zeippen, *J. Phys. B* **14**, 2125 (1981).
27. P. Faucher, *J. Phys. B* **8**, 1886 (1975).
28. P. Faucher and D. A. Landman, *Astron. Astrophys.* **54**, 159 (1977).
29. P. L. Dufton, K. A. Berrington, P. G. Burke, and A. E. Kingston, *Astron. Astrophys.* **62**, 111 (1978).
30. M. Malinovsky, *Astron. Astrophys.* **43**, 101 (1975).
31. A. K. Pradhan, *Phys. Rev. Lett.* **47**, 79 (1981).
32. O. Bely, *Ann. Astrophys.* **30**, 953 (1967).
33. A. H. Gabriel and C. Jordan, *Nature* **221**, 947 (1969).
34. A. H. Gabriel, *Mon. Not. R. Astron. Soc.* **160**, 99 (1970).
35. J. Dubau and S. Volonté, *Rep. Prog. Phys.* **43**, 199 (1980).
36. G. A. Doschek, U. Feldman, and R. D. Cowan, *Astrophys. J.* **245**, 315 (1981).
37. Tokamak de Fontenay aux Roses, J. Dubau, and M. Loulergue, *J. Phys. B* **15**, 1007 (1982).
38. K. Alder, A. Bohr, T. Huus, B. Motteson, and A. Winther, *Rev. Mod. Phys.* **28**, 432 (1956).
39. M. J. Seaton and P. J. Storey, in *Atomic Processes and Applications*, P. G. Burke and B. L. Moiseivitch, Eds., North Holland, Amsterdam, 1976.

40. P. C. W. DAVIES and M. J. SEATON, *J. Phys. B* **2**, 757 (1969).
41. L. P. PRESNYAKOV, and A. M. URNOV, *J. Phys. B* **8**, 1280 (1975).
42. B. SHORE, *Rev. Mod. Phys.* **39**, 439 (1967).
43. I. C. PERCIVAL and M. J. SEATON, *Proc. Camb. Philos. Soc.* **53**, 654 (1957).
44. P. A. M. DIRAC, *The Principle of Quantum Mechanics*, 4th ed., Oxford University Press, London, 1958, see chapter on Collision Problems.
45. U. FANO, *Phys. Rev.* **124**, 1866 (1961).
46. F. H. MIES, *Phys. Rev.* **175**, 164 (1968).
47. W. EISSNER and M. J. SEATON, *J. Phys. B* **5**, 2187 (1972).
48. A. BURGESS, D. G. HUMMER, and J. A. TULLY, *Philos. Trans. R. Soc. Lond. A* **266**, 225 (1970).
49. D. H. SAMPSON, A. D. PARKS, and R. E. N. CLARK, *Phys. Rev. A* **17**, 1619, (1978).
50. M. GAILITIS, *Sov. Phys.-JETP* **17**, 1328 (1963).
51. M. R. C. McDOWELL, L. A. MORGAN, V. P. MYERSCOUGH, and T. SCOTT, *J. Phys. B* **10**, 2727 (1977).
52. M. R. C. McDOWELL, V. P. MYERSCOUGH, and U. NARAIN, *J. Phys. B* **7**, L195 (1974).
53. H. E. SARAPH, M. J. SEATON, and J. SHEMMING, *Philos. Trans. R. Soc. Lond. Ser. A* **264**, 77 (1969).
54. W. EISSNER, in *Physics of Electronic and Atomic Collisions*, T. R. Govers and F. J. de Heer, Eds., North Holland, Amsterdam, 1972, p. 40.
55. D. R. FLOWER and J. M. LAUNAY, *J. Phys. B* **5**, L207 (1972).
56. M. A. HAYES and M. J. SEATON, *J. Phys. B* **10**, L573 (1977).
57. J. DUBAU, *J. Phys. B* **11**, 4095 (1978).
58. M. A. HAYES and M. J. SEATON, *J. Phys. B* **11**, L79 (1978).
59. M. D. HERSHKOWITZ and M. J. SEATON, *J. Phys. B* **6**, 1176 (1973).
60. A. K. PRADHAN, D. W. NORCROSS, and D. G. HUMMER, *Phys. Rev. A* **23**, 619 (1981).
61. A. K. PRADHAN, D. W. NORCROSS, and D. G. HUMMER, *Astrophys. J.* **246**, 1031 (1981).
62. H. VAN REGEMORTER, *Mon. Not. R. Astron. Soc.* **121**, 213 (1960).
63. L. A. VAINSTEIN, *Opt. Spektrosk.* **11**, 301 (1961).
64. S. M. YOUNGER and L. W. WIESE, *J. Quant. Spectrosc. Radiat. Transfer* **22**, 161 (1979).
65. O. BELY and D. PETRINI, *Phys. Lett.* **23**, 442 (1966).
66. D. PETRINI, *Astron. Astrophys.* **9**, 392 (1970).
67. A. BURGESS, *Culham Conference on Atomic Collisions, A.E.R.E. Report 4818*, p. 63 (1964). See also A. Burgess, in *Third Conference on Electron–Atom Collisions*, M. R. C. McDowell, Ed., North Holland, Amsterdam, 1964 or reference 69.
68. A. BURGESS, *Astrophys. J.* **141**, 776 (1964).
69. S. SAHAL-BRÉCHOT, *Astron. Astrophys.* **1**, 91 (1969).
70. HOANG-BINH DY and H. VAN REGEMORTER, *J. Phys. B* **14**, L329 (1981).
71. A. BURGESS, *J. Phys. B* **7**, L364 (1974).
72. J. P. DESCLAUX, *Comput. Phys. Commun.* **13**, 71 (1977).
73. I. P. GRANT, B. J. McKENZIE, P. M. NORRINGTON, *Comput. Phys. Commun.* **21**, 207 (1980).
74. M. KLAPISCH and E. LUC-KOENIG, *Comput. Phys. Commun.*, submitted.
75. D. W. WALKER, *J. Phys. B* **7**, 97 (1974).
76. C. FROESE-FISCHER, *Comput. Phys. Commun.*, **4**, 107 (1972).
77. R. D. COWAN and D. C. GRIFFIN, *J. Opt. Soc. Am.* **166**, 1010 (1976).
78. W. EISSNER, M. JONES, and H. NUSSBAUMER, *Comput. Phys. Commun.* **8**, 270 (1974).
79. H. E. SARAPH, *Comput. Phys. Commun.* **3**, 256 (1972).
79. D. H. SAMPSON and L. B. GOLDEN, *J. Phys. B* **11**, 54 (1978).
81. M. JONES, *Philos. Trans. R. Soc. Lond. Ser. A* **277**, 587 (1975).
82. F. BELY-DUBAU, A. H. GABRIEL, and S. VOLONTÉ, *Mon. Not. R. Astron. Soc.* **189**, 801 (1979).
83. J. DUBAU, M. LOULERGUE, and L. STEENMAN-CLARK, *Mon. Not. R. Astron. Soc.* **190**, 125 (1980).
84. L. A. VAINSTEIN and U. I. SOFRONOVA, *At. Data Nucl. Data Tables* **21**, 49 (1978).

8

THE UNIVERSITY COLLEGE COMPUTER PACKAGE FOR THE CALCULATION OF ATOMIC DATA

ASPECTS OF DEVELOPMENT AND APPLICATION

H. Nussbaumer and P. J. Storey

1. Introduction

In 1955 M. J. Seaton[1] indulged in a short flirtation with Fe II. He did not go further than was considered appropriate at the time, but just offered an estimate of the upper limit for a crucial cross section. The title of that paper is "The kinetic temperature of the interstellar gas in regions of neutral hydrogen," and the abstract of that paper begins:

The paper is mainly concerned with the cooling resulting from excitation by electron impact of low-lying levels in C^+, Si^+, and Fe^+. The collision cross sections have been calculated by quantal methods for C^+ and have been estimated for Si^+ and Fe^+; the precision of the results obtained should be adequate for the present problem. The chemical composition of the gas is assumed to be similar to typical stellar compositions; this assumption is shown to be reasonable but it is not established with certainty.

It shows how closely interwoven in Seaton's work atomic physics and astrophysics have always been. That article is also a revealing example of his sound judgment about the efforts worth investing in a given problem.

Although Seaton was often approached, he never returned to Fe II; but 25 years later two of his pupils, Nussbaumer and Storey, succumbed to its daunting attraction.[2,3] The theoretical foundation to Seaton's 1955 article

H. NUSSBAUMER ● Institute of Astronomy, ETH Zentrum, CH 8092 Zürich, Switzerland
P. J. STOREY ● Department of Physics and Astronomy, University College London, Gower Street, London WC1 E6BT, England.

was laid in 1953. This was not only an excellent year for both claret and burgundy but it also saw the appearance of two important articles by Seaton,[4,5] where he further developed the electron–ion collision theory and gave results in particular for [O I] and [O III] collision strengths. We can easily appreciate the astrophysical relevance of these two papers. The first collision strengths calculated to explain the famous "nebulium" and auroral lines were those of Hebb and Menzel[6] for [O III], which was paper X in a series entitled "Physical Processes in Gaseous Nebulae," and Yamanouchi et al.[7] for [O I].

However, ten years later Aller[8] had to warn the astronomers that these cross sections were probably in error by an order of magnitude. Aller acknowledges Bates for the essence of this information.

Bates et al.[9] had analyzed the collision theory employed by former workers, finding inconsistencies that even violated the principle of detailed balance. Aller thus had to question former interpretations of observed nebular spectra. For example, he could only give lower limits to the O/H abundances based on upper limits of the cross sections. Seaton's calculations were done to remove this order of magnitude uncertainty, and in his work on these crucial cross sections he was confident to have achieved a 40% accuracy.[5] In the paper on the forbidden lines in gaseous nebulae[5] he also introduced for the collision parameter Ω the name collision strength, in analogy to the name line strength introduced by Condon and Shortley[10] for radiative processes.

2.   The Growth of Astronomical Observations

In Bowen's[11] list of nebular lines there were 74 lines between 3313 and 7325 Å; 31 of them belonged to H or He. Thus there were about 40 lines due to complex atoms. The most intense of these lines are due either to the collisionally excited forbidden transitions within the $2p^q$ or $3p^q$ configurations or the Bowen lines excited by fluorescence effects. With the advance in atomic physics heralded by Seaton's 1953 publications, it became a manageable task to calculate the atomic data needed for the interpretation of the most important of these lines. General purpose computer programs would not have been essential for this task.

It was a different matter for the forbidden lines in the solar corona. [Fe X] λ6375 and [Fe XIV] λ5303 were lines observed not only during solar eclipses but daily on coronographs at Kislovodsk (USSR), Kitt Peak (USA), and Pic du Midi (France). The first successful observations of the outer solar atmosphere were performed in 1868 by Lockyer in England and by Janssen in India during a solar eclipse. They observed emission lines of prominences. The pages relating these events in Volumes 67 and 68 of the

Comptes Rendus Hebdomadaires des Seances de l'Academie des Sciences still make interesting reading. The forbidden coronal lines began to be systematically observed in the 1890s. Their identification was the work of Edlén[12,13] after a tentative identification by Grotrian,[14] who was inspired by the identification of [Fe VII] lines in the spectrum of Nova Pictoris by Bowen and Edlén.[15] [An anecdotal history of the coronal identifications was given by Swings[16] at the 1968 Liège meeting.] The coronal lines are fine-structure transitions within the ground term: $3p^5\,^2P^\circ$ for [Fe x], $3p\,^2P^\circ$ for [Fe xIV]. In coronal conditions with $T > 10^6$ K, the excitation of the upper level is no longer a matter of collisional excitation from the ground level alone, by both electrons and protons. Cascades from higher-lying configurations are equally important. Thus it was realized that excitation cross sections for allowed transitions were essential even for the interpretation of forbidden line spectra if the electron temperature is sufficiently high. The $\bar{g}$ approximation for the calculation of collision strengths as given by Van Regemorter[17] was then very widely used.

The situation changed dramatically when observations from space became a reality. The rocket flights by the NRL group in the late 1950s inaugurated this era. Pottasch[18] reaped this first harvest and gave a list of about 150 of the most intense solar lines between 100 and 1900 Å. He also outlined methods of interpretation that would be followed for at least a decade. On the atomic physics side his interpretation was based on the $\bar{g}$ approximation. This allowed him to include a large number of transitions in his analysis. Of course, he was aware that this simplification implied a reduced accuracy.

The British Skylark program which started in the early 1960s allowed acquisition of spatially resolved solar spectra that were adequate for observation of the chromospheric and coronal spectra without disturbance from the photosphere. The emission line list published by Burton, Ridgeley, and Wilson[19] contains several hundred lines between 977 and 2803 Å. That same paper also indicates how close the interaction between solar observations and laboratory plasma physics was: Indeed the same scientists were usually involved in both projects.

The primary problem to be solved was identification; such work was done, for example, at Culham.[20,21] For the low and medium ionized atoms the tables of Moore[22] were invaluable. For the more highly ionized atoms data were much more sparse, and new laboratory investigations were required, which in turn were often looking to the theoretical atomic physicist for guidance or confirmation.

The boost received by solar physics from the rocket and satellite spectra whetted the appetite of other branches of astronomy. Thus Osterbrock[23] sublimated his impatience in an article where he calculated the ultraviolet emission spectrum expected for gaseous nebulae with physical conditions similar to planetary nebulae. Osterbrock relied for his collision cross sections mainly on those given by Seaton.[24,25] Osterbrock's paper would later serve in preparation for space observations of gaseous nebulae. Observations of QSOs in the visual had already brought a glimpse of the intrinsic UV spectra of extra-solar objects. Indeed, Schmidt[26] reported several types of observed transitions like the allowed Mg II doublet at 2798 Å, the C IV doublet at 1550 Å or the C III intercombination line at 1909 Å, in addition to forbidden spectra such as [O II], [Ne III], and [Ne V]. However, because of the dearth of lines in QSO spectra as opposed to the overabundance in the chromospheric or coronal spectra, it was also evident that a lack of accuracy in atomic data could not be compensated for by taking a mean over a large number of lines.

Thus it became evident that astronomical observations were growing into a formidable challenge for atomic physicists in both the quantity and quality of data required. Seaton was the man to respond to that challenge. That the computer would have to take over much of the former tedious calculations was self-evident. But where would the computer program take over? How much should the task be predigested before input? And most important of all, which physical problems should be tackled? Seaton's main concern was the interpretation of solar observations, which by sheer number precluded manual calculations. The bulk of the lines represented electric dipole transitions in highly ionized atoms. The corresponding collision strengths might with sufficient accuracy be calculated in the distorted-wave approximation, provided good bound-state functions could be produced. These functions would then also allow transition probabilities and oscillator strengths to be calculated. In 1967, Seaton outlined such a project to Belling, Eissner, Nussbaumer, and Tully, planning for about a two year effort. After about six months the interests of Belling and Tully shifted to other projects.

3. Some Aspects of the Genesis of the Programs

As the long-term aims, (a) bound-state term structure for assistance in line identification, (b) oscillator strengths, and (c) collision strengths were envisaged. Thus functions of the type

$$\Psi = \mathcal{C} \sum_i \Phi_i \Theta_i \tag{1}$$

were required, where Φ_i is the wave function of the target ion, Θ_i represents the colliding electron, and $\mathcal{C}$ is an antisymmetrizing operator. The evaluation of Φ was seen as the first major step. Matters of principle had to be resolved such as the type of radial functions to be employed: Analytic (hydrogenic), Hartree–Fock self-consistent field and simple central potential functions were considered. In view of the medium to highly ionized atoms that were the main target, scaled Thomas–Fermi functions were accepted as the appropriate compromise between good accuracy over the whole radial range and multipurpose applicability.

When the question of angular momentum coupling arose, it had to be decided whether to preserve the parentage system and work with Racah algebra,[27] or to follow Condon and Shortley[10] and build the eigenfunctions with the help of step operators. Step operators were preferred as being more readily adaptable to the computer, and there was no compelling reason to retain the concept of fractional parentage. Godfredsen[28] had already calculated term energies for many ions with a similar approach except that he employed hydrogenic radial functions. The UCL (University College London) project was primarily concerned not with deriving good energies, but with radiative and collisional transition probabilities. Thus the eigenfunctions themselves were the aim of the atomic structure calculations. For an atom with a nuclear charge Z and N electrons, a potential with the asymptotic form

$$V(r) = \begin{cases} -Z/r, & \text{for} \quad r \to 0 \\ -(Z-N+1)/r, & \text{for} \quad r \to \infty \end{cases} \qquad (2)$$

was considered essential to reproduce the physical charge distribution of the bound states. The nonrelativistic Hamiltonian operator has the form

$$H = \sum_{i=1}^{N} \left(h_i + \sum_{j<i} \frac{2}{r_{ij}} \right) \qquad (3)$$

$$h_i = -\frac{d^2}{dr^2} + \frac{l(l+1)}{r^2} - \frac{2Z}{r} . \qquad (4)$$

Radial functions which are solutions of

$$\left[\frac{d^2}{dr^2} - \frac{l(l+1)}{r^2} - 2V(r) + \varepsilon_{nl} \right] P_{nl}(r) = 0 \qquad (5)$$

should therefore provide a sound basis for a one-electron function expansion. In Eq. (5) n, l are the principal and orbital angular momentum quantum numbers, and ε_{nl} is the one-electron eigenenergy.

That this approach to the calculation of radial waves was justified became clear when Eissner and Nussbaumer[29] compared energy separations in the Na I sequence with results obtained by Godfredsen.[28] As more work was done it also became evident that the use of Hartree–Fock self-consistent field functions would not have brought decisive improvements in the eigenenergies—it being much more important to choose an adequate configuration basis.

Seaton had taken a very active interest in the atomic structure program. He was particularly pleased with the results for the low ionization behavior as shown in Figs. 1 and 2 of Eissner and Nussbaumer.[29] He had always been confident that the chosen approach would be well suited to the coronal ions, but with these results it could be hoped that the Thomas–Fermi wave functions would also serve for the lowly ionized atoms as observed in planetary nebulae. A partial trial was immediately made,[30] in view of an intriguing problem in planetary nebulae, in which radio measurements gave electron temperatures that were different from those determined from the [O III] line ratios.[31] After his 1953 paper, Seaton had reinvestigated, with various collaborators, the [O III] collision strengths. However, collisional coupling to terms of other configurations than $2s^2 2p^2$ had always been neglected. In Reference 30 it was shown that resonances due to collisional coupling between $2s^2 2p^2$ and $2s 2p^3$ could strongly alter the collision strengths. This work incorporated term energies and algebraic coefficients from the new STRUCTURE program.[29] (Although the collision strengths of [O III] changed markedly, the temperature-sensitive emissivity ratios only changed marginally. In fact, the temperature discrepancy was due to the interpretation of the radio measurements.) Full confidence in the Thomas–Fermi radial waves, however, was not yet established, as this excerpt from Eissner et al.[30] shows:

> ... the $1s$, $2s$, and $2p$ functions obtained using the statistical potential may not be very accurate ... we have therefore used the Hartree–Fock $1s$, $2s$, and $2p$ functions.

The first publications giving atomic data based on this new computer package alone, dealt with a particular identification problem in Wolf Rayet stars.[32] The first publication that used the distorted-wave program dealt with [Fe VII] lines[33] in Seyfert galaxies. The first publications demonstrated the importance of the multiconfiguration basis and that two-electron transitions could be dealt with in a satisfactory way. The latter gave the first collision strengths ever calculated for [Fe VII]. The distorted-wave program was to be described by Eissner[34] and Eissner and Seaton.[35] This program calculated reactance matrices in LS coupling. For their transformation to collision strengths, Saraph[36] had written the computer program SIMMEG. The Hamiltonian operator of STRUCTURE did not contain relativistic effects, and the obvious next extension was to include fine-structure terms. This was mainly pursued by Jones.[37] A further extension allowed calculation of transition probabilities, and the resulting program has become known as SUPERSTRUCTURE.[38] In the meantime, Saraph[39] had developed the program JAJOM which allows the calculation of collision strengths in intermediate coupling from reactance matrices calculated in LS coupling.

During the development of SUPERSTRUCTURE, Nussbaumer[40] studied probabilities for forbidden transition in the C I sequence. Such probabilities had already been calculated for a large number of ions. The development of that field is well documented in the reviews of Garstang,[41,42] who was also the most active provider. Garstang's method consisted basically of treating the LS term energies and the spin–orbit and spin–spin interactions as adjustable parameters. He obtained the overlap integral $\int P_{nl} r^2 P_{n'l'} \, dr$ with the best available radial functions. The calculations were mainly done in a one-configuration approximation. Apart from the errors arising from a one-configuration approximation, there is also an inherent inconsistency in this method. The calculation of term energies needs Slater integrals which involve the radial functions. The Slater integrals are treated as free parameters. The spin–orbit parameters, which again involve radial integrals, are also treated as free parameters, thus implying a different set of radial functions. Still another set is finally employed to calculate the transition integrals. Whether or not the combined actions of these shortcomings would cause serious errors still had to be investigated. Garstang[43] gave astrophysical and laboratory evidence "for the basic correctness of the results." His conviction has been vindicated by the bulk of later calculations.

Configuration interaction and its importance for energy levels, transition probabilities, and collision strengths had to be investigated. Layzer[44] had shown that due to the l degeneracy in the hydrogenic limit strong interaction had to be expected between all configurations having the same set of principal quantum numbers and the same parity. When Nussbaumer[40] treated the forbidden transitions in O III he not only included the complex $2s^2 2p^2 + 2p^4$ but also configurations like $2s2p^2 3d$. The adjustable Thomas–Fermi potential, as employed by Eissner and Nussbaumer,[29] was chosen to give the best energy separations within the $2s^2 2p^2$ configuration, disregarding the effects on the eigenenergies of the other configurations. It was discovered that the adjustable potential distorted the $3d$ radial function such that the mean radius of $3d$ came close to the mean radius of the $n = 2$ functions. After a first glance at the computer output, what appeared to be unacceptable behavior of the adjustable potential was, after further reflection, welcomed as a confirmation of the physical soundness of the adjusting procedure with the Thomas–Fermi potential.

The exploratory work[40] on contracted orbitals was followed by a more systematic investigation[45] directed towards C^{2+}. Because of the orthogonality constraints, only orbitals with angular momentum quantum numbers not represented among the spectroscopic orbitals were called in as contracted orbitals. The necessity for this constraint was later lifted by the generalization introduced by Nussbaumer and Storey,[46] who orthogonalized the radial functions numerically. In their investigation of the Na I sequence, Eissner and Nussbaumer did not hit upon contracted orbitals because they used a unique potential for all nl. The contracted orbitals are not unique to the Thomas–Fermi potential. By studying the literature, plenty of useful hints about configuration interaction and contracted orbitals could have been collected [e.g., Weiss[47]], but, as indicated by the brief reference list, the authors of "A programme for calculating atomic structures"[29] evidently preferred to start from first principles and learn for themselves. Indeed, Fig. 1 of Nussbaumer[40] is virtually the same as Fig. 3 of Weiss[47] which already shows $3d$ radial functions contracted to the $n = 2$ volume. Weiss in turn acknowledges his debt to Edmiston and Krauss.[48]

The usefulness of contracted orbitals for the representation of the ground state of He had been shown in several publications by Jucys.[49] However, it is true that all these investigations were based on Hartree–Fock functions, and it could not have been automatically assumed that the Thomas–Fermi potential would produce a similar effect. Also, the applications concentrated on the configurations of the lowest complex. It is therefore unlikely that the UCL program would have been strongly influenced even if these effects had been positively known to the UCL group.

Although the programs were designed as general purpose tools, particular cases had special influence on their final shapes. The C III collision strengths calculated by Osterbrock[50,51] at UCL are an example. C III $\lambda1909$ is one of the stronger lines observed in QSOs. In his Liège review on emission lines in galaxies and QSOs, Osterbrock[52] gave an estimate of $\Omega = 0.3$ for this transition. In the ensuing discussion his attention was drawn to a calculation by Beigman and Vainshtein[53] which gave an observationally implausibly small value of 0.0015 for that collision strength. (It later turned out[50] that there was an error in the scale factor in the figure of Beigman and Vainshtein.) During a visiting period at UCL Osterbrock set out to recalculate that cross section with the close-coupling methods. During the finishing stage of these calculations the UCL distorted-wave (DW) program was nearing completion, and it was planned to repeat Osterbrock's calculations in the DW approximation. Practical problems such as program internal data handling and storage and time requirements could thus be foreseen much more easily. Critical points became apparent early and could be dealt with in a preventative rather than curative fashion.

As mentioned earlier, the atomic structure and distorted-wave programs were originally intended to provide atomic data necessary for the interpretation of the coronal spectrum. Indeed, a large number of such calculations have been, and are still, performed; some examples can be found in the reference list.[54-60] Parallel to the work on DW, Seaton had also taken an active part in further developing the close-coupling method which was mainly intended to deal with electron–ion collisions in lowly ionized systems.[61,62] This program, which later became known as IMPACT,[63] was employed by Jackson[64] to calculate intercombination collision strengths much needed for the interpretation of QSO spectra.

Seaton's original concept of the UCL computer package was conceived as a kind of grand design which would eventually provide the tools to solve all the atomic physics problems presented by astrophysical spectroscopy. This broad outlook has not been abandoned. Programs were, and are, built to solve a certain type of problem and not a particular case. There is one point where Seaton was probably too optimistic. He intended these programs as black boxes for the astrophysicist to run when he needed particular atomic data. In principle, this has been achieved for some programs; in practice, each program still demands considerable insight and feeling for the physics involved.

4. The C III Challenge

4.1. Collision Strengths and Transition Probabilities for the Interpretation of the Solar Spectrum

With its prominent emission lines in planetary nebulae, O^{2+} has captured Seaton's attention over a long period and has been the object of many of his efforts in atomic physics. C^{2+} is another ion that has stimulated many investigations in which UCL computer programs were involved. We shall first mention some of the past involvements, and then describe in Section 4.2 an ongoing project.

Whereas the first impulse for treating C III at UCL came from QSOs,[50,51] the second impulse came from the Sun. Both the $2s^2\,{}^1S$–$2s2p\,{}^1P^\circ$ ($\lambda977$) and the $2s2p\,{}^3P^\circ$–$2p^2\,{}^3P$ ($\lambda1176$) transitions are prominent in the solar spectrum, and their relative intensities provided a welcome diagnostic tool for the chromosphere–corona transition region.[65,66] However, it soon became apparent that the observed C III intensity ratios did not agree with the expected electron pressure, particularly if it was assumed that C III was emitted at the temperature of maximum fractional abundance in a collisional ionization equilibrium. Loulergue and Nussbaumer[67,68] questioned the model, suggesting that the emission originated at a temperature that did not correspond to the temperature of maximum fractional abundance of C^{2+}. This view was contested by Jordan[69] and Dupree et al.,[70] who thought that the standard model was basically correct but that the fault was with the atomic data. This controversy stimulated a new round of C III calculations. Transition probabilities had been calculated by Loulergue and Nussbaumer.[67] A set of collision strengths in the distorted-wave approximation had been provided by Eissner[34]; Flower and Launay[71] had verified their validity by doing a three-configuration close-coupling calculation, and the resonance contributions to the $2s^2\,^1S-2s2p\,^3P^\circ$ collision strength had been investigated by Hershkowitz and Seaton.[72] It was generally assumed that transition probabilities and cross sections had an uncertainty of approximately 25%.

A first indication that the cross sections should indeed be corrected in the direction suggested from the traditional transition region model[69,70] came from a comparison[73] of oscillator strengths calculated in different approximations and assuming the Bethe approximation of $\Omega \propto gf$. Thus in the second paper of Loulergue and Nussbaumer,[68] the discrepancy against the traditional model expectation was much less severe than in their first paper, but it had not disappeared. The controversy inspired the group working with Burke to recalculate all the relevant electron–ion collision strengths with the R-matrix method developed by Burke and Robb.[74] These exhaustive calculations[75] showed again that results of high accuracy (they estimate 10% for most cases) still demand an effort that has to be measured in human years per cross section. Applying their results to the solar observations, they concluded[76] that "the theoretical calculations for C III seem to require an electron pressure lower than the 6×10^{14} cm^{-3} K value, usually associated with the transition region." Obviously the discrepancy between observations and the traditional model was much less severe than had been claimed originally,[67] but it was not simply a matter of low quality atomic data. Additional work showed that observations could be explained if C III was emitted below its temperature of maximum fractional abundance[68,77] or with departures from the ionization equilibrium because of mass flows through a temperature gradient.[78] The $2s^2\,^1S-2s2p\,^3P^\circ$ intercombination probability is also of crucial importance in the solar problem. This probability was first calculated by Garstang and Shamey[79] in a two-configuration approximation. Their probability of 190 s^{-1} is reduced by more than a factor of 2 as soon as the $2s^2\,^1S, 2p^2\,^1S$ interaction is taken into account. In order to calculate this probability, Nussbaumer and Storey[46] extended the UCL atomic structure program allowing each radial wave P_{nl} to be calculated in a different potential with subsequent orthonormalizations. But even with a 56-configuration basis they attach an uncertainty of approximately 10% to their result.

4.2. Excitation of C III by Recombination

4.2.1. Observations. Emission lines seen in the spectra of planetary nebulae are known to be due to collisional excitation, radiative recombination, and fluorescence excitation. The UV spectrum of C III revealed an additional mechanism: dielectronic recombination. Due to the success of Burgess and Seaton[80] in explaining the ionization structure of the solar corona, dielectronic recombination is usually associated with coronal conditions where collisions dominate ionization, and where the electron temperature corresponds approximately to the temperature of maximum abundance of the emitting ion. Only recently have calculations been performed,[81] with the purpose of investigating the contribution of dielectronic recombination to line excitation and ionization balance, under conditions where the temperature is far too low to cause collisional ionization. The calculation of dielectronic recombination coefficients, bringing together electron scattering, bound–free, and bound–bound radiative processes, embodies in one process all the problems that the computer package was designed to solve.

The ultraviolet observation of planetary nebulae with the IUE satellite means that many of the strongest resonance and intercombination lines of the ions of C, N, and O are routinely accessible. Since most of the important ionization stages are observed, abundance determinations can be improved, in principle. However, the determination of ionic abundance ratios relative to H, resting on the relative intensity of a collisionally excited UV line to a hydrogen recombination line, is very sensitive to electron temperature. In the first of a series of papers dealing with the UV spectra of planetary nebulae, Harrington *et al.*[82] derived abundances for carbon and magnesium. They compare their deduced ionization fraction for C^{2+}/H^{+} with that of Torres-Peimbert and Peimbert[83] and find that their value is four times smaller. Torres-Peimbert and Peimbert derive their C^{2+} abundance from the intensity of the C II $\lambda4267$ line, which they take to be populated by recombination; they use hydrogenic recombination coefficients. Harrington *et al.* also deduce the C^{+}/C^{2+} ratio from the intensities of C II] $\lambda2326$ and C III] $\lambda1909$. They find this ratio to be more than a factor of 5 larger than that obtained from models of the ionization structure. To resolve these anomalies, Harrington *et al.* suggest that there may be an additional $C^{2+} \rightarrow C^{+}$ recombination process not included in the model calculations which might also enhance $\lambda4267$. They put forward dielectronic recombination as a possibility, pointing out that earlier estimates of the rate for this process rely on general formulas, which are not valid at nebular temperatures for the ions in question. Individual autoionizing states near the ionization threshold must be considered.

At the same time as these problems were becoming apparent in planetary nebulae, work was in progress in Seaton's group on the spectra of the first nova to be observed systematically with the IUE satellite, Nova Cygni 1978. The nebular stage spectra showed very strong collisionally excited lines, particularly of C and N. Also present, but much weaker, were C II λ1335, C III λ2297, N IV λ1718, and O V λ1371. By analogy with the interpretation of solar transition region spectra, pairs of lines in the same ion were at first used to determine the temperature of formation of the lines.[84] Thus, the pair of lines C III λ2297 and C III] λ1909, both assumed to be collisionally excited, yield temperatures in the range $(4-5) \times 10^4$ K in the C^{2+} region of the nova shell. Similar temperatures were obtained from C II λ2326 and C II λ1335. The physical processes in nova shells during the nebular stage are thought to be essentially the same as those in planetary nebulae, but the temperatures deduced from the nova were far higher than those determined for planetary nebulae (10^4 K). However, there are many differences in detail between novae and planetary nebulae; the source of ionizing photons is probably hotter and certainly more luminous in novae, the density in the shell is higher in the early nebular phase, and the abundances of the major coolants, CNO, are enhanced by more than a factor of ten. These differences are so great that it was not immediately realized that the discrepancy between the nova temperatures and those of planetary nebulae might be due to atomic processes neglected in the interpretation. The identification of C III λ2297 by Harrington and co-workers[85] in the planetary nebula NGC 7009 made it clear that some other mechanism than collisional excitation populates the $2p^2\,^1D$ state, the upper term of this line. For NGC 7009, temperatures between 8900 and 10800 K have been deduced from [N II] and [O III] lines. At such temperatures the rate of collisional excitation is several orders of magnitude too slow to account for the observed λ2297 intensity. Following a suggestion of Seaton, Storey[81] made calculations of dielectronic recombination coefficients for $C^{3+} + e$.

   4.2.2. *Rate Coefficients from Detailed Balance Arguments.* Dielectronic recombination is the result of a radiationless capture from a continuum state $X^+ + e$ to an autoionizing state X_a

$$X^+ + e \rightarrow X_a \tag{6}$$

followed by a radiative decay to some bound state, X_b,

$$X_a \rightarrow X_b + h\nu. \tag{7}$$

We define the probability per unit time for the inverse process in (6), autoionization, to be $\Gamma_a^{(A)}$, and the probability for (7) to be $\Gamma_{ab}^{(R)}$. The recombination coefficient for X_b via X_a is then given by[86,87]

$$\alpha_{ab} = \frac{\omega_a}{2\omega_+} \left(\frac{h^2}{2\pi mkT} \right)^{3/2} \exp\left(-\frac{E_a}{kT} \right) b(X_a)\Gamma_{ab}^{(R)}, \tag{8}$$

where ω_a and ω_+ are the statistical weights of X_a and X^+, E_a is the energy of X_a relative to the ion X^+, and T is the temperature. The dimensionless parameter $b(X_a)$ is a measure of the departure of the population of X_a from its thermodynamic equilibrium and is given by

$$b(X_a) = \frac{\Gamma_a^{(A)}}{\Gamma_a^{(A)} + \sum_b \Gamma_{ab}^{(R)}}, \tag{9}$$

where we have assumed that there is only one continuum channel, $X^+ + e$, but many bound states, X_b. The total dielectronic recombination coefficient is obtained on summing α_{ab} over a and b.

At nebular temperatures, the exponential factor in (8) restricts the summation to those states within about 0.1 Ry of the ionization limit. In the case of $C^{3+} + e$ recombinations, this range includes states belonging to the configurations $2p4l$, $l > 0$. For these terms, the dominant radiative decays are not the core electron decays associated with dielectronic recombination at higher temperatures [Burgess[88]], but outer electron decays. It is shown by Seaton and Storey[87] [Eq. (169)] that the autoionization probability is given by $\Gamma^{(A)} = c/\nu^3$ in atomic units, where c is of order unity and ν is the effective quantum number of the autoionizing state. $\Gamma^{(A)}$ is independent of the ion effective charge z. As a function of increasing orbital angular momentum of the free electron, c generally at first increases, then decreases rapidly. Autoionization probabilities are generally greater for the case where orbital angular momentum is lost in the capture process. For the radiative probability we have, also in atomic units,

$$\Gamma^{(R)} = \alpha^3 R z^p \tag{10}$$

for a core decay, and

$$\Gamma^{(R)} = \alpha^3 R' z^p / \nu^3 \tag{11}$$

for the dominant outer electron decay, where α is the fine-structure constant, R and R' are of order unity and $p = 1$ if there is no change of principal quantum number in the core decay, and $p = 4$ if there is a change. Hence, if z and ν are small, $\Gamma^{(A)} \gg \Gamma^{(R)}$ due to the presence of the α^3 factor in Eqs. (10) and (11). It follows that to obtain α_{ab} at low temperatures where only low-lying resonances are accessible, only knowledge of $\Gamma_{ab}^{(R)}$ is necessary.

Thus the problem of dielectronic recombination $C^{3+} + e$ can be reduced, in a first approximation, to the calculation of radiative transition probabilities for C^{2+}, including the states $2p4l$ and all energetically lower states to which cascading might occur. The program SUPERSTRUCTURE was designed to enable just such large amounts of radiative data to be readily obtained to reasonable accuracy. Storey[81] included in the calculation for C^{2+} all configurations of the type $2snl$, $2pnl$, $2 \leqslant n \leqslant 7$, and $l \leqslant 6$ to ensure a good representation of the important $n = 4$ states. Total recombination coefficients and recombination coefficients for individual bound states can then be calculated using Eq. (8) and solving the capture-cascade equations. The results for $C^{3+} + e$ showed that, at planetary nebula temperatures, dielectronic recombination provides an important additional contribution to the ionization balance of C^{2+}/C^{3+} and also leads to a sufficient population of C^{2+} $2p^2 \, ^1D$ to explain the observed intensity of $\lambda 2297$. Further calculations were performed for the ions observed in Nova Cygni 1978. It was then possible to explain the observed intensities of C II $\lambda 1335$, N IV $\lambda 1718$, and O V $\lambda 1371$ in the nova spectra in terms of population by dielectronic recombination, at temperatures comparable to those of planetary nebulae.[89] This interpretation of the spectrum is also consistent with the intensities of optical forbidden lines of [N II] and [O III].

The calculations also showed that, as conjectured by Harrington et al.,[82] the $C^{2+} \rightarrow C^+$ recombination rate is increased. However, the expected enhancement of C II $\lambda 4267$ was not found.

4.2.3. Improved Low-Temperature Coefficients. We wish to examine, in the specific context of $C^{3+} + e$ recombination and the $\lambda 2297$ line, some of the uncertainties in the calculation of these recombination coefficients[81] with a view to obtaining more reliable values. These uncertainties are as follows:

1. The transition probabilities for the autoionizing states have been calculated in an approximation that makes no allowance for the admixture of the wave functions of these states with continuum states.

2. Relativistic interactions could lead to significant autoionization probabilities for terms that have been assumed to be nonautoionizing.

3. The assumption is made that $b(X_a) = 1$ will become invalid for some sufficiently large value of l.

4. As the temperature increases, captures become possible on increasingly high autoionizing states.

In addition, SUPERSTRUCTURE calculations of oscillator strengths among the bound states show that the perturbations of the $2snl$ series by $2pn'l'$ terms lead to significant radiative probabilities for two-electron transtitions between the series. These probabilities, which may influence the cascading, are difficult to calculate accurately and need further examination.

All that follows refers to LS coupling and low temperatures ($< 1.5 \times 10^4$ K). We do not address points 2 and 4. To be of interest, resonance states must in general lie within an energy $E \simeq kT$ above the ionization limit. In addition, for some sufficiently low T, $E \simeq kT \simeq \Gamma$, where Γ is the width of the resonance. In this case the Maxwellian distribution of electron velocities may vary significantly across the resonance, and the derivation of the recombination coefficient in terms of detailed balancing arguments for narrow autoionizing states is no longer adequate.

A more complete treatment of the theory of dielectronic recombination has been given by Davies and Seaton.[90] This work has been reviewed by Seaton and Storey,[87] whose notation we adopt. Davis and Seaton solve the time-dependent equations for the interaction of an ion with the quantized radiation field. For the case of one initial channel and one final state they obtain for the probability that a colliding electron of energy E relative to X^+ is captured,

$$P(E) = \frac{4\pi^2 |\mathcal{K}(E)|^2}{|1 + Z(E)|^2}, \tag{12}$$

where

$$\mathcal{K}(E) = (\Psi_b | H(\mu) | \Psi_\alpha(E)). \tag{13}$$

Ψ_b and Ψ_α are bound and continuum state eigenfunctions, respectively, of the $X^+ + e$ system neglecting interaction with the radiation field, $H(\mu)$ is related to the dipole length operator, and

$$Z(E) = \lim_{\varepsilon \to 0} \frac{\pi}{i} \int \frac{|\mathcal{K}(W)|^2 \, dW}{W - E - i\varepsilon}. \tag{14}$$

Seaton and Storey show that if $\mathcal{K}(E)$ contains a pole of the form $\mathcal{K}(E) = C/[E - E_0 - i\Gamma^{(A)}/2]$, then

$$Z(E) = \frac{2\pi^2 i |C|^2}{\Gamma^{(A)} \left[(E - E_0 + i\Gamma^{(A)})/2 \right]}, \tag{15}$$

where $|C|^2 = \Gamma^{(A)} \Gamma^{(R)}/4\pi^2$, and $\Gamma^{(A)}$ and $\Gamma^{(R)}$ are identified with the autoionization and radiative widths of the resonance. It can be shown that if $\Gamma^{(A)} \gg \Gamma^{(R)}$, $|1 + Z(E)|^2 = 1$ so that

$$P(E) = 4\pi^2 |\mathcal{K}(E)|^2. \tag{16}$$

Thus, provided $\Gamma^{(A)} \gg \Gamma^{(R)}$, the radiative capture coefficient can be obtained from the matrix element (13) which is directly related to the photoionization cross section from X_b to the channel α.

We use the program IMPACT to obtain wave functions for both the bound states X_b and the continuum states $X^+ + e$ of the $C^{3+} + e$ system. From these wave functions, we can obtain information both about the autoionization probabilities and the amplitudes for photoionization. The C^{3+} system is described by the three states $2s$, $2p$, and $3d$. Radial wave functions for these orbitals are obtained from the program SUPERSTRUC-TURE. The techniques for solving the coupled equations for the ion + electron problem are described elsewhere in this volume. We obtain R-matrix elements and wave functions (single channel in the energy range of interest) for the continuum states $2skl(^1L)$, $l = L = 0, 1, 2, 3, 4$ for a range of energies up to 0.3 Ry above the $2s$ threshold. These R-matrix elements contain poles corresponding to $2pnl$ states with $n = 4, 5$. We fit the R-matrix elements to an expression which is a sum of a quadratic function in energy and simple poles, using the program RANAL. This program, originally written by Seaton, was designed to enable fits to be made to multichannel R-matrices so that resonance structures could be delineated without the need to solve the coupled equations at many points around each resonance. These fits can also be used in combination with the quantum defect method of Seaton, to obtain information about the behavior of the R matrix below a threshold, where closely packed resonances preclude direct calculation. The single-channel scattering matrix is given by

$$S = \frac{1 + iR}{1 - iR}. \tag{17}$$

Near to a pole in R, we can put, approximately,

$$R = (a + b)/(E - E_0), \tag{18}$$

where E_0 is the pole energy and a and b are constants. The pole in the S-matrix is at the complex energy derived from $(1 - iR) = 0$. The auto-ionization probability is related to the position of the pole in the S matrix, E, by

$$E = (E_1 - i\Gamma^{(A)})/2 \tag{19}$$

TABLE 1

Low-Lying Resonances for $C^3 + e$ (^1L): Positions, Autoionization Widths, and Radiative Widths

L	Resonance	Calculated energy (Ry)	$\Gamma^{(A)}$ (Ry)	$\Gamma^{(A)}$ (s^{-1})	$\Gamma^{(R)}$ (s^{-1})
0	$2p4p$	0.04118	0.01686	3.49(+14)	8.66(+8)
	$2p5p$	0.23489	0.007422	1.53(+14)	6.98(+8)
1	$2p5s$	0.18063	0.01082	2.24(+14)	1.45(+9)
	$2p4d$	0.04010	0.002762	5.71(+13)	4.40(+9)
	$2p5d$	0.23820	0.001348	2.79(+13)	3.06(+9)
2	$2p4p$	−0.00001	0.001040	2.15(+13)	1.63(+9)
	$2p5p$	0.21441	0.000575	1.19(+13)	1.14(+9)
	$2p4f$	0.03368	0.0001055	2.18(+12)	1.43(+9)
	$2p5f$	0.23266	0.0000734	1.52(+12)	1.04(+9)
3	$2p4d$	0.02898	0.01648	3.41(+14)	6.88(+9)
	$2p5d$	0.23305	0.007965	1.65(+14)	4.39(+9)
	$2p5g$				7.02(+8)
4	$2p4f$	0.02775	0.008516	1.76(+14)	1.20(+9)
	$2p5f$	0.22868	0.004250	8.79(+13)	2.68(+8)

where E_1 is real, so in terms of the R-matrix fit

$$\Gamma^{(A)} = \frac{-2b}{1 + a^2} \qquad (20)$$

$$E_1 = E_0 - \frac{ab}{1 + a^2} . \qquad (21)$$

In Table 1 we give the positions, widths, and autoionization probabilities of the resonances in the $C^{3+} + e$ singlet continua for $0 < E < 0.3$ Ry. We also give total radiative probabilities for these states calculated with SUPER-STRUCTURE. With the exception of the $2p5g\,^1F^\circ$ resonance, the assumption that $\Gamma^{(A)} \gg \Gamma^{(R)}$ [or $b(X_a) = 1$] is well justified for the $2pnl$ resonances, $n = 4, 5, l < 4$.

We have also obtained photoionization cross sections for the C^{2+} states which are most important for the population of $2p^2\,^1D$, i.e., $2p^2\,^1D, ^1S$ and all terms of the $2p3l$ configurations. The bound and continuum state wave functions are combined in the program PHOTUC written by Saraph, which computes the electric dipole matrix elements. The angular parts of the matrix elements and those radial integrals involving only C^{3+} orbitals are calculated by the program RADALG. The underlying theory has been described by Jones.[91] The photoionization amplitudes are fitted in the same way as the R-matrix elements described above, also using the program RANAL. If we express the photoionization cross section $\sigma_{PI}(E)$ in terms of the differential oscillator strength,

$$\sigma_{PI}(E) = \frac{\pi h e^2}{mc} \frac{df}{dE},$$ (22)

then the radiative capture coefficient α_b, for the state X_b can be obtained from

$$\alpha_b = \frac{\omega_b}{2\omega_+} \left(\frac{h^2}{2\pi mkT} \right)^{3/2} \frac{8\pi^2 e^2}{mc^3} \int_0^\infty \exp\left(\frac{-E}{kT} \right) \nu^2 \frac{df}{dE} \, dE,$$ (23)

where ν is the frequency of the photon emitted in the recombination, and the integal in practice will extend to $E = 0.3$ Ry.

Bound–bound radiative data are obtained from the program package in the same way as outlined for bound–free transitions. Where possible, we replace the SUPERSTRUCTURE radiative probabilities used by Storey[81] with those calculated by Nussbaumer and Storey.[92] They have used the same methods as described here, but a more elaborate C^{3+} target. The effective recombination coefficient for C^{2+} $2p^2\,{}^1D$ can now be calculated by numerically integrating Eq. (23) for those states for which we have calculated photoionization data and solving the capture-cascade equations. For all remaining states in the series $2snl$, $2pnl$, $n \leqslant 5$, we use SUPER-STRUCTURE data and Eq. (8). We also use this source for those bound states of C^{2+} for which it proved impossible to obtain a bound-state wave function with IMPACT due to their proximity to the C^{3+} $(2s)$ limit (e.g., $2p4s\,{}^1P^\circ$).

The resulting recombination coefficients are shown in Fig. 1 (curve c), in the temperature range 100–15000 K. Also shown are the results of Storey (curve a). Curve b shows the result of simply replacing the bound–bound radiative data used by Storey with the more accurate values of Nussbaumer and Storey[92] described above. This curve has been continued down to lower temperatures to show the exponential fall in the coefficient when kT is less than the energy of the lowest autoionizing state. The present calculation leads to a finite recombination coefficient as $T \to 0$, since the photoionization cross section is not zero as $E \to 0$. The difference between curves b and c at 10^4 K is due to two factors:

1. Differences in oscillator strengths; an oscillator strength for an isolated resonance f_a^{CC} may be obtained from the photoionization

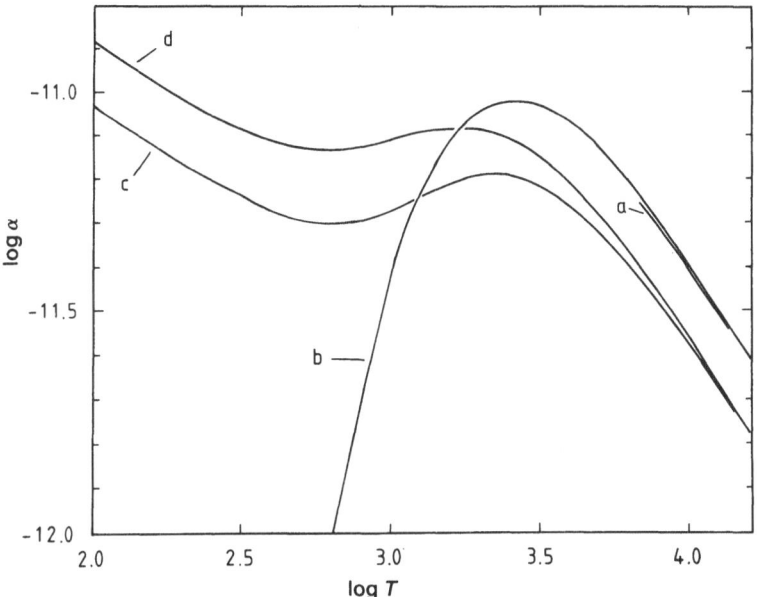

FIGURE 1. Dielectronic recombination coefficients, α (cm^3/s), for C^{2+} (λ2297) as a function of electron temperature T(K). (a) Storey; (b) method of Section 4.2.2 with improved bound–bound transition probabilities; (c) recombination coefficients calculated from photoionization cross sections for the most important terms. Bound–bound radiative data as in (b). (d) Same as (c) but with pole positions corrected.

cross section using

$$f_a^{CC} = \int_{\text{resonance}} \frac{df}{dE}\, dE. \qquad (24)$$

In Table 2 we show resonance oscillator strengths defined in this way compared to values calculated with SUPERSTRUCTURE, f_a^{SS}, for some transitions from $2p^2\,^1D$. The recombination coefficient for this state is dominated by the contribution from the $2p4d\,^1F^o$ resonance. The agreement between f^{CC} and f^{SS} is best for the $2p^2\,^1D$–$2pnl\,^1D^o$ transitions, which is to be expected since the $^1D^o$ series is an unperturbed series of bound states.

2. The $2p4d\,^1F^o$ resonance is the only $2p4l$ singlet autoionizing state for which we found an observed energy. This energy is 0.006 Ry lower than our calculated position. In Fig. 1, curve d, we show the recombination coefficient obtained when we adjust the fits to the photoionization amplitudes and R matrices by correcting the pole

TABLE 2

Absorption Oscillator Strengths for Transitions from C^{2+} $2p^2$ ^1D to $2pnl$ $^1P^\circ$, $^1D^\circ$, $^1F^\circ$ States.

Transition	f^{SSa}	f^{CCb}
$2p^2$ ^1D–$2p3s$ $^1P^\circ$	0.0138	0.0574
$2p^2$ ^1D–$2p4s$ $^1P^\circ$	0.0057	—
$2p^2$ ^1D–$2p5s$ $^1P^\circ$	0.0040	0.0062
$2p^2$ ^1D–$2p3d$ $^1P^\circ$	0.0095	0.0044
$2p^2$ ^1D–$2p4d$ $^1P^\circ$	0.0023	0.0014
$2p^2$ ^1D–$2p5d$ $^1P^\circ$	0.0017	0.0007
$2p^2$ ^1D–$2p3d$ $^1D^\circ$	0.2498	0.2204
$2p^2$ ^1D–$2p4d$ $^1D^\circ$	0.0445	0.0432
$2p^2$ ^1D–$2p5d$ $^1D^\circ$	0.0155	0.0167
$2p^2$ ^1D–$2p3d$ $^1F^\circ$	0.2621	0.4062
$2p^2$ ^1D–$2p4d$ $^1F^\circ$	0.2120	0.1658
$2p^2$ ^1D–$2p5d$ $^1F^\circ$	0.1083	0.0754

[a] f^{SS} denotes SUPERSTRUCTURE calculations.
[b] f^{CC} denotes values obtained from close-coupling wave functions.

position such that it lies at its observed energy.

In Table 3, we give the recombination coefficients for $2p^2$ ^1D corresponding to curve d, which we consider to be the most reliable result. At 10^4 K the values in Table 3 are approximately 25% lower than the results of Storey.[81] The inclusion of the breakdown effects of LS coupling, referred to above, can only increase the recombination coefficient at these temperatures.

TABLE 3

Effective Recombination Coefficients α for C^{2+} (λ2297) as a Function of Temperature T

$T(10^4$ K)	$\alpha \times 10^{12}$ (cm^3/s)	$T(10^4$ K)	$\alpha \times 10^{12}$ (cm^3/s)
0.01	13.0	0.5	5.23
0.02	9.62	0.6	4.49
0.03	8.31	0.7	3.90
0.04	7.68	0.8	3.43
0.05	7.39	0.9	3.04
0.07	7.35	1.0	2.73
0.10	7.73	1.1	2.47
0.15	8.19	1.2	2.26
0.20	8.11	1.3	2.07
0.25	7.72	1.4	1.91
0.30	7.21	1.5	1.77
0.40	6.15	—	—

The incorporation of resonance effects in the process of radiative recombination is important for the study of nebular plasmas. Further work is required to deal with the other ions of interest and to refine rate coefficients for those ions for which first estimates have already been made. The realization of the importance of these effects grew, like the computer package itself, out of the continual interplay between atomic physics and astrophysics, which has always guided the activities of Seaton and his group.

References

1. M. J. SEATON, The kinetic temperature of the interstellar gas in regions of neutral hydrogen, *Ann. Astrophys.* **18**, 188–205 (1955).
2. H. NUSSBAUMER and P. J. STOREY, Atomic data for Fe II, *Astron. Astrophys.* **89**, 308–313 (1980).
3. H. NUSSBAUMER, M. PETTINI, and P. J. STOREY, Sextet transitions in Fe II, *Astron. Astrophys.* **102**, 351–358 (1981).
4. M. J. SEATON, The Hartree–Fock equations for continuous states with applications to electron excitation of the ground configuration terms of O I, *Philos. Trans. R. Soc. Lond. Ser. A* **245**, 469–499 (1953).
5. M. J. SEATON, Electron excitation of forbidden lines occurring in gaseous nebulae, *Proc. Roy. Soc. A* **218**, 400–416 (1953).
6. M. H. HEBB and D. H. MENZEL, Collisional excitation of nebulium, *Astrophys. J.* **92**, 408–423 (1940).
7. T. YAMANOUCHI, T. INUI, and A. AMEMIYA, Excitation of metastable states of oxygen atom by electron impact, *Proc. Phys. Math. Soc. Jpn. Ser. 3*, **22**, 848–854 (1940).
8. L. A. ALLER, Target areas for the collisional excitation of nebular lines, *Astrophys. J.* **111**, 609–610 (1950).
9. D. R. BATES, A. FUNDAMINSKY, J. W. LEECH, and H. S. W. MASSEY, Excitation and ionization of atoms by electron impact—the Born and Oppenheimer approximations, *Philos. Trans. R. Soc. Lond. Ser. A* **243**, 93–141 (1950).
10. E. U. CONDON and G. H. SHORTLEY, *Theory of Atomic Spectra*, Cambridge University Press, London, 1935.
11. I. S. BOWEN, The origin of the nebular lines and the structure of planetary nebulae, *Astrophys. J.* **67**, 1–15 (1928).
12. B. EDLÉN, An attempt to identify the emission lines in the spectrum of the solar corona, *Ark. Mat. Astron. Fys.* **28B**, 1–4 (1941).
13. B. EDLÉN, Die Deutung der Emissionslinien im Spektrum der Sonnenkorona, *Z. Astrophys.* **22**, 30–64 (1942).
14. W. GROTRIAN, Zur Frage der Deutung der Linien im Spektrum der Sonnenkorona, *Naturwissenschaften* **27**, 214 (1939).
15. I. S. BOWEN and B. EDLÉN, Forbidden lines of Fe VII in the spectrum of Nova Pictoris (1925), *Nature* **143**, 373 (1939).
16. P. SWINGS, Finale in *Mem. Soc. R. Sci. Liege Ser. 5* **17**, 407 (1969).
17. H. VAN REGEMORTER, Rate of collisional excitation in stellar atmospheres, *Astrophys. J.* **136**, 906–915 (1962).
18. S. R. POTTASCH, On the interpretation of the solar ultraviolet emission line spectrum, *Space Sci. Rev.* **3**, 816–855 (1964).
19. W. M. BURTON, A. RIDGELEY, and R. WILSON, The ultraviolet emission spectrum of the solar chromosphere and corona, *Mon. Not. R. Astron. Soc.* **135**, 207–223 (1967).
20. A. H. GABRIEL, B. C. FAWCETT, C. JORDAN, Classification of iron lines in the spectrum of the sun and zeta in the range 167 to 220 Å, *Nature* **206**, 390–393 (1965).

21. A. H. GABRIEL and B. C. FAWCETT, Identification of the solar spectrum in the region 60 to 170 Å, *Nature* **206**, 808–809 (1965).
22. C. E. MOORE, *An Ultraviolet Multiplet Table*, Circular No. 488, National Bureau of Standards, Washington, D.C., 1950.
23. D. E. OSTERBROCK, Expected ultraviolet emission spectrum of gaseous nebula, *Planet. Space Sci.* **11**, 621–632 (1963).
24. M. J. SEATON, Thermal inelastic collision processes, *Rev. Mod. Phys.* **30**, 979–991 (1958).
25. M. J. SEATON, The Theory of Excitation and Ionization by Electron Impact, in *Atomic and Molecular Processes*, D. R. Bates, Ed., Academic Press, New York, pp. 374–420.
26. M. SCHMIDT, Large redshifts of five quasistellar sources, *Astrophys. J.* **141**, 1295–1300 (1965).
27. G. RACAH, Theory of complex spectra, *Phys. Rev.* **62**, 438–462 (1942).
28. E. GODFREDSEN, Atomic term energies for atoms and ions with 11 to 28 electrons, *Astrophys. J.* **145**, 308–332 (1966).
29. W. EISSNER and H. NUSSBAUMER, A programme for calculating atomic structures, *J. Phys. B* **2**, 1028–1043 (1969).
30. W. EISSNER, H. NUSSBAUMER H. E. SARAPH, and M. J. SEATON, Resonances in cross sections for excitation of forbidden lines in O^{2+}, *J. Phys. B* **2**, 341–355 (1969).
31. D. R. FLOWER and M. J. SEATON, Forbidden line radiation from gaseous nebulae, *Mém. Soc. R. Sci. Liège Ser. 5* **17**, 251–268 (1969).
32. H. NUSSBAUMER, Oscillator strengths in complex atoms: applications to N IV, *Mon. Not. R. Astron. Soc.* **145**, 141–150 (1969).
33. H. NUSSBAUMER and D. E. OSTERBROCK, On the forbidden emission lines of iron in Seyfert galaxies, *Astrophys. J.* **161**, 811–820 (1970).
34. W. EISSNER, Computer Methods and Packages in Electron–Atom Collisions, in *Physics of Electronic and Atomic Collisions*, North-Holland, Amsterdam, 1972, pp. 460–478.
35. W. EISSNER and M. J. SEATON, Computer programs for the calculation of electron–atom collision cross sections, *J. Phys. B* **5**, 2187–2198 (1972).
36. H. E. SARAPH, Collision strengths from reactance matrices, *Comput. Phys. Commun.* **1**, 232–240 (1970).
37. M. JONES, Relativistic corrections to atomic energy levels, *J. Phys. B* **3**, 1571–1592 (1970).
38. W. EISSNER, M. JONES, and H. NUSSBAUMER, Techniques for the calculation of atomic structures and radiative data including relativistic corrections, *Comput. Phys. Commun.* **8**, 270–306 (1974).
39. H. E. SARAPH, Fine structure cross sections from reactance matrices, *Comput. Phys. Commun.* **3**, 256–268 (1972).
40. H. NUSSBAUMER, Forbidden transitions in the C I sequence, *Astrophys. J.* **166**, 411–422 (1971).
41. R. H. GARSTANG, Forbidden Transitions, in *Atomic and Molecular Processes*, D. R. Bates, Ed., Academic Press, New York, 1962, pp. 1–46.
42. R. H. GARSTANG, Theoretical and experimental forbidden atomic transition probabilities, *Mém. Soc. R Sci. Liège Ser. 5* **17**, 35–44 (1969).
43. R. H. GARSTANG, Discussion on Transition Probabilities for Forbidden Lines, *IAU Symposium 34 on Planetary Nebulae*, D. E. Osterbrock and C. R. O'Dell, Eds., Reidel, Dordrecht, 1968, p. 151.
44. D. LAYZER, On a screening theory of atomic spectra, *Ann. Phys.* (*N.Y.*) **8**, 271–296 (1959).
45. H. NUSSBAUMER, Improved bound wave functions for complex atoms, *J. Phys. B* **5**, 1837–1843 (1972).
46. H. NUSSBAUMER and P. J. STOREY, The C III transition probabilities, *Astron. Astrophys.* **64**, 139–144 (1978).
47. A. W. WEISS, Superposition of configurations and atomic oscillator strengths—carbon I and II, *Phys. Rev.* **162**, 71–80 (1967).
48. C. EDMISTON and M. KRAUSS, Pseudonatural orbitals as a basis for the superposition of configurations. I. He_2^+, *J. Chem. Phys.* **45**, 1833–1839 (1966).

49. A. P. JUCYS, On the Hartree–Fock method in multiconfiguration approximation, *Adv. Chem. Phys.* **14**, 191–206 (1969).

50. D. E. OSTERBROCK, Excitation of semiforbidden $2s^2\,{}^1S-2s2p\,{}^3P^\circ$ lines observed in quasars and nebulae, *J. Phys. B* **3**, 149–160 (1970).

51. D. E. OSTERBROCK, Excitation of C III] λ1909 and other semiforbidden lines in QSO's and nebulae, *Astrophys. J.* **160**, 25–30 (1970).

52. D. E. OSTERBROCK, Forbidden emission lines in galaxies and quasars, *Mém. Soc. R. Sci. Liège Ser. 5*, **17**, 391–403 (1969).

53. I. L. BEIGMAN and L. A. VAINSHTEIN, Effective cross sections for the exchange excitation of atoms and ions by electron impact, *Sov. Phys.-JETP* **25**, 119–123 (1967).

54. D. R. FLOWER, Collision strengths for elelctron excitation of highly ionized complex atoms, *J. Phys. B* **4**, 697–705 (1971).

55. D. R. FLOWER and G. PINEAU DES FORETS, Excitation of the Fe XIII spectrum in the solar corona, *Astron. Astrophys.* **24**, 181–192 (1973).

56. M. LOULERGUE and H. NUSSBAUMER, Fe XVII emission from the solar corona, *Astron. Astrophys.* **24**, 209–213 (1973).

57. M. LOULERGUE and H. NUSSBAUMER, A study of Fe XVII and Ni XIX coronal lines, *Astron. Astrophys.* **45**, 125–134 (1975).

58. H. MASON, The excitation of several iron and calcium lines in the visible spectrum of the solar corona, *Mon. Not. R. Astron. Soc.* **170**, 651–689 (1975).

59. K. P. DERE, H. E. MASON, K. G. WIDING, and A. K. BHATIA, XUV electron density diagnostics for solar flares, *Astrophys. J. Suppl. Ser.* **40**, 341–364 (1979).

60. H. E. MASON, G. A. DOSCHEK, U. FELDMAN, and A. K. BHATIA, Fe XXI as an electron density diagnostic in solar flares, *Astron. Astrophys.* **73**, 74–81 (1979).

61. W. EISSNER and M. J. SEATON, Computer programs for the calcualtion of electron–atom collision cross sections I: General formulation, *J. Phys. B* **5**, 2187–2198 (1972).

62. M. J. SEATON and P. M. H. WILSON, A frozen cores approximation for atomic structure calculations, *J. Phys. B* **5**, L1–L3 (1972).

63. M. A. CREES, M. J. SEATON, and P. M. H. WILSON, IMPACT, a program for the solution of the coupled integro-differential equations of electron–atom collision theory, *Comput. Phys. Commun.* **15**, 23–83 (1978).

64. A. R. G. JACKSON, Excitation of C II] λ2326 O III] λ1664, and other semiforbidden emission lines in quasars, *Mon. Not. R. Astron. Soc.* **165**, 53–60 (1973).

65. C. JORDAN, "The Relative Intensities of Lines from Be I-like Ions in the Solar Spectrum," *Highlights in Astronomy*, C. de Jager, Ed., Reidel, Dordrecht, 1971, pp. 519–526.

66. R. H. MUNRO, A. K. DUPREE, and G. L. WITHBROE, Electron densities derived from line intensity ratios: beryllium isoelectronic sequence, *Solar Phys.* **19**, 347–355 (1971).

67. M. LOULERGUE and H. NUSSBAUMER, The chromosphere–corona transition region as seen in C III, *Astron. Astrophys.* **34**, 225–233 (1974).

68. M. LOULERGUE and H. NUSSBAUMER, The C III problem, *Astron. Astrophys.* **51**, 163–170 (1976).

69. C. JORDAN, The measurement of electron densities from beryllium-like ion line ratios, *Astron. Astrophys.* **34**, 69–73 (1974).

70. A. K. DUPREE, P. V. FOUKAL, and C. JORDAN, Plasma diagnostic techniques in the ultraviolet: the C III density-sensitive lines in the sun, *Astrophys. J.* **209**, 621–632 (1976).

71. D. R. FLOWER and J. M. LAUNAY, Electron collisional excitation of C^{+2}, *Astron. Astrophys.* **29**, 321–326 (1973).

72. M. D. HERSHKOWITZ and M. J. SEATON, The calculation of resonances in electron-ion scattering using the distorted wave approximation, *J. Phys. B* **6**, 1176–1187 (1973).

73. H. P. MUHLETHALER, and H. NUSSBAUMER, Transition probabilities within $2s^2-2s2p-2p^2$ in the Be I sequence, *Astron. Astrophys.* **48**, 109–114 (1976).

74. P. G. BURKE and W. D. ROBB, "The *R*-Matrix Theory of Atomic Processes," in *Advances in Atomic and Molecular Physics*, Vol. 11, D. R. Bates and B. Bederson, Eds., Academic, New York, 1975, pp. 143–214.

75. K. A. BERRINGTON, P. G. BURKE, P. L. DUFTON, and A. E. KINGSTON, Electron collisional excitation of C III and O V, *J. Phys. B* **10**, 1465–1475 (1977).

76. P. L. DUFTON, K. A. BERRINGTON, P. G. BURKE, and A. E. KINGSTON, The interpretation of C III and O V emission line ratios in the sun, *Astron. Astrophys.* **62**, 111–120 (1978).

77. A. MEYER and H. NUSSBAUMER, Formation of emission lines in a shock heated solar atmosphere, *Astron. Astrophys.* **57**, 431–436 (1977).

78. J. C. RAYMOND and A. K. DUPREE, C III density diagnostics in nonequilibrium plasmas, *Astrophys. J.* **222**, 379–383 (1978).

79. R. H. GARSTANG and L. J. SHAMEY, Intercombination line oscillator strengths in the helium and beryllium isoelectronic sequences, *Astrophys. J.* **148**, 665–666 (1967).

80. A. BURGESS and M. J. SEATON, The ionization equilibrium for iron in the solar corona, *Mon. Not. R. Astron. Soc.* **127**, 355–358 (1964).

81. P. J. STOREY, Dielectronic recombination at nebular temperatures, *Mon. Not. R. Astron. Soc.* **195**, 27P–31P (1981).

82. J. P. HARRINGTON, J. H. LUTZ, M. J. SEATON, and D. J. STICKLAND, Ultraviolet spectra of planetary nebulae I: The abundance of carbon in IC 418, *Mon. Not. R. Astron. Soc.* **191**, 13–22 (1980).

83. S. TORRES-PEIMBERT and M. PEIMBERT, Photoelectric photometry and physical conditions of planetary nebulae, *Rev. Mex. Astron. Astrofis.* **2**, 181–207 (1977).

84. D. J. STICKLAND, C. J. PENN, M. J. SEATON, M. A. J. SNIJDERS, P. J. STOREY, and C. R. KITCHIN, Ultraviolet Observations of Nova Cygni 1978, *The First Year of IUE*, A. J. Willis, Ed., University College London, 1979, pp. 63–77.

85. J. P. HARRINGTON, J. H. LUTZ, and M. J. SEATON, Ultraviolet spectra of planetary nebulae IV: The C III 2297 dielectronic recombination line and dust absorption in the C IV λ1549 resonance doublet, *Mon. Not. R. Astron. Soc.* **195**, 21P–26P (1981).

86. D. R. BATES and H. S. W. MASSEY, The negative ions of atomic and molecular oxygen, *Philos. Trans. R. Soc. Lond. Ser. A* **239**, 269–304 (1943).

87. M. J. SEATON and P. J. STOREY, "Dielectronic Recombination," in *Atomic Processes and Applications*, P. G. Burke and B. L. Moisewitsch, Eds., North Holland, Amsterdam, 1976, pp. 134–197.

88. A. BURGESS, Dielectronic recombination and the temperature of the solar corona, *Astrophys. J.* **139**, 776–780 (1964).

89. D. J. STICKLAND, C. J. PENN, M. J. SEATON, M. A. J. SNIJDERS, and P. J. STOREY, Nova Cygni 1978 I: The nebular phase, *Mon. Not. R. Astron. Soc.* **197**, 107–138 (1981).

90. P. C. W. DAVIES and M. J. SEATON, Radiation damping in the optical continuum, *J. Phys. B* **2**, 757–765 (1969).

91. M. JONES, A method for calculating the algebra of matrix elements for photoionization and line radiation, *Comput. Phys. Commun.* **7**, 353–367 (1974).

92. H. NUSSBAUMER and P. J. STOREY, Oscillator strengths for excited states of C^{2+}, *Astron. Astrophys.*, to be submitted.

9

PLANETARY NEBULAE

D. R. Flower

1. Introduction

The importance of planetary nebulae in astronomy is well-known. Stars are believed to evolve from red giants into white dwarfs through ejection of a planetary nebula shell. This phase of stellar evolution is relatively rapid (10,000 yr), implying dramatic changes in the nucleus of the nebula which are still poorly understood. Since they have much of their visual emission concentrated in a few narrow lines (e.g., [O III] $\lambda\lambda4959$, 5007), planetary nebulae are also valuable tracers of Galactic structure.

Studies of planetary nebulae have had an important influence on developments in certain branches of atomic physics. Planetaries might be regarded as extraterrestrial laboratories for the analysis, for example, of forbidden line radiations, which are strongly suppressed at the much higher densities encountered in laboratory plasmas on Earth.

The interplay between atomic physics and astronomical studies of planetary nebulae has proved highly fruitful for both disciplines. For a generation, this field of activity has been dominated by the man to whom this volume is dedicated. His contributions have been prodigious, and I hope to do at least partial justice to their extent and importance in this chapter.

2. Observations

2.1. Optical

Older observations of planetary nebulae were photographic and suffered from the disadvantages of this technique, particularly the nonlinear response and limited intensity range of the photographic plate. The more recent photoelectric observations are to be preferred, when available,

D. R. FLOWER ● Department of Physics, University of Durham, Durham DH1 3LE, England.

although recourse to photographic measurements may still be necessary in order to resolve blends.

As an illustration of the caution that must be exercised when employing photographic measurements, we refer to the discrepancy between the observed and calculated intensities of the Balmer lines in NGC 7027. Photographic measurements by Aller *et al.*[1] indicated that the higher members of this series were systematically more intense than predicted by theory. Seaton[2] reexamined the theory of recombination lines and suggested that the discrepancy arose from errors in the observations. Subsequently, Aller *et al.*[3] used photoelectric observations of the stronger lines to calibrate photographic observations and confirmed the original measurements.[1] Sophisticated calculations of recombination line intensities by Brocklehurst[4] showed that the discrepancies could not be explained by full allowance for the collisional redistribution of angular momentum and energy. The problem was finally resolved by the photoelectric observations of Miller,[5] which showed the older measurements to have systematically overestimated the intensities of the higher members of the Balmer series.

Photoelectric spectra of the brighter northern planetary nebulae were obtained by Liller and Aller,[6] O'Dell,[7] and Peimbert and Torres-Peimbert.[8] Torres-Peimbert and Peimbert[9] extended the observations to the southern hemisphere, using the Cerro Tololo Observatory. Barker[10] has observed additional objects. In Table 1, we compare the observations of O'Dell[7] with those of Peimbert and Torres-Peimbert[8] for NGC 7662 and IC 418, where we also include the more recent measurements of Torres-Peimbert and Peimbert.[9]

The largest discrepancy in Table 1 is between the measurement of [O III] λ4363 by O'Dell and the Peimberts. The Peimberts have taken care to avoid contamination of λ4363 by C II λ4368, which may be the reason for the discrepancy. The line intensities appear to be reproducible to within ± 15%, although larger deviations are present.

Reliable observations of weak lines can be crucial to our understanding of physical conditions in planetary nebulae. We have already mentioned the measurements of the intensities of the higher members of the Balmer series in NGC 7027. The weak λ4363 line, through the ratio λ4363/λλ4959, 5007, provides a measure of the electron temperature in the [O III] emitting region which is invaluable in the studies of elements abundances. The weak recombination line C II λ4267 also provides information on the abundance of carbon in planetary nebulae and has been the subject of several observational and theoretical studies.

The λ4267 line arises from the C II $4f \rightarrow 3d$ transition. Assuming the line to be excited by radiative recombination,

$$C^{2+} + e \rightarrow C^+ + h\nu,$$

TABLE 1

Comparison of Photoelectric Observations, log $F(\lambda)/F(H\beta)$, of NGC 7662 and IC 418 by O'Dell, Peimbert and Torres-Peimbert (PTP), and Torres-Peimbert and Peimbert (TPP)

λ(Å)	Identification	NGC 7662		IC 418		
		O'Dell[a]	PTP[b]	O'Dell[a]	PTP[b]	TPP[c]
3726, 3729	[O II]	− 1.13	− 0.93	0.17	0.14	0.10
3835	H9	− 1.36	—	− 1.26	− 1.26	—
3869	[Ne III]	− 0.20	—	− 1.71	− 1.79	− 1.65
4026	He I	− 1.67	—	− 1.71	− 1.78	—
4102	Hδ	− 0.65	− 0.61	− 0.69	− 0.66	− 0.66
4340	Hγ	− 0.39	− 0.38	− 0.39	− 0.36	− 0.39
4363	[O III]	− 0.84	− 0.82	− 1.78	− 2.18	− 2.09
4472	He I	− 1.77	− 1.61	− 1.48	− 1.51	− 1.44
4686	He II	− 0.24	− 0.37	—	—	—
4861	Hβ	0.00	0.00	0.00	0.00	0.00
4959	[O III]	0.61	—	− 0.37	− 0.38	—
5755	[N II]	—	—	− 1.41	− 1.54	− 1.62
5876	He I	− 1.10	− 1.10	− 0.86	− 0.92	− 0.89
6563	Hα	0.53	0.48	0.60	0.52	0.57
6584	[N II]	—	− 1.31	0.40	0.30	0.29
7320, 7330	[O II]	—	− 1.78	− 0.40	− 0.47	− 0.55
9229	H9	− 1.31	—	− 1.24	− 1.50	—
9532	[S III]	—	—	− 0.12	− 0.09	—
10049	H7	− 1.02	—	− 0.97	− 1.05	—
10830	He I	0.01	—	0.16	0.10	—
10938	H6	− 0.79	—	− 0.75	− 0.81	—

[a] See reference 7.
[b] See reference 8.
[c] See reference 9.

Torres-Peimbert and Peimbert[9] use the observed λ4267/Hβ ratio in the compilation of Kaler[11] to obtain the C^{2+}/H^+ abundance ratio given in Table 2. On applying an ionization correction, based upon model calculations, the total carbon abundance is then derived.

The ultraviolet spectrum of IC 418 exhibits the strong collisionally excited lines, C II] λ2326 and C III] λ1908. From the observed intensities of these lines relative to Hβ and measurements[9] of the [O III] and [N II] electron temperatures, Harrington et al.[12] derive C^+ and C^{2+} abundances listed in Table 2. The C^{2+} abundance is seen to be four times smaller than deduced by Torres-Peimbert and Peimbert[9] from the λ4267/Hβ ratio.

Torres-Peimbert et al.[13] have analyzed their own ultraviolet observations of IC 418 and obtained ionic abundances similar to Harrington et al.[12] and an identical value of the total carbon abundance. Torres-Peimbert et al.[13] also remeasured the intensity of λ4267 photoelectrically and obtained a line flux, relative to Hβ, which is smaller by a factor 1.5

TABLE 2

Abundance of Carbon and Carbon Ions in IC 418 as Determined by Torres-Peimbert and Peimbert, Harrington et al., and Torres-Peimbert et al.[a]

	TPP[b]	HLSS[c]	TPPD[d]
$\log(C^+)$	—	8.72	8.62
$\log(C^{2+})$	$(8.9)^e$	8.28	$8.46(8.69)^e$
$\log(C)$	$(8.9)^e$	8.85 ± 0.2	$8.85(9.08)^e$

[a] The usual scale, on which $\log(H) = 12$, is employed.
[b] See reference 9.
[c] See reference 12.
[d] See reference 13.
[e] Numbers in parentheses are derived from observations of C II λ4267.

than the photographic measurement tabulated by Kaler[11]; this is a further example of the intensity of a weak line being overestimated by photographic techniques. When this correction has been made, the C^{2+} abundance derived from C II λ4267 is larger by a factor of 1.7 than the value deduced from the intensity of C III] λ1908. Torres-Peimbert et al.[13] attribute the remaining discrepancy to unresolved variations in the electron temperature of the ionized gas. Harrington et al.,[12] on the other hand, suggest that an additional recombination process may contribute to the intensity of λ4267; this latter suggestion is not confirmed by the recent calculations of dielectronic recombination rates by Storey.[14]

Ultraviolet and visual (photoelectric) observations of a number of points on the Ring nebula (NGC 6720) by Barker[15,16] show the abundance of C^{2+}, as derived from the intensity of λ4267 assuming recombination theory, to be up to ten times higher than deduced from the intensity of λ1908. Furthermore, there is a possible correlation with distance from the exciting star which leads Barker[16] to suggest that resonant absorption of the stellar ultraviolet continuum radiation may contribute to the excitation of λ4267. This process has been recognized as the principal mechanism of excitation of weak O III lines in the visual spectrum of NGC 7027[17] but seems unlikely to contribute to the intensity of C II λ4267. No satisfactory solution to this problem can be proposed at the present time.

Photoelectric detectors are also used to obtain monochromatic images of planetary nebulae. For example, observations of NGC 7662[18] and IC 418[19] have been made with an electronographic image tube. The diameter of IC 418 is often quoted as 10 arc sec, but the Spectracon measurements[19] show that the emission from the ionized gas extends to a diameter of approximately 20 arc sec. Monochromatic images are invaluable when comparing spectral observations of extended objects obtained with differing aperture sizes.

While it should be clear from this brief review that optical (especially photoelectric) observations are of paramount importance in studies of planetary nebulae, the information that they provide is incomplete in a number of respects. First, most of the radiation from the central stars and much of the radiation from the nebulae are emitted at shorter wavelengths. Second, lines from important ionization stages of abundant elements, such as carbon and nitrogen, are absent or weak in the visual spectra of planetary nebulae. Determinations of element abundances from visual spectra alone are consequently not possible or subject to large uncertainties. Third, optical observations provide only limited information on dust absorption, both internal and external to the nebula. For these reasons, technical advances enabling other spectral regions to be observed have led to important advances in our knowledge of planetary nebulae.

2.2. Infrared

The most striking feature of early infrared observations of planetary nebulae[20-23] was the strength of the continuum, much greater than expected from recombination and free–free emission. In Fig. 1 is shown the infrared and radio spectrum of the planetary nebula NGC 7027[24]; the infrared "excess" is the most prominent feature of the spectrum. Krishna Swamy and O'Dell[25] suggested that dust absorbs the Ly α radiation produced in the nebula and reradiates the energy in the infrared.

More recent observations have shown that Ly α radiation alone is insufficient to acount for the total infrared continuum emission of some nebulae. The deficit is accounted for by the resonance line radiation of

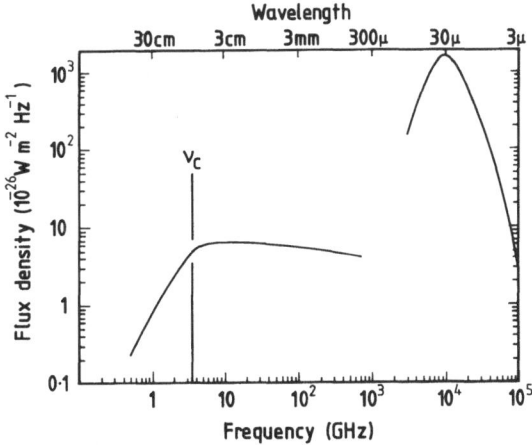

FIGURE 1. Observed radio and infrared spectrum of NGC 7027.[24]

TABLE 3

Infrared Emission Lines Observed in the Spectra of Planetary Nebulae [a]

$\lambda(\mu m)$	Identification	Transition
2.17	H	$n = 7 \to 4$
3.74	H	$8 \to 5$
4.05	H	$5 \to 4$
7.00	[Ar II]	$J = 1/2 \to 3/2$
8.99	[Ar III]	$1 \to 2$
10.52	[S IV]	$3/2 \to 1/2$
12.81	[Ne II]	$1/2 \to 3/2$
24.28	[Ne V]	$1 \to 0$
25.87	[O IV]	$3/2 \to 1/2$

[a] See references 28 and 31–34.

heavy elements, such as C IV $\lambda 1549$, which is significantly attenuated by dust absorption (see Section 2.4). Furthermore, it is probable that stellar ultraviolet radiation undergoes dust absorption directly in some nebulae.[26,27]

In addition to the thermal infrared radiation of the dust, a number of forbidden lines of heavy elements and recombination lines of hydrogen are observed in the spectra of planetary nebulae. The first line to be detected was the [Ne II] 12.8-μm transition in IC 418.[22] The observed flux was $F(12.8\ \mu m) = (2.1 \pm 0.5) \times 10^{-13}$ W m^{-2}; more recent observations[28] give $F(12.8\ \mu m) = (2.8 \pm 0.2) \times 10^{-13}$ W m^{-2}, in agreement with the order measurement to within the combined error bars.

The measurement of [Ne II] $2p^5\ ^2P^o_{1/2} \to\ ^2P^o_{3/2}$ 12.8 μm in IC 418 contributes significantly to our knowledge of this nebula: Ne$^+$ is the predominant stage of ionization of neon in IC 418 and is observable only in the infrared, through the fine-structure transition. From the observed 12.8 μm/$\lambda 4861$ intensity ratio and a model of the nebula, Flower[29] deduced a Ne/H abundance ratio, $N(\text{Ne})/N(\text{H}) = 0.6 \times 10^{-4}$. This determination relies upon the measured 12.8 μm and $\lambda 4861$ fluxes, the reddening constant c to the nebula, the collision strength for electron collisional excitation of the fine-structure transition, and the Ne^{2+}–Ne$^+$ ionization balance in the nebula. Flower[29] used $c = 0.24$, as compared with $c = 0.25$, the value adopted in more recent analyses[12,13] of the spectrum of IC 418. The collision strength employed by Flower,[29] $\Omega(^2P^o_{1/2}, ^2P^o_{3/2}) = 0.39$, is very similar to the effective collision strength at $T_e = 10^4$ K, $\Upsilon(^2P^o_{1/2}, ^2P^o_{3/2}) = 0.37$, recommended by Seaton.[30] As charge exchange is believed to be unimportant in the Ne^{2+}–Ne$^+$ ionization equilibrium (see Section 3.1), the ionization balance computed by Flower[29] should be reliable. Since the

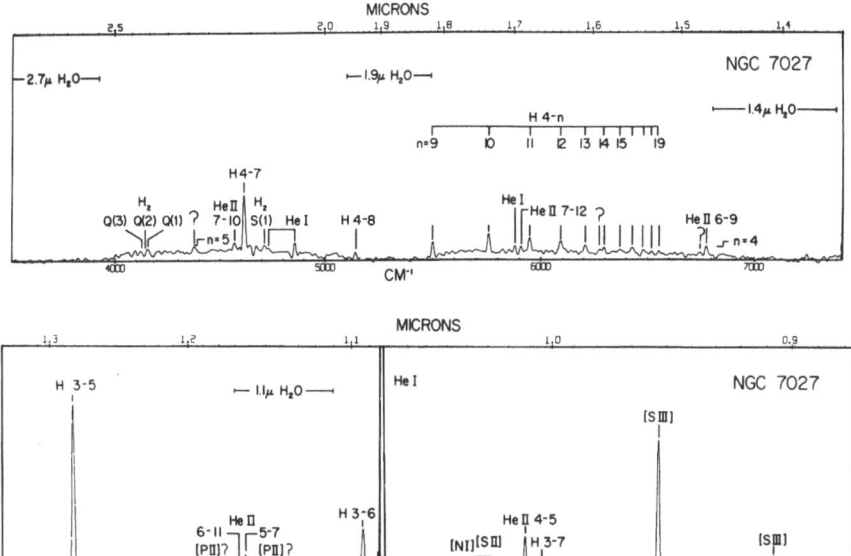

FIGURE 2. Near-infrared spectrum of NGC 7027.[39]

absolute $H\beta$ flux from IC 418 is well determined, it follows that the only significant correction to the value of the Ne/H abundance ratio deduced by Flower[29] arises from the more recent and, in principle, more reliable measurement of the 12.8-μm flux. Applying this correction, we obtain $N(Ne)/N(H) = 0.8 \times 10^{-4}$ in IC 418.

Since the detection of [Ne II] 12.8 μm, a number of other infrared fine-structure transitions and hydrogen recombination lines have been observed in the spectra of planetary nebulae (see Table 3). In addition to the identified atomic transitions, other features are observed at 3.09 and 3.27 μm,[35,36] 6.2 and 7.7 μm,[37] 8.7 and 11.3 μm,[28,33] and at $\lambda > 24$ μm.[38]

The planetary nebula NGC 7027 shows molecular hydrogen emission in the near infrared. Figure 2 shows a Fourier transform spectrum in the 0.9–2.7 μm region.[39] Lines of the $v = 1 \rightarrow 0$ quadrupole spectrum are clearly present, namely, $S(1)$ ($J = 3 \rightarrow 1$), $Q(1)$ ($J = 1 \rightarrow 1$), and $Q(3)$ ($J = 3 \rightarrow 3$), in addition to a number of atomic recombination and forbidden lines. Only ortho-H_2 is observed (odd J). In equilibrium, the relative abundance of ortho- and para-H_2 is

$$\frac{N(J = 1)}{N(J = 0)} = 3\exp(-170.48/T),$$

and the absence of para-H_2 (even J) in the spectrum indicates that the kinetic temperature, T, is at least several hundred degrees Kelvin. In the vicinity of a strong source of ultraviolet radiation, this result is perhaps not surprising. Emission from CO has also been observed from an extended region around NGC 7027.[40]

In view of the widths of the near-infrared features at 3.09 and 3.27 μm, Dabrowski and Herzberg[41,42] suggested that these features might be produced by inverse predissociation and proposed their identification with HeH^+. This process is the inverse of the radiative dissociation of a molecule via a rovibrational level embedded in a continuum; radiative stabilization gives rise to an emission line spectrum characteristic of the molecule. (The analogous atomic process is dielectronic recombination.) However, it was subsequently shown[43] that the rate of inverse predissociation was far too small to support this identification. Allamandola and Norman[44] have proposed that fluorescence of vibrationally excited molecules in grain mantles is responsible for the unidentified emission features.

A number of planetary nebulae, listed in Table 4, display a broad emission feature peaking at 11.2 μm[28,33,45] which is seen in C stars and is attributed to SiC grains. Other nebulae show a broad feature centered at 9.7 μm, seen in late-type oxygen-rich stars and H II regions, which is considered to be silicate emission. Aitken and Roche[45] associate SiC emission with a carbon-rich envelope and silicate emission with an oxygen-rich envelope. Further observations with the International Ultraviolet Explorer satellite will prove crucial in establishing this relationship.

2.3. Radio

Radio frequency observations of planetary nebulae can provide information on the distribution of ionized gas in the nebula, the reddening to

TABLE 4

Planetary Nebulae Displaying SiC or Silicate Emission Features in the 8–13 μm Region[a]

SiC	Silicate
IC 418	Sw St 1
NGC 6572	M 1–26
NGC 6790	Hb 12
M 1–11	IC 4997
IC 2501	Vy 2–2
	He 2–131
	He 2–47

[a] See reference 45.

and across the nebula, the electron temperature of the emitting plasma. The physical process giving rise to the observed radio emission—bremsstrahlung (free–free emission) of the electrons—is well understood. The emission coefficient may be written[46]

$$4\pi j_\nu = 3.77 \times 10^{-38} N_e T_e^{-1/2} \left\{ \left[N(\text{H}^+) + N(\text{He}^+) \right] \left[17.72 + \ln(T_e^{3/2}/\nu) \right] \right.$$

$$\left. + 4N(\text{He}^{2+}) \left[17.03 + \ln(T_e^{3/2}/\nu) \right] \right\}$$

in units of erg cm^{-3} s^{-1} Hz^{-1}. The limits of validity of this expression are discussed by Gayet.[47]

Thermodynamic equilibrium is attained among the free electrons, so the emission coefficient and the opacity, κ_ν, are related through

$$j_\nu = \kappa_\nu B_\nu,$$

where B_ν is the Planck function. At radio frequencies ($h\nu \ll kT_e$),

$$B_\nu \approx 2kT_e(\nu/c)^2.$$

Therefore,

$$\kappa_\nu = 9.8 \times 10^{-3} N_e N(\text{H}^+) T_e^{-3/2} \nu^{-2}$$

$$\times \left\{ (1 + y') \left[17.72 + \ln(T_e^{3/2}/\nu) \right] + 4y'' \left[17.03 + \ln(T_e^{3/2}/\nu) \right] \right\},$$

where $y' = N(\text{He}^+)/N(\text{H}^+)$ and $y'' = N(\text{He}^{2+})/N(\text{H}^+)$.

The optical depth at frequency ν is given by

$$\tau_\nu = \int \kappa_\nu \, dl.$$

If we consider a pure hydrogen nebula at a uniform T_e, then

$$\tau_\nu = 9.8 \times 10^{-3} T_e^{-3/2} \nu^{-2} \left[17.72 + \ln(T_e^{3/2}/\nu) \right] E,$$

where $E = \int N_e^2 \, dl$ is the emission measure. Defining a critical frequency, ν_{crit}, such that $\tau_{\nu_{\text{crit}}} = 1$, we have[24]

$$E = \nu_{\text{crit}}^2 T_e^{3/2} \left\{ 9.8 \times 10^{-3} \left[17.72 + \ln(T_e^{3/2}/\nu_{\text{crit}}) \right] \right\}^{-1},$$

with ν_{crit} in Hz and E in cm^{-5}.

The flux received at the Earth at frequency ν is given by

$$F_\nu = 2k\left(\frac{\nu}{c}\right)^2 \int T_e(1 - e^{-\tau_\nu})\,d\Omega;$$

F_ν is currently expressed in Jansky, equivalent to the former flux unit, f.u. $= 10^{-26}$ W m^{-2} Hz^{-1} $= 10^{-23}$ erg cm^{-2} s^{-1} Hz^{-1}. If the nebula is optically thick at frequency ν, $\tau_\nu \gg 1$, for all lines of sight passing through it, then

$$F_\nu = 2kT_e(\nu/c)^2\Omega, \qquad \nu \ll \nu_{crit}$$

where Ω is the solid angle subtended at the Earth by the nebula. On the other hand, if the nebula is optically thin at frequency ν, $\tau_\nu \ll 1$, then

$$F_\nu = \int j_\nu\,dl\,d\Omega$$

$$= 3.00 \times 10^{-39}ET_e^{-1/2}\left[17.72 + \ln\left(T_e^{3/2}/\nu\right)\right]\Omega, \qquad \nu \gg \nu_{crit}$$

where F_ν is expressed in erg cm^{-2} s^{-1} Hz^{-1}. The variation of F_ν with ν is illustrated in Fig. 1 for the planetary nebula NGC 7027.

If the radio brightness distribution at an optically thin frequency is known, then $E(\Omega)$ and $\tau_\nu(\Omega)$ may be calculated for an assumed value of T_e. The radio frequency flux, F_ν, may then be calculated and compared with the observed spectrum; the best fit determines the value of T_e. As noted by Scott,[48] this method underestimates T_e if the source contains unresolved structure. Scott presents observations of five planetary nebulae at $\nu = 5$ GHz, obtained with an angular resolution of 2 arc sec. For NGC 7662, he obtains $T_e = 5500 \pm 2000$ K, as compared with $T_e = 12,500$ K from observations of the [O III] lines.[9]

Discrepancies between values of T_e derived from forbidden line ratios and by other methods, such as radio continuum observations and measurements of the Balmer lines and continuum, have been recognized for many years. Peimbert[49,50] attributes these discrepancies to spatial variations of T_e, characterized by the mean square fluctuation,

$$t^2 = \frac{\int[T(\mathbf{r}) - T_0]^2 N_e(\mathbf{r})N_i(\mathbf{r})\,dl\,d\Omega}{T_0^2\int N_e(\mathbf{r})N_i(\mathbf{r})\,dl\,d\Omega},$$

where $T(\mathbf{r})$ is the local value of T_e, $N_i(\mathbf{r})$ is the local ion density, and

$$T_0 = \frac{\int T(\mathbf{r})N_e(\mathbf{r})N_i(\mathbf{r})\,dl\,d\Omega}{\int N_e(\mathbf{r})N_i(\mathbf{r})\,dl\,d\Omega}.$$

Apparent discrepancies in the abundance of C^{2+} in the planetary nebula IC 418, as determined from C III] $\lambda1909$ and C II $\lambda4267$, have also been attributed to spatial temperature variations.[13]

Scott[48] relates the supposed temperature fluctuations to the existence of cool, dense filaments within the hot, rarefied ambient gas[51] and invokes the observations of Boeshaar[52] in support of such variations. Boeshaar studied inhomogeneities in planetary nebulae by means of photographic spectrophotometry. Values of the electron temperature were deduced from the [N II] nebular to auroral line ratio, $(\lambda\lambda6548, 6584)/\lambda5755$. The nebular lines, $\lambda\lambda6548, 6584$, are relatively strong and, according to Boeshaar, their fluxes should be accurate to within 10%. The auroral line is weak and the uncertainty is 25%.

The nebular lines, $\lambda\lambda6548, 6584$, are emitted from a common upper level and their relative intensity is determined by a ratio of spontaneous radiative transition probabilities, i.e., a constant. The transition probability calculations of Nussbaumer and Rusca[53] yield $I(\lambda6584)/I(\lambda6548) = 2.9$; the differential reddening may be neglected. The values of this ratio observed by Boeshaar[52] vary from less than 1 to almost 30. We conclude that the observed intensities of the strong lines exceed the estimated 10% error and infer that the errors in the weak lines may be much greater than 25%. The validity of the conclusions regarding temperature fluctuations in nebulae, derived from these observations, must be seriously questioned.

The reddening constant, c, defined as the logarithmic extinction of $H\beta(\lambda4861)$, may be directly determined from a comparison of the $H\beta$ flux and the radio flux at a frequency at which the nebula is optically thin.[24,54] The rate of emission per unit volume at radio frequencies, $4\pi j_\nu$, is given above. The rate of emission in the $H\beta$ line, $4\pi j_\beta$, is given by Brocklehurst[4] for a nebula that is optically thick in Lyman radiation and may be fitted to within 2% over the range $5 \times 10^3 \leqslant T_e \leqslant 2 \times 10^4$ K by

$$4\pi j_\beta = 3.96 \times 10^{-22} T_e^{-0.878} N_e N(H^+)$$

in units of erg cm^{-3} s^{-1}. The values of the numerical coefficients are insensitive to the electron density. Hence, the relative intensity of $H\beta$ and radio radiation at an optically thin frequency is given by

$$\frac{I_\beta}{I_\nu} = 1.05 \times 10^{16} T_e^{-0.378} \left\{ (1 + y') \left[17.72 + \ln(T_e^{3/2}/\nu) \right] \right.$$

$$\left. + 4y'' \left[17.03 + \ln(T_e^{3/2}/\nu) \right] \right\}^{-1}$$

in units of Hz. The intensity ratio at the source is related to the intensity ratio at the Earth through

$$\frac{I_\beta}{I_\nu} = 10^c \frac{F_\beta}{F_\nu} .$$

from which c may be determined. It is clearly important that the radio and Hβ fluxes should refer to the same region, usually the whole nebula.

As mentioned above, the 2.6-mm $J = 1 \rightarrow 0$ rotational transition of $^{12}C \, ^{16}O$ has been observed in association with the planetary nebula NGC 7027[40]; $^{13}C \, ^{16}O$ may also have been detected. The line is emitted from a region that is approximately 1 arc min $\times$ 1 arc min in angular size, much more extended than the 15 arc sec diameter of the ionized part of the nebula.[55] Mufson et al.[40] also report the marginal detection of $^{12}C \, ^{16}O$ emission from the planetary nebulae IC 418 and NGC 6543. The CO column density implied by the observations of IC 418 should have been easily detectable with the IUE satellite through absorption of the stellar ultraviolet continuum radiation in the fourth positive system of CO. No such absorption is seen.[56] It would appear that either the reported detection of CO in IC 418 at 2.6 mm is incorrect or that the CO is not to be found along the line of sight to the exciting star. Further radio observations with higher sensitivity are required to resolve this question.

2.4. Ultraviolet

Interesting and quantitatively useful ultraviolet observation of planetary nebulae had been made prior to the launch of the IUE satellite. We shall briefly review these observations, although they have been largely superseded by more recent measurements with IUE. Early reviews[57,58] of the expected ultraviolet spectra of planetary nubulae were followed by a more detailed discussion of some aspects of the related physics.[59]

Ultraviolet photometric observations of planetary nebulae were made with the OAO-2 satellite.[60] The TD-1 satellite has also been used to observe the central stars of planetary nebulae.[61,62] Intermediate band measurements of planetary nebulae between 1500 and 3300 Å have been made with ANS.[63,64] We shall discuss the interpretation of stellar ultraviolet spectra below, in connection with NGC 7662.

Pottasch et al.[64] deduce the flux in the C IV resonance line doublet from a comparison of wide (150 Å) and narrow (50 Å) band fluxes, centered at λ1550. In Table 5, we compare ANS results for four nebulae with rocket ultraviolet measurements[65,66] and with more recent IUE obervations.[67–70] In these cases, the nebula is larger than the IUE large aperture and it is possible that only part of the line flux from the nebula

TABLE 5
Measured Fluxes of the C IV Resonance Doublet
$\lambda\lambda 1548, 1551$ (in Units of 10^{-12} erg $cm^{-2} s^{-1}$)

NGC	Rocket	ANS[c]	IUE
2392		45 ± 12	8.7[d]
3242		85 ± 6	7.5[e]
7027	10.6[a]	20 ± 3	17[f]
7662	277[b]	330 ± 10	181 ± 3[g]

[a] Bohlin et al.[65]
[b] Bohlin et al.[66]
[c] Pottasch et al.[64]
[d] Aller and Keyes.[67]

[e] Perinotto and Benvenuti.[68]
[f] Boggess et al.[69]
[g] Harrington et al.[70]

has been recorded. However, the discrepancies between the ANS and IUE measurements, particularly for the first two nebulae, would appear to be too large to be explicable in this way. The method used to derive the flux in the C IV doublet from the ANS measurements[64] will be most reliable when the contrast between the line and continuum is greatest, i.e., for NGC 7027, where the stellar continuum is not observed.

The paper by Bohlin et al.[65] is remarkable in a number of respects: it presents the first rocket ultraviolet spectrographic observations of a planetary nebula (NGC 7027); it demonstrates the importance of dielectronic recombination on the $C^{2+}-C^{3+}$ ionization balance; and it notes that internal dust absorption of the C IV resonance doublet is significant in attenuating the observed line intensity. Observations with IUE (see below) and in the far infrared[27] have since shown that this mechanism is generally important and that the observed $\lambda 1550$ flux does not correctly reflect the abundance of C^{3+} in the nebula. Pottasch et al.[64] determined the carbon abundance from their measurements of the flux in the C IV resonance doublet, discussed above, without allowing for internal dust absorption.

The advent of the IUE satellite has led to a vast improvement in the quality and quantity of ultraviolet observations of planetary nebulae and of their central stars. Observations made up to the second year of operation of the satellite have been reviewed by Nussbaumer[71] and Peimbert.[72]

A description of the IUE satellite and its modes of operation is given by Boggess et al.[73] The satellite carries long-wave and short-wave cameras, the short-wave prime (SWP) camera covering the wavelength range 1150–1950 Å and the long-wave redundant (LWR) camera covering the range 1900–3200 Å. Detection is photoelectric, with digitalized read-out of the image (768 × 768 pixels) in terms of data numbers (DN). The raw images may be examined after each exposure to assess the data quality. A

TABLE 6

Dimensions (in arc sec Projected on the Sky) of the IUE Apertures to the Short-Wave and Long-Wave Cameras. The Large Aperture has Width a and Length b; the Small Aperture has Diameter d

	SW	LW
a	10.3	10.2
b	23.0	23.8
d	3.21	3.98

problem encountered when observing emission line objects is the limited DN range of the cameras: exposures that leave strong lines unsaturated are too short to yield a satisfactory signal-to-noise ratio in the weak lines. Several exposures of any given nebula are generally necessary.

Observations may be made in low- or high-dispersion modes, the latter being obtained by interposition of an echelle grating in the light path. For a point source, the resolution is 6 Å for SWP images and 8 Å for LWR images. The resolving power of the echelle spectrograph is 1.2×10^4. Two entrance slots, one circular and the other oval, are available for use with each of the cameras; their dimensions are given in Table 6. For an extended source filling the large aperture, the resolution at low dispersion is degraded to 11 Å for SWP and 17 Å for LWR. The angular sizes of planetary nebulae observed with IUE range from an effective point source (e.g., IC 4997) to an extended source filling the large aperture (e.g., NGC 6720).

The image (set of DN values) is corrected for geometric and photometric distortions and converted into flux numbers (FN) by means of an intensity transfer function (ITF).[74] The net spectrum may then be obtained as the difference between the gross spectrum and the background. Care must be taken, particularly when dealing with the extended sources, to ensure that the full width of the spectrum is included in the extraction slit. The spectrum is then calibrated to yield the observed radiation intensity in the customary units of erg cm^{-2} s^{-1} $Å^{-1}$. The fluxes of emission lines (erg cm^{-2} s^{-1}) are derived by numerical quadrature over the width of the observed line profile. The degree of uncertainty in fixing the level of the continuum provides a measure of the error in the line flux.

Data analysis has been greatly facilitated in the United Kingdom by the introduction of the STARLINK system of VAX computers. Software for the interactive reduction and analysis of the IUE spectra has been

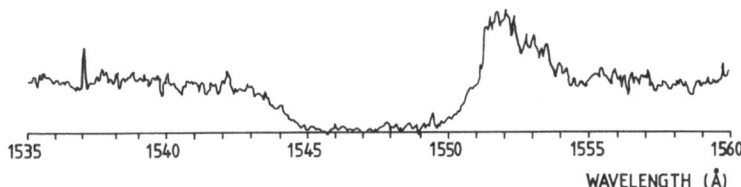

FIGURE 3. The C IV λλ1548, 1551 resonance line doublet observed in IC 418, from IUE image number SWP 5076.[56]

written notably by S. Adams, J. Giddings, and M. A. J. Snijders of University College London.

One of the salient features of ultraviolet spectra of the central stars of planetary nebulae is the evidence of continuing mass loss furnished by the presence of P Cygni profiles. In Fig. 3, we present a portion of a high-dispersion SWP image of IC 418, in the vicinity of $\lambda = 1550$ Å, taken with the large aperture centered on the exciting star. The emission components could be partly nebular and partly circumstellar in origin. As IC 418 is an extended object with angular dimensions comparable to the size of the larger aperture, line profiles obtained with the small and large slots would then be expected to differ, the emission component being stronger in the large aperture spectrum. In fact, comparison of large and small slot observations of the C IV resonance doublet at low dispersion shows the profiles to be identical to the level of the noise,[12] establishing their circumstellar origin. Allowing for the instrumental resolution of 6 Å and the known separation of the components of the doublet (498 km s^{-1}), Harrington et al.[12] derive a stellar wind terminal velocity, $v(\infty) \approx 1000$ km s^{-1}; this result is confirmed by the high-dispersion profiles in Fig. 3.

Other lines in the spectrum of IC 418 exhibit P Cygni profiles, including Si IV λλ1394, 1403, C II λλ1335, 1336, and C III λ1176. Castor et al.[75] have analyzed the P Cygni line profiles in the spectrum of NGC 6543. In Table 7, we list some ultraviolet lines that have been observed with P Cygni profiles in the spectra of planetary nebulae.

It is clearly important to couple studies of the ultraviolet spectra of the central stars with detailed analysis of the nebular spectra. The origin and evolution of planetary nebulae are incompletely understood, and the careful reduction and analysis of IUE observations may be expected to greatly improve this understanding. Much of the effort to date has been directed toward the interpretation of the nebular spectra alone, to the exclusion of the emission from the central stars. It is to be hoped that this tendency will be corrected during the next few years.

One of the expectations of ultraviolet observations of planetary nebu-

TABLE 7

Transitions Observed with P Cygni Profiles in the Spectra of the Central Stars of Some Planetary Nebulae

			λ (Å)
N v	$2s\,^2S_{1/2}-2p$	$^2P^o_{1/2}$	1242.80
	–	$^2P^o_{3/2}$	1238.82
C ii	$2s^2 2p\,^2P^o_{1/2}-2s2p^2$	$^2D_{3/2}$	1334.53
	$^2P^o_{3/2}-$	$^2D_{3/2}$	1335.66
	$^2P^o_{3/2}-$	$^2D_{5/2}$	1335.71
O iv	$2s2p^2\,^2P_{1/2}-2p^3$	$^2D^o_{3/2}$	1338.60
	$^2P_{3/2}-$	$^2D^o_{3/2}$	1342.98
	–	$^2D^o_{5/2}$	1343.51
O v	$2s2p\,^1P^o_1-2p^2$	$^1D_2$	1371.29
Si iv	$3s\,^2S_{1/2}-3p$	$^2P^o_{1/2}$	1402.73
	–	$^2P^o_{3/2}$	1393.73
C iv	$2s\,^2S_{1/2}-2p$	$^2P^o_{1/2}$	1550.77
	–	$^2P^o_{3/2}$	1548.20
He ii	2–3		1640.5
N iv	$2s2p\,^1P^o_1-2p^2$	$^1D_2$	1718.55

lae was improved determination of the abundances of C, N, and O relative to H. This expectation has been largely realized, although not without some unforeseen difficulties. Ultraviolet observations are particularly valuable for determining the C and N abundances, whose evaluations from visual line intensities alone are subject to large uncertainties. The C/O abundance ratio can be deduced from ultraviolet observations with greater precision than the abundances of each of these elements relative to H. The value of this ratio has important implications for the expected grain composition and the possible chemistry of planetary nebulae.

We pursue our review of IUE observations with a discussion of two of the best studied nebulae, NGC 7662 and IC 418.

2.4.1. NGC 7662. Pottasch *et al.*[63] conclude, on the basis of their analysis of ANS observations of planetary nebulae, that there are large discrepancies between determinations of the temperature of the central stars based upon the Zanstra method and upon ultraviolet continuum observations. In an exhaustive study of the planetary nebulae NGC 7662, Harrington *et al.*[70] consider the reasons for these discrepancies. First of all, they show that IUE observations imply values of the total continuum emission from the central star and the whole nebula which are substantially smaller than the ANS measurements and conclude that Pottasch et al.[63] underestimated the errors in these measurements by large factors. Harrington et al.[70] then consider the sensitivity of the ratio,

$$R(T_C) = F_*(\lambda 1550)/F_*(\lambda 3300),$$

of the stellar continuum emission at the extreme wavelengths observed with ANS. They show that this ratio is insensitive to the value adopted for the color temperature of the central star, T_C. Alternatively, a reliable determination of T_C from $R(T_C)$ requires accurate measurements of the ultraviolet stellar continuum fluxes. Furthermore, $R(T_C)$ is shown to be sensitive to the adopted value of the reddening constant, c, and to the subtraction of the nebular continuum flux at the longer wavelengths. Harrington et al.[70] obtain $T_C \approx 100,000$ K from their IUE observations and measurements of the visual continuum fluxes by J. B. Kaler, consistent with the He II Zanstra temperature, $T_Z(\text{He II}) = 113,000$ K.

In Table 8 are presented the results of IUE measurements of emission line fluxes in NGC 7662. The agreement between different observers is generally good, to within the combined probable error bars [N v λ1240 falls in a part of the spectrum where the stellar continuum is rapidly varying[70] and an accurate measurement of its intensity is difficult to achieve].

In Table 9, we list the abundances of C, N, O, and Ne relative to H, as determined from IUE[70,76,77] and rocket ultraviolet[66] observations. There are striking differences between the various determinations. The results of Peña and Torres-Peimbert[77] are systematically higher than those obtained by the other groups. Harrington et al.[70] interpret C III λ2297 as being produced by dielectronic recombination of C^{3+} (see below) and deduce a larger abundance of carbon than Benvenuti and Perinotto[76] and Bohlin et al.[66]; this has the important consequence that C/O > 1 according to Harrington et al.,[70] whereas Benvenuti and Perinotto[76] obtain C/O < 1.

Of the four studies, the results of which are reported in Table 9, that of Harrington et al.[70] is the most comprehensive. These authors pay careful attention to the reduction of the ultraviolet observations of both the nebula and the central star, consider all available observations in other spectral regions, including the far infrared,[27] and base their conclusions on sophisticated models incorporating all physical processes known to be important.

As mentioned above, the observation of C III $2p^2\ ^1D \rightarrow 2s2p\ ^1P^o$ (λ2297) plays a key role in the analysis of Harrington et al.[70] This transition, in C III, N IV (λ1718), and O v (λ1370), was identified in the spectrum of Nova Cygni 1978[78] and recognized as being produced by

TABLE 8
Ultraviolet Line Intensities in NGC 7662 as Measured by Different Observers Using the IUE Satellite

λ (Å)	Identification	Observed flux (10^{-12}erg cm^{-2} s^{-1})		
		Harrington *et al.*[a]	Benvenuti and Perinotto[b]	Peña and Torres-Peimbert[c]
1176	C III	—	4.0	4.7
1240	N V	5 ± 2.5	1.5	< 0.8
1403	O IV], Si IV	8.0 ± 1.0	7.4	6.9
1485	N IV]	8.5 ± 1.5	5.2	7.4
1549	C IV	181 ± 3	—	174
1603	[Ne IV]	—	0.9	0.4
1640	He II	101 ± 2	—	90
1664	O III]	5.5 ± 1.0	6.2	4.8
1751	N III	3.0 ± 1.0	4.0	2.8
1908	C III]	92 ± 2	—	96
2253	He II	0.6 ± 0.2	1.5	—
2297, 2307	C III, He II	2.8 ± 0.6	2.9	3.0
2321, 2326	[O III], C II]	3.6 ± 0.6	3.8	3.5
2385	He II	0.9 ± 0.3	1.6	—
2423	[Ne IV]	27 ± 2	—	—
2511	He II	1.6 ± 0.3	3.0	2.0
2733	He II	4.3 ± 0.5	4.3	4.8
2800	Mg II	≤ 0.3	—	—
2837	O III	4.0 ± 0.7	4.7	3.7
2855	[Ar IV]	0.5 ± 0.5	—	—
3023–3054	O III	7.4 ± 0.5	8.0	7.4
3133	O III	35 ± 1	35	41
3203	He II	8.3 ± 1.5	9.5	11.5

[a] See reference 70.
[b] See reference 76.
[c] See reference 77.

TABLE 9
Element Abundances Relative to H in NGC 7662

Element	Harrington *et al.*[a]	Benvenuti and Perinotto[b]	Peña and Torres-Peimbert[c]	Bohlin *et al.*[d]	Solar[e]
C	6.2(−4)[f]	4.7(−4)	8.5(−4)	3.7(−4)	4.7(−4)
N	6.0(−5)	9.1(−5)	1.9(−4)	8.0(−5)	9.8(−5)
O	3.6(−4)	5.2(−4)	8.1(−4)	3.7(−4)	8.3(−4)
Ne	7.0(−5)	8.9(−5)	2.0(−4)	7.0(−5)	—

[a] See reference 70.   [d] See reference 66.
[b] See reference 76.   [e] See reference 80.
[c] See reference 77.   [f] Figures in brackets are powers of 10.

dielectronic recombination via low-lying autoionizing states. Storey[14] subsequently calculated the dielectronic recombination coefficients. The intensity of $\lambda 2297$ depends on the abundance of C^{3+}, as does the intensity of the collisionally excited resonance line doublet C IV $\lambda 1550$. Harrington et al.[70] derive a C^{3+} abundance from $\lambda 2297$ and a total carbon abundance that is almost a factor of 2 larger than the value deduced from $\lambda 1550$; they conclude that $\lambda 1550$ undergoes dust absorption. The far-infrared measurements[27] confirm this suspicion.

As may be seen from Table 8, only an upper limit to the intensity of Mg II $\lambda 2800$ is established by the IUE observations, implying an abundance of magnesium at least 50 times less than the solar value. Silicon is also depleted, by about a factor of 4. The obvious interpretation[70] is that both Mg and Si have been removed from the gas phase by grain formation. Depletion of gas-phase magnesium has also been found in NGC 6572.[79]

2.4.2. IC 418. This nebula has already been mentioned in Sections 2.1–2.3. The apparent uniformity of the shell, as seen in monochromatic images of the nebula,[19] makes this object particularly attractive to observational and theoretical study.

In Table 10, we list the results of IUE observations of emission line intensities in IC 418,[12,13] dereddened and referred to $I(H\beta) = 100$. The agreement is seen to be good. Practically all of the carbon in IC 418 is in the form of C^+ and C^{2+}, and lines of these ions are prominent features of the ultraviolet spectrum. Good optical[7–9] and ultraviolet[12,13] measurements of O II and O III line intensities are also available. As O^+ and O^{2+} are the predominant ionization stages of oxygen, IC 418 would appear to be a prime candidate for the determination of the C/H and O/H abundance ratios.

Values of the carbon ionic abundances derived by Harrington et al.[12]

TABLE 10

Ultraviolet Line Intensities in IC 418, Corrected for Interstellar Reddening[81]
Using $c = 0.25$; Intensities on the Scale of $I(H\beta) = 100$.

		Intensity	
λ (Å)	Identification	Harrington et al.^{a}	Torres-Peimbert et al.^{b}
1908	C III]	33	36
2326	C II]	90	85
2470	[O II]	21.5	21.4
2800	Mg II	23	18

^{a} See reference 12.
^{b} See reference 13.

and Torres-Peimbert *et al.*[13] from their respective observations are given in Table 2. The agreement in the values of the total carbon abundance is excellent, although there are differences in the abundances of individual ions. Torres-Peimbert *et al.*[13] obtain a higher C^{2+} abundance from the intensity of C II λ4267 and attribute the discrepancy with the value derived from C III] λ1908 to the neglect of spatial temperature variations. With an assumed mean square temperature fluctuation $t^2 = 0.02$, the C/H abundance ratio derived from ultraviolet line measurements is found to agree well with the value obtained using the measured flux of λ4267.

The C/O ratio is less sensitive to the values of T_e and the assumed value of t^2 in the nebula than the C/H and O/H ratios. Harrington *et al.* obtain C/O = 1.8 ± 0.8; Torres-Peimbert *et al.*[13] obtain C/O = 1.9 (t^2 = 0) and C/O = 2.1 ($t^2 = 0.02$). In the Sun, C/O = 0.6.[80] Hence, there is a clear indication of an enhancement of the C/O ratio of IC 418 compared with the Sun.

Both groups of authors[12,13] note that the observed C^+/C^{2+} ratio is significantly greater than predicted by model calculations. It has since been shown[14] that the rate of dielectronic recombination,

$$C^{2+} + e \rightarrow (C^+)^* \rightarrow C^+ + h\nu,$$

is approximately 2.6 times faster than the rate of radiative recombination,

$$C^{2+} + e \rightarrow C^+ + h\nu$$

at $T_e = 10^4$ K. Dielectronic recombination has not been taken into account in model calculations on IC 418. Its inclusion will tend to bring the calculated C^+/C^{2+} ratio into better agreement with observations.

Dielectronic recombination manifests itself directly in the C II spectrum emitted by IC 418. Clavel *et al.*[82] have shown that the $2s2p^2 \, ^2D \rightarrow 2s^2 2p \, ^2P^\circ$ resonance doublet at 1335 Å is excited principally by dielectronic recombination, whereas the intercombination transition, $2s2p^2 \, ^4P \rightarrow 2s^2 2p \, ^2P^\circ$ at 2326 Å is excited by inelastic electron collisions. The profiles of the components of the resonance doublet, λ1334.53 and λλ1335.66, 1335.71 (blended), are shown in Fig. 4. Both components show a pronounced P Cygni profile. The ultraviolet continuum radiation from the central star undergoes resonant scattering by C^+ ions in the nebula. There is interstellar absorption, particularly in $2s^2 2p \, ^2P^\circ_{1/2} - 2s2p^2 \, ^2D_{3/2}$ λ1334.53, which connects with the ground fine-structure state, $^2P^\circ_{1/2}$; and there is emission from the nebula, produced by the dielectronic recombination process discussed above, followed by cascades to $2s2p^2 \, ^2D$.

From a comparison of the fluxes in C II λ1335 and C II λ4267, Clavel *et al.*[82] conclude that the resonance doublet is attenuated by approximately a factor of 2 due to internal dust absorption; the dust optical depth

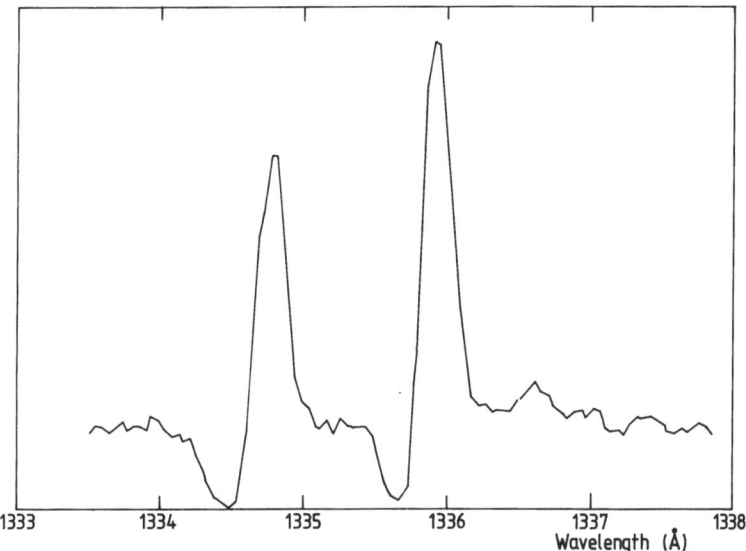

FIGURE 4. Profiles of the C II resonance lines observed in IC 418.[82]

is estimated to be $\tau_D = 0.08$. Analogous conclusions have been reached regarding internal dust absorption of the C IV resonance line doublet ($\lambda 1550$) in NGC 7009 and NGC 7662.[83]

2.4.3. The C/O Ratio in Planetary Nebulae. As mentioned in Section 2.2, Aitken and Roche[45] classify nebulae as carbon-rich (C/O > 1) or oxygen-rich (C/O < 1) according to the presence of "SiC" or "silicate" features, respectively, in their infrared spectra. In Table 11, we summarize determinations of the C/O ratio from ultraviolet observations of nebulae which exhibit SiC or silicate infrared features. The correlation proposed by

TABLE 11
*IUE Measurements of the C/O Abundance Ratio
(by Number) in Nebulae Displaying SiC or Silicate
Features in their Infrared Spectra*

Nebula	IR feature	C/O
IC 418	SiC	1.8[12,13]
NGC 6572	SiC	1.1[79], 1.3[84]
IC 4997	Silicate	0.4[85]
He 2–131	Silicate	0.5[86]
Sw St 1	Silicate	0.5[87]

Aitken and Roche[45] is seen to be followed, but the sample is small and further IUE observations are required.

3.   Models: Atomic Data

Sophisticated computer models of planetary nebulae have been constructed and used in the analysis of their spectra. Elaborate techniques are employed to solve the transfer of stellar ultraviolet continuum radiation and the "diffuse" line and continuum radiation produced in the nebula and to solve the related equations of thermal balance, which determine the temperature distribution in the nebula. Level populations and line intensities are determined by solving the relevant statistical equilibrium equations.

Comparison of computed and observed spectra yields information on the physical conditions in the nebula and can also furnish clues to deficiencies in the atomic data being employed. We mentioned in Section 2.4.2 the discrepancy between model computations of the C^+/C^{2+} abundance ratio in IC 418 and the value determined from IUE observations of this nebula. At least part of the discrepancy arises from the neglect of dielectronic recombination in the models.

On the basis of model computations of the spectrum of NGC 7027, Péquignot et al.[88] proposed that charge-transfer reactions with hydrogen were important for a number of ions, notably O^{2+}. As will be seen later, subsequent calculations of charge-transfer cross sections have confirmed some but not all of their predictions.

3.1.   Charge Transfer

We consider processes of the type

$$X^{+m} = H^0 \rightleftarrows X^{+(m-1)} + H^+ + \Delta E$$

and

$$X^{+m} + He^0 \rightleftarrows X^{+(m-1)} + He^+ + \Delta E,$$

where ΔE is the energy defect. In Table 12, we compile the most recent results of calculations of charge-transfer (CT) rate coefficients. Where calculations have been made by more than one group, the discrepancies between the results are a guide to remaining uncertainties. We append some comments on individual ions.

3.1.1. O^+.   Field and Steigman[93] used the orbiting approximation to calculate a rate coefficient of approximately 1×10^{-9} cm^3 s^{-1} for the

TABLE 12

Calculated Rate Coefficients for Charge Transfer with H^0 (*Unless Otherwise Stated*)

Ion		Rate coefficient (10^{-9} cm^3 s^{-1})		
		$T_e = 5 \times 10^3$ K	10^4 K	2×10^4 K
C^{2+}		$1.00(-3)^g$	$1.00(-3)$	$1.35(-3)^a$
C^{3+}		$\begin{cases} 3.09 \\ 1.6 \end{cases}$	$\begin{array}{c} 3.58 \\ 1.6 \end{array}$	$\begin{array}{c} 4.22^a \\ 1.6^d \end{array}$
C^{4+}		2.16	2.13	2.28^c
N^+		$1.23(-3)$	$1.04(-3)$	$0.84(-3)^b$
N^{2+}		0.78	0.86	0.97^a
N^{3+}		$\begin{cases} 1.54 \\ 1.82 \end{cases}$	$\begin{array}{c} 2.93 \\ 3.41 \\ 3.5^d \end{array}$	$\begin{array}{c} 5.14^a \\ 5.97^c \end{array}$
O^+			1.04^e	
O^{2+}	$\begin{cases} H^0 \\ He^0 \end{cases}$	$\begin{array}{c} 0.60 \\ 0.10 \end{array}$	$\begin{array}{c} 0.77 \\ 0.20 \end{array}$	$\begin{array}{c} 1.03^a \\ 0.39^a \end{array}$
O^{3+}		6.34	8.63	11.81^a
Ne^{3+}		4.00	5.68	8.28^a
Si^{2+}		4.26	5.16	6.09^f

[a] Butler *et al.*[89]   [b] Butler and Dalgarno.[90]
[c] Gargaud *et al.*[91]   [d] Watson and Christensen.[92]
[e] Field and Steigman.[93]   [f] McCarroll and Valiron.[94]
[g] Numbers in parentheses are powers of 10.

reaction

$$O^+ + H^0 \rightarrow O^0 + H^+$$

with only a weak dependence on the kinetic temperature, T. Chambaud *et al.*[95] have since performed quantum mechanical calculations of the cross sections for this reaction and tabulated the rate coefficients in the temperature range $10 \leqslant T \leqslant 1000$ K. Their results are smaller than those of Field and Steigman[93] at lower temperatures but differ by less than 40% at $T = 1000$ K. In planetary nebulae, where $T \approx 10^4$ K, the exact value of the rate coefficient is unimportant: the process is near resonance ($\Delta E \ll kT$) and is sufficiently rapid for detailed balance to be established. Under these conditions

$$\frac{N(O^+)}{N(O^0)} = \frac{8}{9} \frac{N(H^+)}{N(H^0)} ,$$

and the ionization of O^0 is bound to that of H^0. The importance of this process in the planetary nebula IC 418 was first demonstrated by Williams.[96]

   3.1.2 O^{2+}. The possible importance of the CT reaction

$$O^{2+} + H^0 \rightarrow O^+ + H^+ + \Delta E$$

on the ionization equilibrium of oxygen in gaseous nebulae was discussed by Perinotto.[97] Comparison of models of the emission spectrum of the planetary nebula NGC 7027 with observations convinced Péquignot *et al.*[88] that the rate coefficient for this reaction must be large ($\approx 10^{-9}$ cm^3 s^{-1}) at nebular temperatures. Quantum mechanical calculations of the cross section[89] confirmed this expectation.

   3.1.3. Ne^{2+}. On the basis of models of NGC 7027[88] and NGC 7662,[98] the reaction

$$Ne^{2+} + H^0 \rightarrow Ne^+ + H^+ + \Delta E$$

was predicted to be fast, with a rate coefficient $k \approx 10^{-10}$ cm^3 s^{-1}. In this case, quantal calculations have not confirmed the empirical prediction. Butler *et al.*[89] conclude that radiative charge transfer,

$$Ne^{2+} + H^0 \rightarrow Ne^+ + H^+ + h\nu,$$

will proceed more rapidly, with a rate coefficient $10^{-15} \leqslant k \leqslant 10^{-14}$ cm^3 s^{-1}, which is negligibly small. Thus the *ad hoc* introduction of CT processes should not be regarded as a panacea for the weaknesses of the nebular models.

3.2. Dielectronic Recombination

   We consider the process

$$X^{+m} + e \rightarrow [X^{+(m-1)}]^*,$$

where $X^{+(m-1)}$ is formed in an autoionizing state. Application of the principle of detailed balance gives

$$\gamma = \frac{\omega_*}{2\omega_{+m}} \left(\frac{h^2}{2\pi m k T} \right)^{3/2} \exp(-E_*/kT) A_a$$

for the rate coefficient γ. In this expression, ω denotes a statistical weight, E_* the energy of the autoionizing level relative to the ground state of X^{+m}, and A_a the autoionization probability; the other symbols have their usual meanings.

TABLE 13

Dielectronic Recombination Coefficients,[14] α_d, in Units of $10^{-12}\ cm^3$ s^{-1}, for the Process $X^{+m} + e \rightarrow [X^{+(m-1)}]^* \rightarrow X^{+(m-1)} + h\nu$

$X^{+(m-1)}$	$T_e(10^4\ K)$				
	0.7	0.9	1.1	1.3	1.5
C^{2+}	18.1	14.3	11.9	10.2	9.02
N^{3+}	16.6	16.4	15.7	14.8	13.9
O^{4+}	3.06	5.07	6.72	7.94	8.79
C^{+}	7.58	6.39	5.53	4.90	4.42
N^{2+}	22.2	22.1	21.4	20.5	19.6

The second stage of the dielectronic recombination process is radiative stabilization,

$$[X^{+(m-1)}]^* \rightarrow X^{+(m-1)} + h\nu,$$

and the dielectronic recombination coefficient is

$$\alpha_d = \gamma A_r / (A_r + A_a),$$

where A_r is the radiative transition probability. At high temperatures, the full Rydberg series of autoionizing levels must be considered, and core relaxation is the principal mechanism of radiative stabilization. The general formula of Burgess[99] is then applicable. At the lower temperatures ($T_e \approx 10^4$ K) more typical of planetary nebulae, the general formula of Burgess,[99] employed, for example, by Aldrovandi and Péquignot,[100] grossly underestimates the rate of dielectronic recombination.[14] Radiative stabilization through transitions of the captured electron must then be taken into account.

Assuming $A_a \gg A_r$ for the autoionizing levels, Storey[14] calculated rate coefficients for dielectronic recombination of a number of ions of C, N, and O and found that they exceed the corresponding radiative recombination coefficients at $T_e = 10^4$ K. In Table 13, we summarize the results that Storey[14] obtained. It is to be expected that the dielectronic recombination will prove to be important for other ions at nebular temperatures.

3.3. Photoionization and Radiative Recombination

Photoionization and its inverse, radiative recombination,

$$X^{+m} + h\nu \rightleftarrows X^{+(m+1)} + e$$

are fundamental to the ionization equilibrium in nebulae. Various theoretical techniques have been used to evaluate photoionization cross sections for complex ions: the Hartree–Fock method, quantum defect theory, and, more recently, the close-coupling approximation. Quantum defect theory, as embodied in the general formulas of Burgess and Seaton[101] and Peach,[102] has been extensively applied.[103,104] Limited comparison with experimental results can be made for neutral species. The reliability of the theoretical methods should improve with the degree of ionization.

Photoionization cross sections and recombination coefficients for hydrogen have been evaluated by Burgess.[105] A FORTRAN program that computes hydrogenic recombination coefficients is available through Computer Physics Communication.[106] Hydrogenic theory may be applied when evaluating coefficients for recombination to excited states of complex ions. A useful list of references to photoionization cross sections for ions of the more abundant elements in planetary nebulae (H, He, C, N, O, Ne, Mg, Si, and S) is to be found in Table 13 of Harrington et al.[70]

More recent elaborate calculations take account not only of direct photoionization but also of the autoionization process. For example, photoionization of the ground state of atomic nitrogen may proceed directly,

$$N^0(2s^22p^3\,{}^4S^\circ) + h\nu \to N^+(2s^22p^2\,{}^3P) + ks(d),$$

where k denotes the wave number and $s(d)$ the angular momentum state of the ejected electron; or via autoionizing levels in the first ionization continuum ($h\nu \geqslant 14.5$ eV), for example,

$$N^0(2s^22p^3\,{}^4S^\circ) + h\nu \to N^0(2s2p^3({}^5S^\circ)np\,{}^4P)$$

$$\to N^+(2s^22p^2\,{}^3P) + ks(d).$$

The autoionizing levels give rise to resonances in the photoionization cross section,[107,108] as illustrated in Fig. 5. A similar study[109] of photoionization from the metastable states ($2s^22p^3\,{}^2D^\circ$ and ${}^2P^\circ$) of N^0 emphatically illustrates the possible importance of resonances over restricted energy ranges. The process

$$N^0(2s^22p^3\,{}^2D^\circ, {}^2P^\circ) + h\nu \to N^0(2s2p^4\,{}^2P)$$

$$\to N^+(2s^22p^2\,{}^3P, {}^1D) + ks(d)$$

gives rise to a broad resonance in the photoionization cross section, as seen in Fig. 6. Similar results have been obtained[110] for photoionization of the ground ($2s^22p^2\,{}^3P$) and metastable ($2s^22p^2\,{}^1D$ and ${}^1S$) states of C^0. The

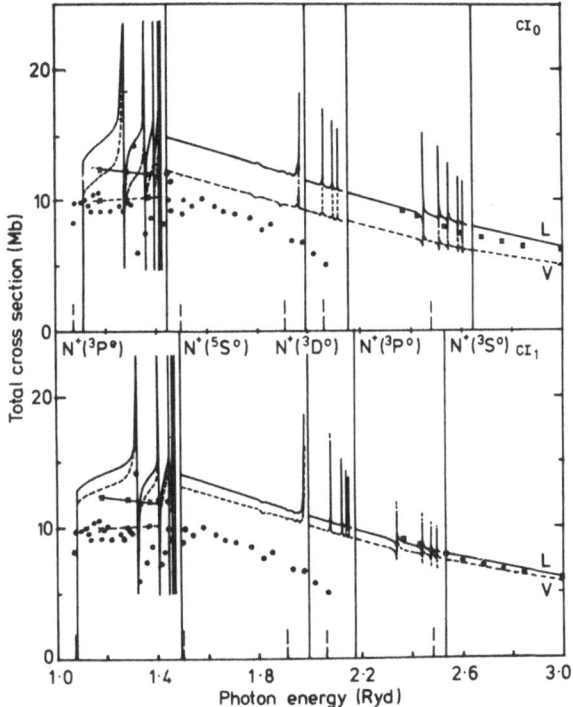

FIGURE 5. Resonances in the calculated[107] cross section for photoionization of N^0-$2s^2 2p^3\,^4S^\circ$ and a comparison with experimental points. (Copyright Institute of Physics).

ground-state photoionization cross section of O^0 has also been computed.[111]

Results of calculations of ground-state photoionization cross sections for neon ions (Ne^+, Ne^{2+}, and Ne^{3+}) in the close-coupling approximation and including autoionizing states have been published by Pradhan.[112,113] Developing techniques for incorporating the results of these and similar calculations in nebular models is a task in itself. The long-term answer probably lies in the use of computer data banks.

Once the photoionization cross section, a_ν, is known as a function of frequency, ν, the radiative recombination coefficient may be evaluated from

$$\alpha_i(T_e) = \frac{1}{c^2}\left(\frac{2}{\pi}\right)^{1/2}(mkT_e)^{-3/2}\frac{\omega_i}{\omega_+}\exp\left[I_i/(kT_e)\right]$$

$$\times \int_{I_i}^{\infty}(h\nu)^2 a_\nu \exp\left[-h\nu/(kT_e)\right]d(h\nu),$$

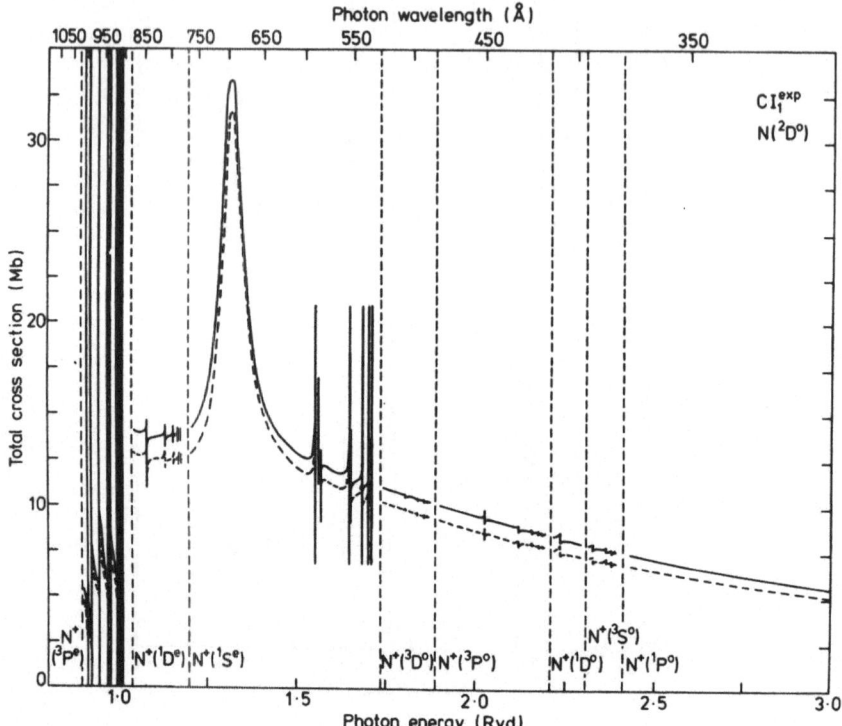

Figure 6. Resonances in the calculated[109] N^0 $2s^2 2p^3$ $^2D^\circ$ photoionization cross section. (Copyright Institute of Physics).

where ω_i is the statistical weight of the (final) state of the recombined ion, ω_+ the statistical weight of the (initial) state of the recombining ion, and I_i the ionization energy of the state i. It should be noted that, if a_ν includes the effects of autoionizing resonances, discussed above, no further correction for dielectronic recombination is necessary, providing that $A_a \gg A_r$.

In planetary nebulae, recombination occurs predominantly with the recombining ion initially in its ground state. The total recombination coefficient may be obtained by summing over the ground and all excited states of the recombined ion. For excited states whose principal quantum number differs from that of the ground state, the hydrogenic approximation is satisfactory.[106] For recombination to the ground states of complex ions, the coefficients should be evaluated from the corresponding photoionization cross sections. Tarter,[114,115] Aldrovandi and Péquignot,[100] and Gould[116] have obtained expressions for radiative recombination coefficients of complex ions. In none of these compilations are resonances in the photoionization cross sections taken into account. Much work remains to be done in this respect. When constructing models of nebulae, it is preferable to use the photoionization cross sections from which the radiative recombination coefficients have been evaluated when solving the equations of ionization equilibrium. In this way, some systematic errors may be avoided.

3.4. Electron Collisional Excitation

Electron collisional excitation is the primary mechanism giving rise to the strong forbidden and intercombination lines observed in planetary nebulae. This process is also responsible for the excitation of some of the permitted lines in the ultraviolet (e.g., C IV $\lambda\lambda1548$, 1551). Table 13 of Harrington et al.[70] contains a useful list of references to the results of recent calculations of collision cross sections.

Let us consider, as an example, electron collisional excitation of the C III $\lambda1909$ intercombination transition. The direct process is

$$C^{2+}(2s^2\,{}^1S) + kl \rightarrow C^{2+}(2s2p\,{}^3P^\circ) + k'l - 1(l+1)$$

in which the incident $k(l)$ electron is inelastically scattered into a $k'(l-1)$ or $k'(l+1)$ state. However, it is also necessary to consider processes of the type

$$C^{2+}(2s^2\,{}^1S) + kp \rightarrow C^+(2s2p({}^1P)Ns(d)\,{}^2P^\circ)$$

$$\rightarrow C^{2+}(2s2p\,{}^3P^\circ) + k's(d)$$

in which an autoionizing state, $2s2p({}^1P)ns(d)^2P^\circ$, is excited during the collision. This series of autoionizing Rydberg states gives rise to resonances in the collisional excitation cross section which are analogous to those encountered in our discussion of photoionization (Section 3.3). In Fig. 7 we present the results of calculations[117] of the collision strength,

$$\Omega(i, j) = k_i^2 \omega_i Q(i \rightarrow j),$$

where the cross section $Q(i \rightarrow j)$ is expressed in units of πa_0^2 and k_i^2 is in atomic units, for the $2s^2\,{}^1S \rightarrow 2s2p\,{}^3P^\circ$ transition in C^{2+}.

In astrophysical applications, one generally requires the rate coefficient, obtained by averaging over the Maxwellian velocity distribution of the electrons. For de-excitation, the rate coefficient (cm^3 s^{-1}) may be written

$$q(j \rightarrow i) = 8.63 \times 10^{-6} \frac{\Upsilon(j,i)}{\omega_j T_e^{1/2}},$$

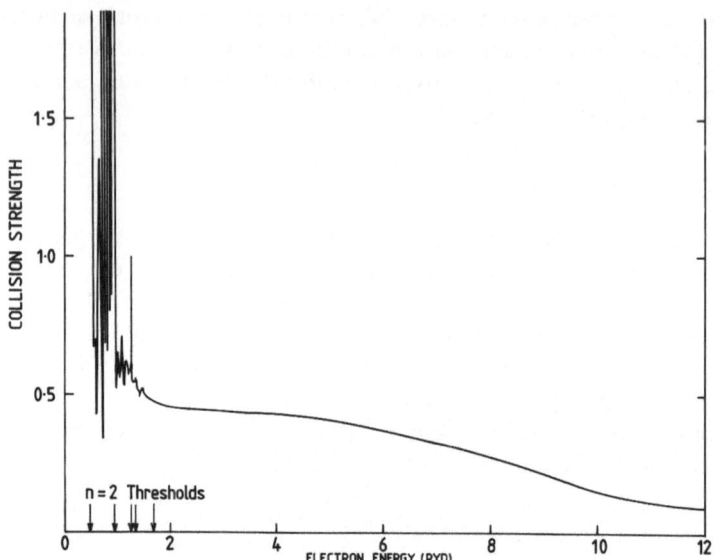

FIGURE 7. Resonances in the cross section for electron collisional excitation of the $2s^2\,{}^1S$–$2s2p\,{}^3P^\circ$ transition in C III.[117] (Copyright Institute of Physics).

where

$$\Upsilon(j,i) = \int_0^\infty \Omega(j,i)e^{-x_j}\,dx_j$$

and

$$x_j = E_j/kT_e,$$

E_j being the collision energy relative to the upper state j. From the principle of detailed balance, the rate coefficient for excitation is

$$q(i \to j) = \frac{\omega_j}{\omega_i}\,q(j \to i)\exp(-\Delta E_{ij}/kT_e),$$

where $\Delta E_{ij} > 0$ is the excitation energy of the transition. Such calculations have been carried out for transitions of C III and O V,[118] O III,[119] Ne V,[120] and Si III.[121]

The importance of resonance contributions to cross sections for forbidden transitions was demonstrated by earlier work[122,123] on [O III]. Seaton[30] has given convenient fits to collision strength calculations for [N II],

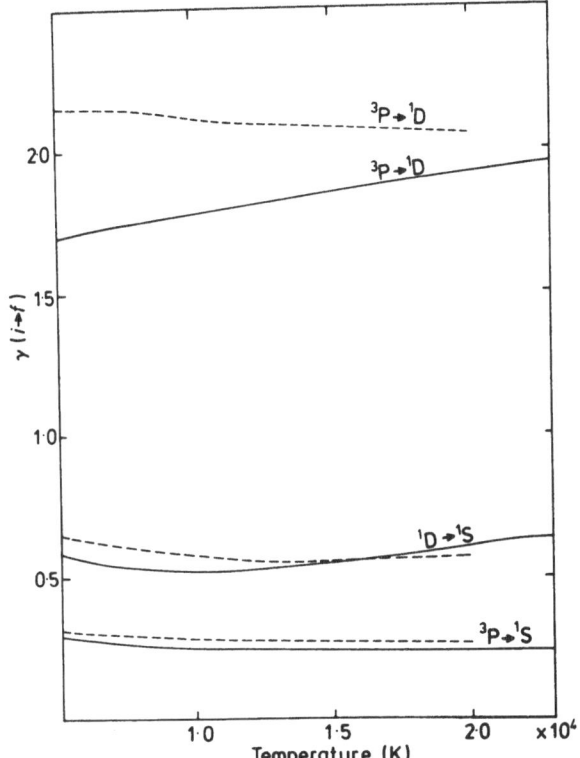

FIGURE 8. Comparison of calculations of the effective collision strengths for transitions within the Ne^{4+} $2s^2 2p^2$ ground configuration: continuous lines, Baluja *et al.*[120]; dashed lines, Giles[125]. (Copyright Institute of Physics).

[O III], [Ne II], and [Ne III]. For [O II] and [S II], reference should be made to Pradhan.[124] Giles has computed effective collision strengths Υ for [Ne V][125] and [Ne IV][126] His results for [Ne V] are compared with those of Baluja *et al.*[120] in Fig. 8; given the complexity of the calculations, the agreement is good.

3.5. Radiative Transition Probabilities

The difficulties attendant on calculations of probabilities for optically forbidden transitions are, in some respects, less severe than those encountered in computations of photoionization and electron collisional excitation cross sections. The compilation of transition probabilities by Garstang[127] is still useful. Multiconfiguration calculations have since been performed for a number of important ions, in particular, those in the carbon isoelectronic sequence.[53]

An outstanding recent achievement in this field has been the resolution of a long-standing discrepancy between the observed and calculated values of the [O II] $\lambda 3729/\lambda 3726$ intensity ratio. In the high density limit ($N_e \rightarrow \infty$), the level populations are determined by a Boltzmann distribution and the intensity ratio is given by[128]

$$r(\infty) = \frac{I(\lambda 3729)}{I(\lambda 3726)} = \frac{3}{2} \frac{A\left(^2D_{5/2} \rightarrow {}^4S_{3/2}\right)}{A\left(^2D_{3/2} \rightarrow {}^4S_{3/2}\right)}.$$

Seaton and Osterbrock[129] calculate $r(\infty) = 0.43$ as compared with the value $r(\infty) = 0.35 \pm 0.04$ observed in high-density planetary nebulae. They surmised that the effects of configuration interaction, which had been neglected, might be responsible for the discrepancy, but later calculations[130] showed that this was not the case. Zeippen[130] suggested that higher-order (relativistic) corrections to the magnetic dipole operator were important, and subsequent work[131] confirmed this suspicion. Zeippen[132] presents comprehensive results of very elaborate calculations of forbidden transition probabilities for ions in the nitrogen isoelectronic sequence up to Fe xx.

Zeippen[132] thanks Professor M. J. Seaton for "his generous hospitality at University College London." Many of the contributors to this volume have good reason to do likewise!

References

1. L. H. ALLER, I. S. BOWEN, and R. MINKOWSKI, *Astrophys. J.* **122**, 62 (1955).
2. M. J. SEATON, *Mon. Not. R. Astron. Soc.* **120**, 326 (1960).
3. L. H. ALLER, I. S. BOWEN, and O. C. WILSON, *Astrophys. J.* **138**, 1013 (1963).
4. M. BROCKLEHURST, *Mon. Not. R. Astron. Soc.* **153**, 471 (1971).
5. J. S. MILLER, *Astrophys. J.* **165**, L101 (1971).
6. W. LILLER, and L. H. ALLER, *Proc. Natl. Acad. Sci. USA* **49**, 675 (1963).
7. C. R. O'DELL, *Astrophys. J.* **138**, 1018 (1963).
8. M. PEIMBERT and S. TORRES-PEIMBERT, *Bol. Obs. Tonantzintla Tacubaya* **6**, 21 (1971).
9. S. TORRES-PEIMBERT and M. PEIMBERT, *Rev. Mex. Astron. Astrofis.* **2**, 181 (1977).
10. T. BARKER, *Astrophys. J.* **219**, 914 (1978).
11. J. B. KALER, *Astrophys. J. Suppl.* **31**, 517 (1976).
12. J. P. HARRINGTON, J. H. LUTZ, M. J. SEATON, and D. J. STICKLAND, *Mon. Not. R. Astron. Soc.* **191**, 13 (1980).
13. S. TORRES-PEIMBERT, M. PEIMBERT, and E. DALTABUIT, *Astrophys. J.* **238**, 133 (1980).
14. P. J. STOREY, *Mon Not. R. Astron. Soc.* **195**, 27P (1981).
15. T. BARKER, *Astrophys. J.* **240**, 99 (1980).
16. T. BARKER, *Astrophys. J.* **253**, 167 (1982).
17. M. J. SEATON, *Mon Not. R. Astron. Soc.* **139**, 129 (1968).

18. C. I. COLEMAN, N. K. REAY, and S. P. WORSWICK, *Adv. Electron. Electron Phys.* **40B**, 817 (1976).
19. N. K. REAY and S. P. WORSWICK, *Astron. Astrophy.* **72**, 31 (1979).
20. F. C. GILLETT, F. J. LOW, and W. A. STEIN, *Astrophys. J.* **149**, L97 (1967).
21. F. J. LOW and D. E. KLEINMANN, *Astron. J.* **73**, 868 (1968).
22. F. C. GILLETT and W. A. STEIN, *Astrophys. J.* **155**, L97 (1969).
23. N. J. WOOLF, *Astrophys. J.* **157**, L37 (1969).
24. Y. TERZIAN, in *Planatary Nebulae—Observations and Theory* Y. Terzian, Ed., IAU Symposium Number 76, Reidel, Dordrecht, 1978, pp. 111–120.
25. K. S. KRISHNA SWAMY and C. R. O'DELL, *Astrophys. J.* **151**, L61 (1968).
26. M. COHEN and M. J. BARLOW, *Astrophys. J.* **238**, 585 (1980).
27. H. MOSELEY, *Astrophys. J.* **238**, 892 (1980).
28. S. P. WILLNER, B. JONES, R. C. PUETTER, R. W. RUSSELL and B. T. SOIFER, *Astrophys. J.* **234**, 496 (1979).
29. D. R. FLOWER, *Mon. Not. R. Astron. Soc.* **147**, 245 (1970).
30. M. J. SEATON, *Mon. Not. R. Astron. Soc.* **170**, 475 (1975).
31. J. Z. HOLTZ, T. R. GEBALLE, and D. M. RANK, *Astrophys. J.* **164**, L29 (1971).
32. T. R. GEBALLE and D. M. RANK, *Astrophys. J.* **182**, L113 (1973).
33. D. K. AITKEN, P. F. ROCHE, P. M. SPENSER, and B. JONES, *Astrophys. J.* **233**, 925 (1979).
34. W. J. FORREST, J. F. MCCARTHY, and J. R. HOUCK, *Astrophys. J.* **240**, L37 (1980).
35. K. M. MERRILL, B. T. SOIFER, and R. W. RUSSELL, *Astrophys. J.* **200**, L37 (1975).
36. G. L. GRASDALEN and R. R. JOYCE, *Astrophys. J.* **205**, L11 (1976).
37. R. W. RUSSELL, B. T. SOIFER, and S. P. WILLNER, *Astrophys. J.* **217**, L149 (1977).
38. W. J. FORREST, J. R. HOUCK, and J. F. MCCARTHY, *Astrophys. J.* **248**, 195 (1981).
39. R. R. TREFFERS, U. FINK, H. P. LARSON, and T. N. GAUTIER, *Astrophys. J.* **209**, 793 (1976).
40. S. L. MUFSON, J. LYON, and P. A. MARIONNI, *Astrophys. J.* **201**, L85 (1975).
41. I. DABROWSKI and G. HERZBERG, *Trans. N.Y. Acad. Sci.* **38**, 14 (1978).
42. I. DABROWSKI and G. HERZBERG, Proceedings of the 21st International Astrophysics Colloquium (Institut d'Astrophysique, Université de Liège) pp. 341–349 (1980).
43. D. R. FLOWER and E. ROUEFF, *Astron. Astrophys.* **72**, 361 (1979).
44. L. J. ALLAMANDOLA and C. A. NORMAN, *Astron. Astrophys.* **63**, L23 (1978).
45. D. K. AITKEN and P. F. ROCHE, *Mon. Not. R. Astron. Soc.* **200**, 217 (1982).
46. L. OSTER, *Astrophys. J.* **134**, 1010 (1961).
47. R. GAYET, *Astron. Astrophys.* **9**, 312 (1970).
48. P. F. SCOTT, *Mon. Not. R. Astron. Soc.* **170**, 487 (1975).
49. M. PEIMBERT, *Astrophys. J.* **150**, 825 (1967).
50. M. PEIMBERT, *Bol. Obs. Tonantzintla Tacubaya* **6**, 29 (1971).
51. D. VAN BLERKOM and T. T. ARNY, *Mon. Not. R. Astron. Soc.* **156**, 91 (1972).
52. G. O. BOESHAAR, *Astrophys. J.* **187**, 283 (1974).
53. H. NUSSBAUMER and C. RUSCA, *Astron. Astrophys.* **72**, 129 (1979).
54. D. K. MILNE and L. H. ALLER, *Astron. Astrophys.* **38**, 183 (1975).
55. Y. TERZIAN, B. BALICK, and C. BIGNELL, *Astrophys. J.* **188**, 257 (1974).
56. J. CLAVEL and D. R. FLOWER, *Mon. Not. R. Astron. Soc.* **190**, 1P (1980).
57. D. E. OSTERBROCK, *Planet. Space Sci.* **11**, 621 (1963).
58. D. R. FLOWER, in *Planetary Nebulae*, D. E. Osterbrock and C. R. O'Dell, Eds., IAU Symposium Number 34, Reidel, Dordrecht (1968), pp. 77–86.
59. D. R. FLOWER, *Mem. Soc. Astron. Ital.* **47**, 313 (1976).
60. A. V. HOLM, in *Scientific Results from OAO-2*, A. D. Code, Ed., NASA SP-310, p. 229, 1972.
61. A. BOKSENBERG, D. CARNOCHAN, J. CAHN, and S. P. WYATT, *Mon. Not. R. Astron. Soc.* **172**, 395 (1975).
62. J. H. LUTZ and D. J. CARNOCHAN, *Mon. Not. R. Astron. Soc.* **189**, 701 (1979).
63. S. R. POTTASCH, P. R. WESSELIUS, C. C. WU, H. FIETEN, and R. J. VAN DUINEN, *Astron. Astrophys.* **62**, 95 (1978).
64. S. R. POTTASCH, P. R. WESSELIUS, and R. J. VAN DUINEN, *Astron. Astrophys.* **70**, 629 (1978).

322 D. R. FLOWER

65. R. C. Bohlin, P. A. Marionni, and T. P. Stecher, *Astrophys. J.* **202**, 415 (1975).
66. R. C. Bohlin, J. P. Harrington, and T. P. Stecher, *Astrophys. J.* **219**, 575 (1978).
67. L. H. Aller and C. D. Keyes, in *The Universe at Ultraviolet Wavelengths*, R. D. Chapman, Ed., NASA CP-2171, p. 649, 1981.
68. M. Perinotto and P. Benvenuti, *Astron. Astrophys.* **100**, 241 (1981).
69. A. Boggess, W. A. Feibelman, and C. W. McCracken, in *The Universe at Ultraviolet Wavelengths*, R. D. Chapman, Ed., NASA CP-2171, p. 663, 1981.
70. J. P. Harrington, M. J. Seaton, S. Adams, and J. H. Lutz, *Mon. Not. R. Astron. Soc.* **199**, 517 (1982).
71. H. Nussbaumer, Proceedings of the Second European IUE Conference, ESA SP-157, p. xliii, ESA Scientific and Technical Publications, Noordwijk (1980).
72. M. Peimbert, in *The Universe at Ultraviolet Wavelengths*, R. D. Chapman, Ed., NASA CP-2171, p. 557, 1981.
73. A. Boggess, F. A. Carr, D. C. Evans, D. Fischel, H. R. Freeman, C. F. Fuechsel, D. A. Klinglesmith, V. L. Krueger, G. W. Longanecker, J. V. Moore, E. J. Pyle, F. Rebar, K. O. Sizemore, W. Sparks, A. B. Underhill, H. D. Vitagliano, D. K. West, F. Macchetto, B. Fitton, P. J. Barker, E. Dunford, P. M. Gondhalekar, J. E. Hall, V. A. W. Harrison, M. B. Oliver, M. C. W. Sandford, P. A. Vaughan, A. K. Ward, B. E. Anderson, A. Boksenberg, C. I. Coleman, M. A. J. Snijders, and R. Wilson, *Nature* **275**, 372 (1978).
74. R. C. Bohlin, A. V. Holm, B. D. Savage, M. A. J. Snijders, and W. M. Sparks, *Astron. Astrophys.* **85**, 1 (1980).
75. J. I. Castor, J. H. Lutz, and M. J. Seaton, *Mon. Not. R. Astron. Soc.* **194**, 547 (1981).
76. P. Benvenuti and M. Perinotto, *Astron. Astrophys.* **95**, 127 (1981).
77. M. Peña, and S. Torres-Peimbert, *Rev. Mex. Astron. Astrofis.* **6**, 309 (1981).
78. D. J. Stickland, C. J. Penn, M. J. Seaton, M. A. J. Snijders, and P. J. Storey, *Mon. Not. R. Astron. Soc.* **197**, 107 (1981).
79. D. R. Flower and C. J. Penn, *Mon. Not. R. Astron. Soc.* **194**, 13P (1981).
80. D. L. Lambert, *Mon. Not. R. Astron. Soc.* **182**, 249 (1978).
81. M. J. Seaton, *Mon. Not. R. Astron. Soc.* **187**, 73P (1979).
82. J. Clavel, D. R. Flower, and M. J. Seaton, *Mon. Not. R. Astron. Soc.* **197**, 301 (1981).
83. J. P. Harrington, J. H. Lutz, and M. J. Seaton, *Mon. Not. R. Astron. Soc.* **195**, 21P (1981).
84. S. Torres-Peimbert and M. Peña *Rev. Mex. Astron. Astrofis.*, **6**, 301 (1981).
85. D. R. Flower, *Mon. Not. R. Astron. Soc.* **193**, 511 (1980).
86. M. J. Seaton, *Q. J. R. Astron. Soc.* **21**, 229 (1980).
87. D. R. Flower, A. Goharji, and M. Cohen, *Mon. Not. R. Astron. Soc.*, to be submitted.
88. D. Péquignot, S. M. V. Aldrovandi, and G. Stasinska, *Astron. Astrophys.* **63**, 313 (1978).
89. S. E. Butler, T. G. Heil and A. Dalgarno, *Astrophys. J.* **241**, 442 (1980).
90. S. E. Butler and A. Dalgarno, *Astrophys. J.* **234**, 765 (1979).
91. M. Gargaud, J. Hanssen, R. McCarroll, and P. Valiron, *J. Phys. B* **14**, 2259 (1981).
92. W. D. Watson and R. B. Christensen, *Astrophys. J.* **231**, 627 (1979).
93. G. B. Field and G. Steigman, *Astrophys. J.* **166**, 59 (1971).
94. R. McCarroll and P. Valiron, *Astron. Astrophys.* **53**, 83 (1976).
95. G. Chambaud, J. M. Launay, B. Levy, P. Millie, E. Roueff, and F. Tran Minh, *J. Phys. B* **13**, 4205 (1980).
96. R. E. Williams, *Mon. Not. R. Astron. Soc.* **164**, 111 (1973).
97. M. Perinotto, *Astron. Astrophys.* **61**, 247 (1977).
98. D. Péquignot, *Astron. Astrophys.* **83**, 52 (1980).
99. A. Burgess, *Astrophys. J.* **141**, 1588 (1965).
100. S. M. V. Aldrovandi and D. Péquignot, *Astron. Astrophys.* **25**, 137 (1973).
101. A. Burgess and M. J. Seaton, *Mon. Not. R. Astron. Soc.* **120**, 121 (1960).
102. G. Peach, *Mem. R. Astron. Soc.* **71**, 13 (1967).
103. D. R. Flower, in *Planetary Nebulae*, D. E. Osterbrock and C. R. O'Dell, Eds., IAU Symposium Number 34, Reidel, Dordrecht, 1968, pp. 205–208.

104. M. B. HIDALGO, *Astrophys. J.* **153**, 981 (1968).
105. A. BURGESS, *Mem. R. Astron. Soc.* **69**, 1 (1964).
106. D. R. FLOWER and M. J. SEATON, *Comput. Phys. Commun.* **1**, 31 (1969).
107. M. LE DOURNEUF, VO KY LAN, and A. HIBBERT, *J. Phys. B* **9**, L359 (1976).
108. M. LE DOURNEUF, VO KY LAN, and C. J. ZEIPPEN, *J. Phys. B* **12**, 2449 (1979).
109. C. J. ZEIPPEN, M. LE DOURNEUF, and VO KY LAN, *J. Phys. B* **13**, 3763 (1980).
110. P. G. BURKE and K. T. TAYLOR, *J. Phys. B* **12**, 2971 (1979).
111. K. T. TAYLOR and P. G. BURKE, *J. Phys. B* **9**, L353 (1976).
112. A. K. PRADHAN, *J. Phys. B* **12**, 3317 (1979).
113. A. K. PRADHAN, *Mon. Not. R. Astron. Soc.* **190**, 5P (1980).
114. C. B. TARTER, *Astrophys. J.* **168**, 313 (1971).
115. C. B. TARTER, *Astrophys. J.* **181**, 607 (1973).
116. R. J. GOULD, *Astrophys. J.* **219**, 250 (1978).
117. K. A. BERRINGTON, P. G. BURKE, P. L. DUFTON, and A. E. KINGSTON, *J. Phys. B* **10**, 1465 (1977).
118. P. L. DUFTON, K. A. BERRINGTON, P. G. BURKE, and A. E. KINGSTON, *Astron. Astrophys.* **62**, 111 (1978).
119. K. L. BALUJA, P. G. BURKE, and A. E. KINGSTON, *J. Phys. B* **14**, 119 (1981).
120. K. L. BALUJA, P. G. BURKE, and A. E. KINGSTON, *J. Phys. B* **13**, 4675 (1980).
121. K. L. BALUJA, P. G. BURKE, and A. E. KINGSTON, *J. Phys. B* **14**, 1333 (1981).
122. W. EISSNER, H. NUSSBAUMER, H. E. SARAPH, and M. J. SEATON, *J. Phys. B* **2**, 341 (1969).
123. W. EISSNER and M. J. SEATON, *J. Phys. B* **7**, 2533 (1974).
124. A. K. PRADHAN, *Mon. Not. R. Astron. Soc.* **177**, 31 (1976).
125. K. GILES, *Mon. Not. R. Astron. Soc.* **187**, 49P (1979).
126. K. GILES, *Mon. Not. R. Astron. Soc.* **195**, 63P (1981).
127. R. H. GARSTANG, in *Planetary Nebulae*, D. E. Osterbrock and C. R. O'Dell, Eds., IAU Symposium Number 34, Reidel, Dordrecht, 1968, pp. 143–152.
128. D. R. FLOWER and M. J. SEATON, *Mem. Soc. R. Sci. Liège, Sér. V*, **17**, 251 (1969).
129. M. J. SEATON and D. E. OSTERBROCK, *Astrophys. J.* **125**, 66 (1957).
130. C. J. ZEIPPEN, *J. Phys. B* **13**, L485 (1980).
131. W. EISSNER and C. J. ZEIPPEN, *J. Phys. B* **14**, 2125 (1981).
132. C. J. ZEIPPEN, *Mon. Not. R. Astron. Soc.*, **198**, 111 (1982).

10

FORBIDDEN ATOMIC LINES IN AURORAL SPECTRA

D. R. Bates

1. Introduction

By way of background information we give in Table 1 the origins of the forbidden multiplets of interest together with representative values of their intensities in an International Brightness Coefficient (IBC) 3 aurora; and we give in Table 2 the excitation energies and the Einstein A transition probabilities. In terms of the intensity of $\lambda5577$ Å, the IBC scale, which is logarithmic, runs from 1 kR for IBC 1 to 1000 kR for IBC 4. It is appropriate to mention here that the earliest calibration of the scale was made by Seaton.[1]

2. Beginnings

Vegard was the first to consider the auroral excitation processes. In a well known review article, written at a time when he thought that allowed O I and N I lines were absent, Vegard[3] not unreasonably dismissed excitation by electron impact as not being sufficiently selective. He took the central problem to be the explanation of how the $O(^1D)$ and $O(^1S)$ levels are selectively populated and suggested that excitation transfer, perhaps from $N_2(A\,^3\Sigma_u^+)$, is involved. Of the transfer processes that might be envisaged he rejected the most obvious

$$N_2(A\,^2\Sigma_u^+) + O(^3P) \rightarrow N_2 + O(^1S) \tag{1}$$

on the grounds that because the number density ratio $n(O)/n(N_2)$ increases with altitude according to Chapman[4] so also should the intensity ratio $I(5577)/I(N_2^+\,1NG)$, which would be in conflict with observation. Much

D. R. BATES ● Department of Applied Mathematics and Theoretical Physics, Queen's University of Belfast, Belfast BT7 1NN, Northern Ireland.

TABLE 1

Intensities of Multiplets in IBC 3 Aurora

Emitter	Wavelength (Å)	Intensity (kR)
$O(^1D)$	6300, 6364	50
$O(^1S)$	5577	100
$O(^1S)$	2972	5
$N(^2D)$	5199, 5201	2
$N(^2P)$	10395, 10404	100
$N(^2P)$	3466	6

stress was laid on an apparent slight decrease of the observed ratio. It is now accepted that in ordinary auroras $I(5577)/I(N_2^+ 1NG)$ is remarkably constant up to at least 200 km [see Omholt[5]].

Although allowed atomic lines were not identified with certitude until strong lines in the near infrared were observed by Meinel[6, 7] and Petrie[8] evidence for their presence began to accumulate much earlier.[9-11] Hence the basis for the original objection to invoking excitation by electron impact disappeared, and when Bates *et al.*[12] were examining possible auroral processes it was natural for them to regard

$$O(^3P) + e \to O(^1D, ^1S) + e \tag{2}$$

as likely to be important. They considered that secondary processes occurring at thermal energy would contribute significantly. In particular, they directed attention to dissociative recombination

$$O_2^+ + e \to O(^1D, ^1S) + O \tag{3}$$

TABLE 2

Forbidden O I and N I Multiplets

	Transition	Excitation energy	Wavelength (Å)	Transition probability A^a (s^{-1})
O I	$^1D \to ^3P$	1.96	6300, 6364	6.7×10^{-3}
	$^1S \to ^1D$	4.17	5577	1.34
	$^1S \to ^3P$	4.17	2972	6.7×10^{-2}
N I	$^2D \to ^4S$	2.37	5199, 5201	1.07×10^{-5}
	$^2P \to ^2D$	3.56	10395, 10404	7.8×10^{-2}
	$^2P \to ^4S$	3.56	3466	5.0×10^{-3}

^{a} Wiese *et al.*[2]

and argued that process (1) should be reinstated—the case against invoking it having no real substance.

Knowing that direct excitation (2) and dissociative recombination (3) *must* occur, and having little information on possible rival processes, it was almost inevitable that auroral scientists should proceed initially on the assumption that these rival processes might be neglected. The assumption proved to be wrong. This is not especially surprising in itself. What is rather surprising is the extent to which the wrong assumption seemed to lead, at more than one stage, to good accord with observation.

3. Seaton's Work

Seaton's research on the forbidden atomic lines in auroras was carried out over quite a brief period, 1953–1958, but together forms a key chapter in the history of the subject. They began with his classic study[13] of the Hartree–Fock equations for continuous states with application to the excitation of the ground configuration terms of O I by electron impact. This was followed by similar calculations on O II and N II[14] and by accurate estimates for N I.[15] Deactivation by an electron gas, of course, was also covered. The results that Seaton provided initiated the slow advance of auroral physicists from the promiscuity of qualitative reasoning to the discrimination of quantitative reasoning.

The current state of knowledge led Seaton[16] to propose that the most straightforward application of the results of his quantal calculations would be to the quenching of the forbidden lines. Following his convenient notation let 1, 2, and 3 denote the terms of the ground configuration in order of increasing excitation energy; let S_3 be the rate of entry into term 3 and S_2 be the rate of entry into term 2 other than by cascading or collisional deactivation from term 3; let d_n be the total rate of collisional deactivation experienced by an atom in term n and d_{32} be the specific rate of $3 \rightarrow 2$ deactivation; and affix subscripts to the Einstein coefficient A to indicate the particular transition concerned. The equations describing the steady state are then

$$(A_{32} + A_{31} + d_3)n(X_3) = S_3 \tag{4}$$

and

$$(A_{21} + d_2)n(X_2) = S_2 + (A_{32} + d_{32})n(X_3). \tag{5}$$

Combination of them gives

$$\frac{n(X_2)}{n(X_3)} = \left(\frac{A_{31} + A_{32} + d_3}{A_{21} + d_2} \right) \left(\frac{S_2}{S_3} + \frac{A_{32} + d_{32}}{A_{31} + A_{32} + d_3} \right). \tag{6}$$

The intensity of a multiplet $I(\lambda_{mn})$ may be expressed

$$I(\lambda_{mn}) = C(E_n - E_m)A_{nm}n(X_n), \tag{7}$$

where the E's are the term energies and C is a constant. Noting that A_{31} and A_{32} are in all instances much greater than A_{21}, Seaton neglected d_3 and d_{32} but retained d_2; thus on substituting numerically in Eqs. (6) and (7) he obtained

$$\frac{I(6300,64)}{I(5577)} = \frac{0.94}{1 + 110d_2} \left(\frac{S_2}{S_3} + 0.94 \right), \qquad \text{for} \quad \text{O I} \tag{8}$$

and

$$\frac{I(5200)}{I(3466)} = \frac{10}{1 + 9.6 \times 10^4 d_2} \left(\frac{S_2}{S_3} + 0.94 \right), \qquad \text{for} \quad \text{N I}. \tag{9}$$

He pointed out that deactivation must occur if $I(6300,64)/I(5577)$ and $I(5200)/I(3466)$ are less than 0.88 and 9.4, respectively. These values would be approximately doubled if S_2/S_3 were unity, that is, if terms 2 and 3 were populated at the same rate. Observations by Petrie and Small[17] showed that $I(6300,64)/I(5577)$ is 0.40 in the region between the base and the level of maximum luminosity of steady homogeneous arcs (100–110 km) and observations by Vegard and Kvifte[11] showed that $I(5200)$ $/I(3466)$ is around 0.5. Seaton[16] deduced that quenching is therefore important. On the assumption that it is caused by electron collisions he sought to obtain information on the electron density in auroras with the aid of his calculated deactivation coefficients. However, as he came to recognize,[18] the assumption is incorrect. This is signaled by the failure of observers to report a dependence of the intensity ratio on the absolute brightness of the aurora. Again, the $\lambda 5577$ nightglow emission from near 100 km is not accompanied by $\lambda\lambda 6300, 64$ emission from the same region, showing that quenching of the red doublet is quite effective even when the electron density is low.[18] Deactivation in collisions with neutral molecules can be much more rapid than aeronomists had expected from chemical kinetics. The transformation that has taken place is especially striking in the case of spin-forbidden reactions like $O(^1D)$–N_2 deactivation. These were not understood until comparatively recently.[19, 20]

Seaton[16] believed the principal source of $\lambda5577$ and $\lambda\lambda6300,64$ to be excitation by electron impact,[2] remarking that it is difficult to see what other mechanism could be operative in the atmospheric region where oxygen is mainly in the atomic form. He recalled that Meinel[21] had found that the intensities of these lines fall off more slowly with decreasing altitude below the level of the peak luminosity than does the N_2^+ 1NG intensity and attributed the effect to an additional source. Dissociative recombination (3) seemed to him the most likely possibility. He presented a plausible case that the observed intensities could be accounted for by the combination of the two sources. In his treatment he adopted the crude model of the auroral electrons introduced by Bates[22] who divided these electrons into two groups: *active* electrons, which have enough energy to excite or ionize the atmospheric constitutents, and *passive* electrons, which have only thermal energy. The source of the N_2^+ 1NG system is[23]

$$N_2 + e \rightarrow N_2^+ (B\,^2\Sigma_u^+) + 2e. \tag{10}$$

Seaton therefore took $q(eA)$, the rate of production of active electrons (by the primaries), to be at least equal to $P(1NG)$, the volume rate of photon emission of this system. From the cross sections involved he reckoned that the number of forbidden line excitations during the thermalization of an active electron is of the order of unity. Because atmospheric extinction at short wavelengths is greatly overestimated he believed that $P(1NG)$ $/P(5577)$ was about 10 and was thus confident that process (2) is an ample source. It is now known that

$$P(1NG)/P(5577) = 1.5, \tag{11}$$

[see Vallance Jones[24]] but compensating this smaller ratio it has been calculated that

$$q(eA)/P(1NG) = 50/3 \tag{12}$$

[see Omholt[5]]. In order to estimate the contribution from (3), Seaton supposed that the average distribution of the resulting two oxygen atoms among the ground configuration terms to be proportional to their statistical weights; thus he took $f(^1S)$, the number of ^1S atoms per recombination, to be $2/15$. Assuming that O_2 and N_2 have equal ionization probabilities, he inferred that in the steady state dissociative recombination (3) gives

$$\frac{P(5577)}{P(1NG)} = \frac{2}{15} \frac{q(eA)}{P(1NG)} \frac{n(O_2)}{n(N_2)} . \tag{13}$$

He concluded that this process may account for practically the whole emission at the lower border of auroras.

Turning to the forbidden nitrogen lines, Seaton,[16] with the aid of his cross sections, proved that direct excitation

$$N + e \rightarrow N(^2D, {}^2P) + e \tag{14}$$

would require the degree of dissociation of the nitrogen in the auroral region to be almost complete, which he rightly judged to be quite unacceptable. He reckoned dissociative recombination,

$$N_2^+ + e \rightarrow N(^2D, {}^2P) + N, \tag{15}$$

must be the process involved.

The sources that Seaton favored—direct excitation and dissociative recombination—are often referred to as the classical (or historical) sources.

4. Refinement of Classical Theory

Chamberlain[25] recognized that the emission rate ratio $P(5577)$ $/P(1NG)$ presented a problem: it would be far greater than observed unless the active electrons had a distribution with a maximum at an energy much higher than 4 eV. He suggested that a distribution function which Omholt[26] had calculated could be modified in the desired manner if electron exchange made the e–N_2 excitation cross sections peak closer to threshold than assumed.

In the same context Dalgarno[27] pointed to another degradation process: electrons of energy $E(eV)$ lose energy in elastic collisions with ambient thermal electrons at a rate

$$\frac{dE}{dx} = -\frac{2 \times 10^{-12}}{E} n(e) \qquad \text{eV cm}^{-1}, \tag{16}$$

whereas they lose energy in exciting $O(^1S)$ at a rate

$$\frac{dE}{dx} = -2 \times 10^{-17} n(O) \qquad \text{eV cm}^{-1} \tag{17}$$

[Dalgarno et al.[28]]. Detailed calculations were carried out by Dalgarno and Khare[29] who also improved the treatment of the contribution from dissociative recombination by computing $n(O_2^+)$ using the conventional reaction scheme [Donahue[30]] which incorporates many thermal processes including

$$N_2^+ + O_2 \rightarrow N_2 + O_2^+ \tag{18}$$

and

$$O^+ + O_2 \rightarrow O + O_2^+. \tag{19}$$

The near invariance of $P(5577)/P(3914)$ which had been confirmed by rocket measurements [Cummings et al.[31]] was convincingly reproduced for IBC 1 to 4 auroras, and moreover the deduced magnitude of this ratio is approximately correct.

Further calculations were done by Rees et al.[32] on an aurora for which photographic triangulation from two ground stations had provided well determined altitude profiles. These calculations covered $\lambda6300$ and took into account excitation by the ambient electron gas the temperature of which was raised above that of the neutral gas. Allowance was also made for deactivation

$$O(^1D) + N_2 \rightarrow O(^3P) + N_2 \tag{20}$$

with β_{20} set at 3×10^{-11} cm^3 s^{-1}—a value deduced by Hunten and McElroy[33] from an analysis of observational data. The results, which are in fair agreement with observations, are given in Table 3. According to them, direct excitation and dissociative recombination are the main sources of the green line in the respective regions below and above about 200 km; the predicted green line emission is about half that measured. In the case of the red line, direct excitation predominates, the importance of the ambient electron gas relative to the secondary electrons increasing with altitude. Due to quenching at low altitudes and diffusion at high altitudes the emission rates are less than the $O(^1D)$ excitation rates. The diffusion within the lifetime of the metastable atom broadens the $\lambda6300$ profile but not to the extent that is observed. Rees et al.[32] suggested that motion of the arc may have been responsible for the effect.

The consensus on the role of impact excitation was broken by Romick and Belon[34] and by Murcray.[35] They observed that the horizontal volume emission rate profiles of $\lambda5577$ and $\lambda3914$ [the N_2^+ 1NG (0, 0) band] differ, the former being concentrated to the center of the latter. There is no doubt that source (10) is responsible for the $\lambda3914$ emission. The nonlinear (nearly square) relationship between the horizontal emission rate profiles was therefore judged to imply that $\lambda5577$ arises from a secondary process in which both interacting systems are auroral products. Romick and Belon mentioned dissociative recombination as a possibility. However, as they

TABLE 3

Green and Red Line Excitation and Volume Emission Rates [a]

| Altitude (km) | $\lambda5577$ Rates[b] (cm^{-3} s^{-1}) | | | $O + e$ | | $\lambda6300$ Rates (cm^{-3} s^{-1}) | | | |
	$O + e$	$O_2^+ + e$	$P(5577)$	Secondary electrons	Thermal electrons	$O_2^+ + e$	$^1S \to {}^1D$	Total	$P(6300)$
120	1.3^4	5.2^3	1.8^4	4.8^5	2.6^{-2}	2.6^4	1.8^4	5.2^5	4.0^2
130	9.7^3	3.8^3	1.3^4	3.7^5	2.7^1	1.9^4	1.3^4	4.0^5	7.3^3
140	7.2^3	2.8^3	9.9^3	2.7^5	3.2^2	1.4^4	9.9^3	3.0^5	1.1^3
150	5.3^3	2.1^3	7.4^3	2.0^5	1.3^3	1.0^4	7.4^3	2.2^5	1.5^3
160	4.0^3	1.7^3	5.7^3	1.5^5	2.6^3	8.4^3	5.7^3	1.7^5	1.9^3
170	3.0^3	1.3^3	4.3^3	1.1^5	3.7^3	6.8^3	4.3^3	1.3^5	2.1^3
180	2.2^3	1.1^3	3.3^3	8.2^4	4.3^3	5.6^3	3.3^3	9.5^4	2.2^3
190	1.5^3	9.2^2	2.5^3	5.8^4	4.9^3	4.6^3	2.5^3	7.0^4	2.2^3
200	1.1^3	7.7^2	1.8^3	4.1^4	5.6^3	3.8^3	1.8^3	5.1^4	2.1^3
250	1.6^2	3.0^2	4.6^2	5.8^3	5.4^3	1.5^3	4.6^2	1.3^4	1.1^3
300	3.5^1	8.0^1	1.1^2	1.3^3	4.3^3	4.0^2	1.1^2	6.0^3	5.8^2
350	—	—	—	3.5^2	3.2^3	9.0^1	—	3.6^3	3.1^2
400	—	—	—	1.1^2	2.0^3	1.9^1	—	2.1^3	1.4^2
500	—	—	—	1.2^1	6.7^2	—	—	6.8^2	2.3^1

[a] Rees *et al.*[32]
[b] Notation 1.3^4 means 1.3×10^4.

recalled, Omholt[26] has objected to dissociative recombination as a major source because the $\lambda5577$ emission follows rapid variations in the $\lambda3914$ emission with a delay no longer than would be expected from the $O(^1S)$ radiative lifetime (0.75 s).

5. Advent of In Situ Measurements

Analysis of photometric and electron flux data obtained during two sounding rocket auroral experiments led Donahue *et al.*[36] to the conclusion that direct excitation (2) is responsible for only 10% of the green line emission. Using the calculated steady-state value of $n(O_2^+)$ they inferred that if dissociative recombination (3) had a branching ratio $f(^1S)$ for $O(^1S)$ production equal to 0.2 it would be a strong enough source. However, 0.2 is a high value for $f(^1S)$ and, furthermore, later composition measurements with a rocket-borne mass spectrometer [Donahue *et al.*[37]] revealed $n(O_2^+)$ to be a factor 10 less than had been calculated from the ion chemistry so that dissociative recombination was precluded from being a major source of green line emission except in the upper part of the aurora [Parkinson *et al.*[38]]. Similar composition results, manifest as a high $n(NO^+)/n(O_2^+)$ ratio, were obtained and discussed by Swider and Narcisi[39] and have been discussed further by Jones and Rees.[40] Odd nitrogen atoms (N and NO) have to be taken into account.

Little seemed to be left of the classical theory of the green line. Starting almost afresh Parkinson et al.[38] proposed the dissociative excitation process

$$O_2 + e \rightarrow O(^1S) + O. \tag{21}$$

A cross section as large as 10^{-16} cm^2 is required, whereas the measured cross section is less than 2×10^{-17} cm^2 [Zipf[41]]. Because of this the hypothesis did not arouse sustained interest. Zipf[42] proposed instead that the intense eUV radiation, which Stolarski and Green[43] had predicted would ensue from precipitating electrons impacting on N$_2$, would contribute to the green line through photodissociative excitation

$$O_2 + h\nu \rightarrow O(^1S) + O \tag{22}$$

[Lawrence and McEwan[44]]. This hypothesis had to be abandoned when *in situ* measurements [Park et al.[45]] showed that the total eUV intensity is small compared with that of the green line because of radiation entrapment and predissociation [Zipf and McLaughlin[46]].

Meanwhile, the work of theorists [Judge,[47] Rees and Luckey,[48] and Roble and Rees[49]] emphasized that the relative contributions from the various sources is not invariant from aurora to aurora. In particular, the contribution from direct excitation (2) to the green line is greatest for a soft (0.1 keV) primary electron spectrum. It is then of major importance but is not if the primary electron spectrum is medium (1 keV) or hard (10 keV). This explains at least some of the apparent discrepancies in the results.

6. $N_2(A\,^3\Sigma_u^+)$–O Excitation Transfer

The excitation transfer process (1) remained speculative until Meyer et al.[50] measured $\lambda5577$ emission in an apparatus that coupled two discharge-flow systems—one producing $A\,^3\Sigma_u^+$ nitrogen molecules and the other producing ground-state oxygen atoms. The rate coefficient for

$$N_2(A\,^3\Sigma_u^+) + O \rightarrow N_2(X\,^1\Sigma_g^+) + O, \tag{23}$$

which includes (1) with the branching ratio $f(^1S)$ judged to lie between 0.16 and 0.5, was found by Meyer et al.[51] to be about 3.5 times that for

$$N_2(A\,^3\Sigma_u^+) + O_2 \rightarrow N_2(X\,^1\Sigma_g^+) + O_2, \tag{24}$$

which is 4×10^{-12} cm^3 s^{-1} [Young et al.[52], Callear and Wood[53], and Meyer et al.[54]]. The corresponding value of β_{23} is thus 1.4×10^{-11} cm^3 s^{-1}. Deactivation by nitrogen,

$$N_2(A\,^3\Sigma_u^+) + N_2 \rightarrow N_2 + N_2, \tag{25}$$

is relatively slow with β_{25} less than 10^{-18} cm^3 s^{-1} [Noxon[55]]. The radiative transition probability for $N_2(A\,^3\Sigma_u^+)$ is about 0.5 s^{-1} [Shemansky[56]], making quenching in auroras important.

The system intensity of the N$_2$1PS system gives the rate of formation of $N_2(A\,^3\Sigma_u^+)$ by cascading alone. In an IBC 3 aurora this intensity is 880 kR, while the system intensity of N$_2$VK is 55 kR [Vallance Jones[24]]. The result of the experiment of Meyer et al.[50] therefore makes it seem likely that (1) gives a significant contribution to the 5577-Å emission (the intensity of which is 100 kR in a IBC 3 aurora, Table 1).

There have been a number of determinations of β_{23} from auroral data. Beiting and Feldman[57] have collected together the values obtained which range from 1.7×10^{-11} cm^3 s^{-1} to as high as 4×10^{-10} cm^3 s^{-1}. If β_{23} is indeed much greater than 1.4×10^{-11} cm^3 s^{-1}, then presumably Meyer et al.[50,51] must have seriously overestimated the amount of atomic oxygen in their apparatus. Their determination of the branching ratio $f(^1S)$ would not be affected by this. It is $f(^1S)$ rather than β_{23} which controls the contribution of (1) to the green line emission. Thus, because (23) is the main $N_2(A\,^3\Sigma_u^+)$ sink, a steady state is established with

$$\beta_{23}n(N_2,A\,^3\Sigma_u^+)n(O) = F(N_2,A\,^3\Sigma_u^+), \tag{26}$$

$F(N_2,A\,^3\Sigma_u^+)$ being the volume rate of formation of $N_2(A\,^3\Sigma_u^+)$; and consequently the contribution of (1) to the volume rate of formation of O(^1S) is

$$F_{VK}(O\,^1S) = f(^1S)F(N_2,A\,^3\Sigma_u^+). \tag{27}$$

Clearly β_{23} influences the effective lifetime of $N_2(A\,^3\Sigma_u^+)$ molecules and, therefore, how rapidly $F_{VK}(O\,^1S)$ can change. It was initially taken to be 1×10^{-11} cm^3 s^{-1} [Parkinson and Zipf[58] and Parkinson[59]]. The results obtained did not encourage acceptance of (1) as a major source of the green line. However, later calculations [Henriksen[60] and Brekke[61]] showed that there would be no difficulty if β_{23} were 1×10^{-10} cm^3 s^{-1}.

Recent measurements[98] show that for vibrational levels 0 and 1, β_{23} is 2.8 and 3.4×10^{-11} cm^3 s^{-1}; and that $F(^1S)$ is 0.75.

Attention has been drawn to the possibility of another excitation transfer process

$$O_2(x) + O \rightarrow O_2 + O(^1S), \tag{28}$$

where x represents the $c\,^1\Sigma_u^-$ state [Solheim and Llewellyn[62]] or the $A\,^3\Sigma_u^+$ or $C\,^3\Delta_u$ states [Yau and Shepherd[63]]. In favor of $O_2(x)$ being formed by electron impact it was recalled that measurements by Trajmar et al.[64] on energy loss in molecular oxygen showed a peak at 6.1 eV. Gattinger and Vallance Jones[65] have discounted the relevance of this, pointing out that almost all the energy loss occurs above the dissociation limit at 5.1 eV.

7. Quenching

Over the decade following 1966 the generally accepted rate coefficient for

$$O(^1S) + O(^3P) \rightarrow O(^3P) + O(^3P,^1D) \tag{29}$$

increased 100-fold to become

$$\beta_{29} = 5 \times 10^{-11}\exp(-305/T)\ \text{cm}^3\ \text{s}^{-1} \tag{30}$$

[Slanger and Black[66]] so that (29) was considered to be significant in auroras. A dramatic reversal then occurred. Slanger and Black[67] announced that the laboratory studies of (29) were completely vitiated by the rapid deactivation process

$$O(^1S) + O_2(a\,^1\Delta_g) \rightarrow \text{products} \tag{31}$$

[see also Slanger and Black[68]], and that contrary to what had been supposed there was no experimental evidence against β_{29} being more than 2×10^{-14} cm^3 s^{-1}—the upper limit set by the calculations of Krauss and

Neumann.[69] This leaves

$$O(^1S) + O_2 \rightarrow O + O_2 \tag{32}$$

as the main quenching process for the green line. Its rate coefficient β_{32} is only $4 \times 10^{-12}\exp(-865/T)$ cm^3 s^{-1} [Slanger et al.[70]] so that there is little quenching above 95 km.

The red doublet $\lambda\lambda6300, 64$ is quenched by

$$O(^1D) + N_2 \rightarrow O + N_2. \tag{33}$$

Measurements by Streit et al.[71] show that the rate coefficient decreases with increasing temperature: thus

$$\beta_{33} = 2.0 \times 10^{-11}\exp(107 \text{ K}/T) \text{ cm}^3 \text{ s}^{-1},$$

$$\tag{34}$$

$$104 \text{ K} < T < 354 \text{ K}.$$

In good agreement with (34) a determination by Aminoto et al.[72] gives β_{33} to be 2.4×10^{-11} cm^3 s^{-1} at 295 K. A value of 3×10^{-11} cm^3 s^{-1} [the same as the early estimate of Hunten and McElroy[33]] was deduced from a series of measurements using Atmosphere Explorer [Hays et al.[73]].

Deactivation of $N(^2D)$ may take place in collisons with atomic oxygen

$$N(^2D) + O(^3P) \rightarrow N(^4S) + O(^3P) \tag{35}$$

or in reactive collisions with molecular oxygen

$$N(^2D) + O_2 \rightarrow NO + O \tag{36}$$

[Lin and Kaufman[74]]. A laboratory measurement by Davenport et al.[75] gives β_{35} to be $(1.8 \pm 0.6) \times 10^{-12}$ cm^3 s^{-1} at 300 K with an activation energy of (1 ± 0.5)kcal/mol, but a smaller value of 4×10^{-13} cm^3 s^{-1} at 800 K has been deduced by Frederick and Rusch[76] from $\lambda5200$ nightglow studies using Atmosphere Explorer. There have been several determinations of β_{36} at 300 K: $(6 \pm 2) \times 10^{-12}$ cm^3 s^{-1} [Lin and Kaufman[74]], $(7.4 \pm 0.7) \times 10^{-12}$ cm^3 s^{-1} [Slanger et al.[77]] and $(5.2 \pm 0.4) \times 10^{-12}$ cm^3 s^{-1} [Husain et al.[78]]. Slanger et al.[77] find that the activation energy is zero.

The main $N(^2P)$ sink is

$$N(^2P) + O \rightarrow N(^4S,^2D) + O(^3P,^1D). \qquad (37)$$

Golde and Thrush[79] have reported a provisional measurement of β_{27}: 7×10^{-11} cm^3 s^{-1} at 300 K. A smaller value, 1×10^{-11} cm^3 s^{-1} has been obtained by Young and Dunn.[80] Reactive deactivation

$$N(^2P) + O \rightarrow NO + O(^3P,^1D) \qquad (38)$$

is relatively unimportant, β_{38} being 2.6×10^{-12} cm^3 s^{-1} [Husain et al.[78]].

8. Coordinated Rocket and Satellite Measurements

The most notable series of coordinated measurements in the history of auroral science was carried out on an IBC 1 aurora occurring on March 20, 1974 [Rees et al.[81]]. They were made from an Atmosphere Explorer satellite, a rocket, and a ground station. Information was obtained on the neutral density $n(N_2), n(O)$, the primary electron flux (0.2–25 keV), the secondary electron flux (0.2–500 eV), the electron density, the electron temperature, the positive ion densities, and the $\lambda 3914$ (N_2^+ 1NG), $\lambda 3220$ (N_2VK), $\lambda 3371$ (N_2PS), $\lambda 5577$, and $\lambda 6300$ altitude profiles. A major objective was to test auroral models by determining all relevant parameters simultaneously. As far as $\lambda 5577$ and $\lambda 6300$ are concerned it was concluded that the classical processes (Section 4) suffice. Thus in the case of $\lambda 5577$ the measured intensity was 4.5 kR, while electron impact (2) was calculated to give 1.6 kR and dissociative recombination (3) to give 5.5 kR [so that the N_2(A)–O excitation transfer process (23) did not seem to be required, although, as was noted, it would give to 3.4 kR if $f(^1S)$ were as high as unity]. In the case of $\lambda 6300$ the measured intensity was 0.69 kR, while electron impact was calculated to give 0.59 kR and dissociative recombination (3) to give 0.1 kR. The measurements had shown the primary electron spectrum to be very hard so that (Section 5) this reinstatement of the classical processes was puzzling. However, it was ephemeral in that it was due to the use of calculated values of $n(e)$, $n(O_2^+)$, and the secondary electron flux. The measured value of $n(O_2^+)$ is much less [Sharp et al.[82]]. In order to account for the discrepancy, a large amount of nitric oxide was invoked within the auroral arc. Again, the measured electron flux below 10 eV is also much less than that calculated. This difference was not explained.

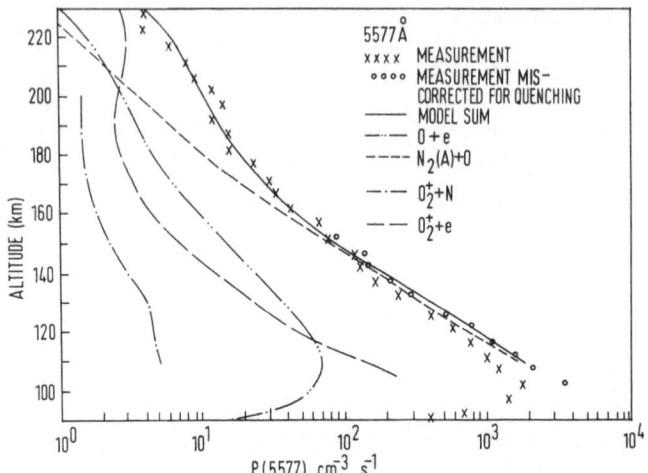

FIGURE 1. Comparison of green line volume emission rate deduced from photometric measurements and calculated from model. [After Sharpe et al.[82]].

Sharp et al.[82] gave further consideration to the excitation processes in the aurora of March 20, 1974 with emphasis on the altitude profiles. Figure 1 summarizes their results on the green line. Excitation transfer (1) from $N_2(A)$ is the dominant source up to 160 km and the greatest source up to 200 km; above this dissociative recombination (3) is the greatest source. Electron impact excitation (2) gives only a minor contribution, while

$$N + O_2^+ \rightarrow NO^+ + O(^1S) \tag{39}$$

which was introduced in connection with the dayglow [Frederick et al.[83]] is quite inappreciable. When correcting the measured volume emission rate for quenching the erroneous high rate coefficient (30) was used. The true correction is negligible. In consequence, the proposed total source function is too strong at low altitudes.

Sharp and Torr[84] analyzed the data from a somewhat different viewpoint. They subtracted the sum of the contributions from impact excitation (2), dissociative recombination (3), and the ion–molecule reaction (39) from the apparent green line to obtain a residual source function $F_r(5577)$. They then sought to identify the source concerned from the altitude profile of the ratio $F_r(5577)/P(3914)$, where $P(3914)$ is the volume emission rate of the N_2^+ 1NG (0, 0) band. Figure 2 shows the derived profile and also the profiles for the $N_2(A)$ excitation transfer process (1) and for

$$N(^2D) + NO \rightarrow N_2 + O(^1S) \tag{40}$$

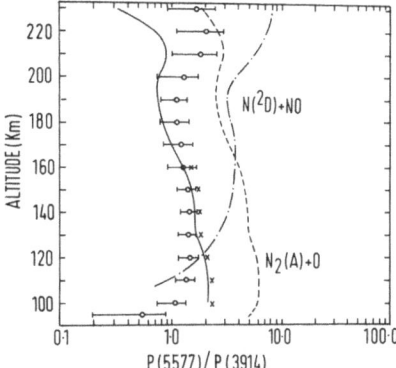

FIGURE 2. Apparent missing green line source function (full line and see text) and two possible source functions: O, original data points; ×, data points adjusted to allow for (29) by use of (30) which Slanger and Black[67] have since shown to be much too high. [After Sharp and Torr.[84]]

which Rusch et al.[85] introduced as a possibility. Black et al.[86] measured the rate coefficient β_{40} to be $1.8 \times 10^{-10} f(^1S)$ cm^3 s^{-1}, where $f(^1S)$ is the unknown branching ratio. Sharp and Torr[84] judged from Fig. 2 (in essence, the same as Fig. 1) that the $N_2(A)$ excitation transfer process (1) meets the profile criterion satisfactorily. However, they were making comparison with a residual curve calculated with overcorrection for quenching due to (30) being used again.

The $P(5577)/P(3914)$ data points may be fitted accurately (see Fig. 3) by including a significant contribution from (40) and a reduced contribution from (1) 0.1 and 0.2 times the respective contributions depicted in Fig. 2, and by taking the $f(^1S)$ branching ratio in dissociative recombination to be 0.05 instead of the original 0.08 [see Bates and Zipf[87]]. These are acceptable adjustments.

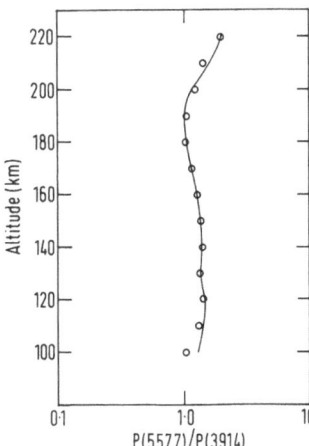

FIGURE 3. Comparison of green line volume emission rate deduced from photometric measurements (O) and calculated from model (———) (see text).

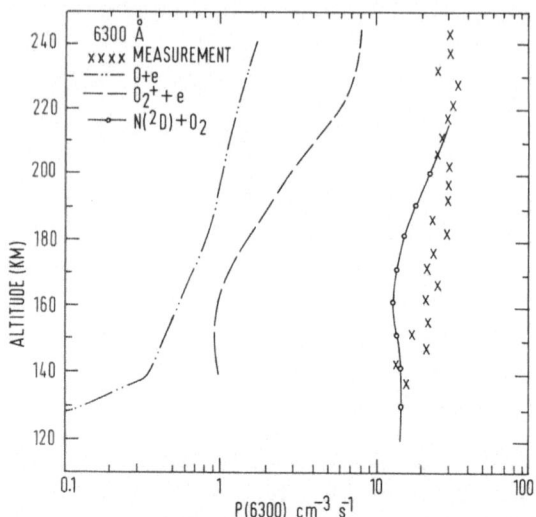

FIGURE 4. Comparison of red line volume emission rate deduced from photometric measurements and calculated from model. [After Rusch et al.[88]]

When they turned their attention to λ6300 Sharp et al.[82] discovered a truly surprising deficiency: the sum of the contributions from impact excitation (2) and dissociative recombination (3) falls far short of what is required.

A proposal that the missing source is

$$N(^2D) + O_2 \rightarrow NO + O(^1D) \tag{41}$$

soon came from Rusch et al.[88] Measurements (Section 7) have given β_{41} to be around $6 \times 10^{-12} f(^1D)$ cm^3 s^{-1} but have left the branching ratio $f(^1D)$ undetermined. Rusch et al.[88] suggested that $f(^1D)$ may be nearly unity and showed that the auroral results would then be explicable (Fig. 4). An understanding of the $N(^2D)$ problem was naturally entailed.

Impact excitation of atomic nitrogen may be dismissed as inappreciable [Seaton[16]]. When the high abundance of NO$^+$ ions was discovered it became evident that the dissociative recombination process concerned would not be (15) but

$$NO^+ + e \rightarrow N(^2D) + O, \qquad f(^2D) = 0.8. \tag{42}$$

In their study of the dayglow, Wallace and McElroy[89] considered also

$$N_2^+ + O \rightarrow N(^2D) + NO^+, \qquad f(^2D) = 1. \tag{43}$$

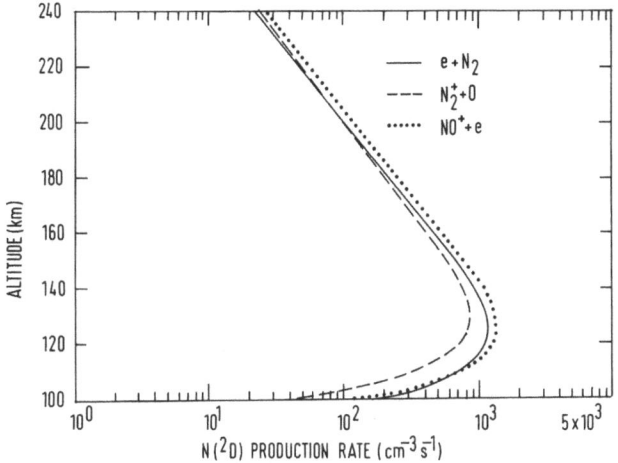

FIGURE 5. Calculated $N(^2D)$ production rates. [After Rusch and Gérard.[91]]

While investigating the diurnal variation of the nitric oxide in the atmosphere Strobel[90] saw the need to introduce

$$N_2 + e \rightarrow N(^2D) + N, \qquad f(^2D) = 0.8. \qquad (44)$$

It is now believed that (42), (43), and (44) are the main λ5200 dayglow sources [see Frederick and Rusch[76]]. The dayglow work provided a sound basis for the auroral calculations of Rusch *et al.*[88] Taking the $f(^2D)$ branching ratio to be as indicated in each equation it is found [Rusch and Gérard[91]] that the contributions from (42), (43), and (44) are nearly equal (Fig. 5).

The λ5200 and λλ6300,64 auroral emissions are closely related. Equating the rate at which $O(^1D)$ is formed by (41) to the rate at which it is removed by (33) and by radiative transitions gives

$$n(N,^2D)n(O_2)\beta_{41} = n(O\,^1D)\big[n(N_2)\beta_{33} + A(6300,64)\big] \qquad (45)$$

and hence that

$$\frac{P(6300,64)}{P(5200)} = \frac{A(6300,64)n(O_2)\beta_{41}}{A(5200)\big[n(N_2)\beta_{33} + A(6300,64)\big]} \qquad (46)$$

$$\approx 1.2 \times 10^2 n(O_2)/n(N_2) \qquad (47)$$

[Rusch *et al.*[88]]. This rationalizes neatly the observation by Gérard and

Harang[92] that under steady auroral conditions the $\lambda\lambda6300,64$ to $\lambda5200$ intensity ratio is about 20, irrespective of the incident energy flux.

Rusch et al.[88] pointed out that the chemical lifetime of $N(^2D)$ is long, 200 s at 180 km, and that due to diffusion (41) is therefore a source of $O(^1D)$ which is operative at a considerable distance from the center of an auroral arc. They advanced this as the explanation of broadening of the spatial profile of the $\lambda\lambda6300,64$ emission (Section 4).

Contemporaneously and consistent with the studies on the data obtained from the March 20, 1974 aurora, Arnoldy and Lewis[93] made extensive in situ measurements on two auroras. Their primary electron spectra were quite hard. Arnoldy and Lewis[93] confirmed the close correlation between the $\lambda5577$ and $\lambda3914$ emissions. They found also that the former could not be accounted for by direct excitation (2) and dissociative recombination (3)—these together giving only about a third of the observed intensity. The contributions from the other suggested sources were not determined.

9. $\lambda3466$ and $\lambda10,400$ of N I

The $\lambda3466$ and $\lambda10,400$ emissions have received little attention until recently. Ignoring the endothermic process

$$N^{O+} + e \rightarrow N(^2P) + O, \tag{48}$$

Rees and Jones[94] took the sources to be

$$N_2^+ + e \rightarrow N(^2P) + N \tag{49}$$

and

$$N_2 + e \rightarrow N(^2P) + N^+ + 2e. \tag{50}$$

The model seemed to meet with some success. However, it is incorrect. This was not recognized initially because there was a serious delay before two laboratory results came to the notice of auroral scientists. These results are firstly that $N(^2P)$–O deactivation (37) is quite rapid [Golde and Thrush[79] and Young and Dunn[80]] and secondly that the cross section for dissociative ionization through the $N(^2P)$ channel (50) is very small [Ehrhardt and Kresling[95]].

Another source,

$$N_2 + e \rightarrow N(^2P) + N + e, \tag{51}$$

was introduced by Zipf et al.[96] and Gérard and Harang.[97]

Having uncovered the overlooked work of Ehrhardt and Kresling[95]

TABLE 4
$N(^2P)$ Production and Loss[a]

	Production[b] (cm^{-3}s^{-1})		Loss[b] (s^{-1})	
Altitude	$N_2 + e$	$N_2^+ + e$	$N(^2P) + O$	$N(^2P) + O_2$
100	6.4^3	1.6^0	5.0^1	5.2^0
120	2.9^4	2.7^2	7.6^0	2.0^{-1}
140	1.0^4	3.5^2	2.2^0	2.5^{-2}
160	3.7^3	2.4^2	9.8^{-1}	6.8^{-3}
180	1.5^3	1.4^2	5.4^{-1}	2.3^{-3}
200	6.7^2	7.4^1	3.2^{-1}	9.4^{-4}
250	1.3^2	1.5^1	1.1^{-1}	1.4^{-4}

[a] Gérard and Harang.[97]
[b] Notation 6.4^3 means 6.4×10^3.

on (50), Zipf *et al.* carried out an analysis of rocket data of λ3466 N I and and λ3914 N_2^+ 1NG taking into account the results of contemporary laboratory studies of e–N_2 collisions and N_2^+ dissociative recombination. They concluded that (51) is the dominant source. Gérard and Harang[87] reached the same conclusion from measurements on the intensities of λ3466 and λ3426 $N_2(1, 10)$ VK which showed the ratio of these intensities to be independent of the brightness of the aurora. The conclusions follows because the VK system is excited in e–N_2 collisions at a rate proportional to that of (51) and is quenched in $N_2(A)$–O collisions. Table 4 shows the position as deduced by Gérard and Harang.[97] It remains possible that there is another significant production process at high altitudes. Zipf *et al.*[96] have tentatively suggested

$$N_2^+(v' \geqslant 2) + O \to N(^2P) + NO^+. \tag{52}$$

The process has not been studied in the laboratory.

ACKNOWLEDGMENT

I am indebted to Dr. M. H. Rees and Dr. A. I. F. Stewart for helpful correspondence.

References

1. M. J. SEATON, *J. Atmos. Terr. Phys.* **4**, 285 (1954).
2. W. L. WIESE, M. W. SMITH, and B. M. GLENNON, *Atomic Transition Probabilities*, Vol. I, *Hydrogen through Neon*, National Bureau of Standards Data Series, Washington D.C., U.S. Government Printing Office, 1966.

3. L. VEGARD, in *Terrestrial Magnetism and Electricity*, J. A. Fleming, Ed., McGraw-Hill, New York, 1939, Chap. 11.
4. S. CHAPMAN, *Proc. R. Soc. Lond. Ser. A* **132**, 353 (1931).
5. A. OMHOLT, *The Optical Aurora*, Springer-Verlag, Berlin, 1971, Sections 2.2 and 4.2.2.
6. A. B. MEINEL, *Publ. Astron. Soc. Pacific* **60**, 373 (1948).
7. A. B. MEINEL, *Astrophys. J.* **113**, 583 (1951).
8. W. PETRIE, *J. Geophys. Res.* **55**, 143 (1950).
9. L. VEGARD, *Geophys. Publ.* **12**, No. 8 (1938).
10. J. DUFAY and MAO-LIN TCHENG, *Cahiers Phys.* **2** (8), 51 (1942).
11. L. VEGARD and G. KVIFTE, *Geofys. Publ.* **16**, No. 7 (1945).
12. D. R. BATES, H. S. W. MASSEY, and R. W. B. PEARCH, in *Emission Spectra of the Night Sky and Aurora* (Reports of the Gassiot Committee) The Physical Society, London, 1948, p. 97.
13. M. J. SEATON, *Phil. Trans. R. Soc. Lond. Ser. A* **245**, 469 (1953).
14. M. J. SEATON, *Proc. R. Soc. Lond. Ser. A* **218**, 400 (1953).
15. M. J. SEATON, in *The Airglow and Aurorae*, E. B. Armstrong and A. Dalgarno, Eds., Pergamon, London, 1956, p. 289.
16. M. J. SEATON, *J. Atmos. Terr. Phys.* **4**, 295 (1954).
17. W. PETRIE and R. SMALL, *Can. J. Phys.* **31**, 911 (1953).
18. M. J. SEATON, *Astrophys. J.* **127**, 67 (1958).
19. J. C. TULLY, *J. Chem. Phys.* **61**, 61 (1974).
20. G. E. ZAHR, R. K. PRESTON, and W. H. MILLER, *J. Chem. Phys.* **62**, 1127 (1975).
21. A. B. MEINEL, *Mem. Soc. R. Sci. Liège* **12**, 203 (1952).
22. D. R. BATES, *Proc. R. Soc. Lond. Ser. A* **196**, 217 (1949).
23. D. R. BATES, *Proc. R. Soc. Lond. Ser. A* **196**, 562 (1949).
24. A. VALLANCE JONES, *Aurora*, Reidel, Dordrecht, 1974, p. 88.
25. J. W. CHAMBERLAIN, *Physics of the Aurora and Airglow*, Academic, New York, 1961, p. 297.
26. A. OMHOLT, *Geofys. Publ.* **20**, No. 5 (1959).
27. A. DALGARNO, *Ann. Geophys.* **20**, 65 (1964).
28. A. DALGARNO, M. B. McELROY, and R. J. MOFFETT, *Planet. Space Sci.* **11**, 463 (1963).
29. A. DALGARNO and S. P. KHARE, *Planet. Space Sci.* **15**, 938 (1967).
30. T. M. DONAHUE, *Planet. Space Sci.* **14**, 33 (1966).
31. W. D. CUMMINGS, R. E. LA QUEY, and B. J. O'BRIEN, *J. Geophys. Res.* **71**, 1399 (1966).
32. M. H. REES, J. C. G. WALKER, and A. DALGARNO, *Planet. Space. Sci.* **15**, 1097 (1967).
33. D. M. HUNTEN and M. B. McELROY, *Rev. Geophys.* **4**, 303 (1966).
34. G. J. ROMICK and A. E. BELON, *Planet. Space Sci.* **15**, 1695 (1967).
35. W. B. MURCRAY, *J. Geophys. Res.* **72**, 1047 (1967).
36. T. M. DONAHUE, T. D. PARKINSON, E. C. ZIPF, J. P. DOERING, W. G. FASTIE, and R. E. MILLER, *Planet. Space Sci.* **16**, 737 (1968).
37. T. M. DONAHUE, E. C. ZIPF, and T. D. PARKINSON, *Planet. Space Sci.* **18**, 171 (1970).
38. T. D. PARKINSON, E. C. ZIPF, and T. M. DONAHUE, *Planet. Space Sci.* **18**, 187 (1970).
39. W. SWIDER and R. S. NARCISI, *Planet. Space Sci.* **18**, 379 (1970).
40. R. A. JONES and M. H. REES, *Planet. Space Sci.* **21**, 537 (1973).
41. E. C. ZIPF, private communication to A. Vallance Jones [24], p. 143.
42. E. C. ZIPF, *EOS Trans. Am. Geophys. Union* **54**, 403 (1973).
43. R. S. SIOLARSKI and A. E. S. GREEN, *J. Geophys. Res.* **72**, 3967 (1967).
44. G. M. LAWRENCE and M. J. McEWAN, *J. Geophys. Res.* **78**, 8314 (1973).
45. H. PARK, P. D. FELDMAN, and W. G. FASTIE, *Geophys. Res. Lett.* **4**, 41 (1977).
46. E. C. ZIPF and R. W. McLAUGHLIN, *Planet. Space Sci.* **26**, 449 (1978).
47. R. J. R. JUDGE, *Planet. Space Sci.* **20**, 2081, (1972).
48. M. H. REES and D. LUCKEY, *J. Geophys. Res.* **79**, 5181 (1974).
49. R. G. ROBLE and M. H. REES, *Planet. Space Sci.* **25**, 991 (1977).
50. J. A. MEYER, D. W. SETSER, and D. H. STEDMAN, *Astrophys. J.* **157**, 1023 (1969).
51. J. A. MEYER, D. W. SETSER, and D. H. STEDMAN, *J. Phys. Chem.* **74**, 2238 (1970).
52. R. A. YOUNG, G. BLACK, and T. G. SLANGER, *J. Chem. Phys.* **50**, 303 (1969).

53. A. B. CALLEAR and P. M. WOOD, *Trans. Faraday Soc.* **67**, 272 (1971).
54. J. A. MEYER, D. W. SETSER, and W. G. CLARK, *J. Phys. Chem.* **76**, 1 (1972).
55. J. F. NOXON, *J. Chem. Phys.* **36**, 926 (1962).
56. D. E. SHEMANSKY, *J. Chem. Phys.* **51**, 689 (1969).
57. E. J. BEITING and P. D. FELDMAN, *J. Geophys. Res.* **84**, 1287 (1979).
58. T. D. PARKINSON and E. C. ZIPF, *Planet. Space Sci.* **18**, 895 (1970).
59. T. D. PARKINSON, *Planet. Space Sci.* **19**, 251 (1971).
60. K. HENRIKSEN, *Planet. Space Sci.* **21**, 863 (1973).
61. A. BREKKE, *Planet. Space Sci.* **21**, 698 (1973).
62. B. H. SOLHEIM and E. J. LLEWELLYN, *Planet. Space Sci.* **27**, 473 (1979).
63. A. W. YAU and G. G. SHEPHERD, *Planet. Space Sci.* **27**, 481 (1979).
64. S. TRAJMAR, W. WILLIAMS, and A. KUPPERMANN, *J. Chem. Phys.* **56**, 3759 (1972).
65. R. L. GATTINGER and A. VALLANCE JONES, *Planet. Space Sci.* **27**, 169 (1979).
66. T. G. SLANGER and G. BLACK, *J. Chem. Phys.* **64**, 3763 (1976).
67. T. G. SLANGER and G. BLACK, *Geophys. Res. Lett.* **8**, 535 (1981).
68. T. G. SLANGER and G. BLACK, *J. Chem. Phys.* **75**, 2247 (1981).
69. M. KRAUSS and D. NEUMANN, *Chem. Phys. Lett.* **36**, 372 (1975).
70. T. G. SLANGER, B. J. WOOD, and G. BLACK, *Chem. Phys. Lett.* **17**, 401 (1972).
71. G. E. STREIT, J. H. CARLETON, A. L. SCHMELTEKOPF, J. A. DAVIDSON, and H. I. SCHIFF, *J. Chem. Phys.* **65**, 4761 (1976).
72. S. T. AMIMOTO, A. P. FORCE, R. G. GULOTTY, and J. R. WIESENFELD, *J. Chem. Phys.* **71**, 3640 (1979).
73. P. B. HAYS, D. W. RUSCH, R. G. ROBLE, and J. C. G. WALKER, *Rev. Geophys. Space Phys.* **16**, 255 (1978).
74. C. L. LIN, and F. KAUFMAN, *J. Chem. Phys.* **55**, 3760 (1971).
75. J. E. DAVENPORT, T. G. SLANGER, and G. BLACK, *J. Geophys. Res.* **81**, 12 (1976).
76. J. E. FREDERICK and D. W. RUSCH, *J. Geophys. Res.* **82**, 3509 (1977).
77. T. G. SLANGER, B. J. WOOD, and G. BLACK, *J. Geophys. Res.* **76**, 8430 (1971).
78. D. HUSAIN, S. K. MITRA, and A. N. YOUNG, *J. Chem. Soc. Faraday 2*, **70**, 1721 (1974).
79. M. F. GOLDE and B. A. THRUSH, *Discuss. Faraday Soc.* **53**, 233 (1972).
80. R. A. YOUNG and O. J. DUNN, *J. Chem. Phys.* **63**, 1150 (1975).
81. M. H. REES, A. I. STEWART, W. E. SHARP, P. B. HAYS, R. A. HOFFMAN, L. H. BRACE, J. P. DOERING, and W. K. PETERSON, *J. Geophys. Res.* **82**, 2250 (1977).
82. W. E. SHARP, M. H. REES, and A. I. STEWART, *J. Geophys. Res.* **84**, 1977 (1979).
83. J. E. FREDERICK, D. W. RUSCH, G. A. VICTOR, W. E. SHARP, P. B. HAYS, and H. C. BRINTON, *J. Geophys. Res.* **81**, 3923 (1976).
84. W. E. SHARP and D. G. TORR, *J. Geophys. Res.* **84**, 5345 (1979).
85. D. W. RUSCH, D. G. TORR, W. E. SHARP, T. M. DONAHUE, and K. HENRIKSEN, *J. Atmos. Terr. Phys.* **37**, 1173 (1975)
86. G. BLACK, T. G. SLANGER, G. A. ST. JOHN, and R. A. YOUNG, *J. Chem. Phys.* **51**, 116 (1969).
87. D. R. BATES and E. C. ZIPF, *Planet. Space Sci.* **28**, 1081 (1980).
88. D. W. RUSCH, J.-C. GÉRARD, and W. E. SHARP, *Geophys. Res. Lett.* **5**, 1043 (1978).
89. L. WALLACE and M. B. McELROY, *Planet. Space Sci.* **14**, 677 (1966).
90. D. F. STROBEL, *J. Geophys. Res.* **76**, 2441 (1971).
91. D. W. RUSCH and J.-C. GÉRARD, *J. Geophys. Res.* **85**, 1285 (1980).
92. J.-C. GÉRARD and O. E. HARANG, IN *Physics and Chemistry of Upper Atmospheres*, B. M. McCormac, Ed., Reidel, Dordrecht, 1973, p. 267.
93. R. L. ARNOLDY and P. B. LEWIS, *J. Geophys. Res.* **82**, 5563 (1977).
94. M. H. REES and R. A. JONES, *Planet. Space. Sci.* **21**, 1213 (1973).
95. H. EHRHARDT and A. KRESLING, *Z. Naturforsch. A* **22**, 2036 (1967).
96. E. C. ZIPF, P. J. ESPY, and C. F. BOYLE, *J. Geophys. Res.* **85**, 687 (1980).
97. J.-C. GÉRARD and O. E. HARANG, *J. Geophys. Res.* **85**, 1757 (1980).
98. L. G. PIPER, G. E. CALEDONIA, and J. P. KENNEALY, *J. Chem. Phys.* **75**, 2847 (1981); L. G. PIPER, *J. Chem. Phys.* **77**, 2373 (1982).

INDEX